POLYSACCHARIDE-PROTEIN COMPLEXES IN INVERTEBRATES

POLYSACCHARIDE-PROTEIN COMPLEXES IN INVERTEBRATES

by

Stephen Hunt

Department of Biology, University of Lancaster

1970

Academic Press · London · New York

ACADEMIC PRESS INC. (LONDON) LTD.
Berkeley Square House
Berkeley Square,
London, W1X 6BA

U.S. Edition published by
ACADEMIC PRESS INC.
111 Fifth Avenue,
New York, New York 10003

Library of Congress Catalog Card Number: 79—107931
SBN: 12–362050–3

Printed in Great Britain by
J. W. Arrowsmith Ltd., Bristol

Foreword and Acknowledgements

I wish here to thank my colleagues in the Departments of Biological Chemistry, Medical Biochemistry and Zoology at Manchester University who have always been ready with advice and willing to answer my most naive questions. It would also be impossible and ungrateful not to express my gratitude to those who taught me biochemistry and zoology and who made it unthinkable, for me at least, to study one without the other.

My thanks are due to my wife for her continued patience and encouragement during the preparation of this book and also to her and the secretaries of the Biological Chemistry department who painstakingly deciphered my writing and maintained the integrity of the English language.

I am indebted to the British Biochemical Society; the Japanese Biochemical Society; the Royal Society; the American Chemical Society; Pergamon Press, Oxford; Elsevier, Amsterdam; Macmillan (Journals), London; John Wiley and Sons, New York; and the Marine Biological Laboratory, Woods Hole for permission to use original material from their publications as well as to all the authors concerned who also gave their permission.

I would also like to express my appreciation of the diligence of the editorial staff of Academic Press, London in preparing the manuscript for publication.

Finally I would like to thank Professor G. R. Barker for the provision of facilities in the Department of Biological Chemistry at Manchester both for the writing of this book and such new research as has been included in it.

Lancaster
October, 1969   Stephen Hunt

Contents

CHAPTER 1

Introduction

Ten years ago the author of a review which dealt with the chemistry of protein-polysaccharide complexes felt compelled to write, "Much of the lamentation which served as introduction to the review of 'Mucoids and Glycoproteins' by Meyer in 1945 still seems apt today. Our knowledge of the structure and metabolism of protein-carbohydrate complexes stills lags seriously behind that which has been obtained about proteins and polysaccharides proper; the relevant literature remains scattered and by present day standards not voluminous" (Jevons, 1958). Jevons largely had in mind, of course, the protein-carbohydrate complexes of vertebrates, and at that time, with the vertebrate field in such straitened circumstances, it was natural that the more comparative aspects should take second place in the priority list. Happily, many of the wrongs voiced by Jevons have been redressed for vertebrate biochemistry, and during the past decade ample interest has been paid. The time now seems ripe to turn to the invertebrates and to examine the comparative biochemistry of these complex molecules in detail. It is unfortunate therefore that everything which Jevons said ten years ago one would wish now to repeat in the invertebrate context. The relatively few publications in the invertebrate field which were quoted in 1958 admittedly have had their ranks considerably swollen since then (there would be little point in writing the present review had they not), but the flow of new literature is small and the whole field of study still seems to lack the direction and purpose which has characterized work with vertebrate glycoproteins and mucopolysaccharides in the last few years.

The reason for this neglect is in part easy to see and in a sense under-standable; the proper study of man, in the medical and related sciences, is after all man, or at least some closely related animal. Much of the support for the study of glycoproteins and mucopolysaccharides has come from bodies whose interest lies in the diseases of man which involve these molecules either directly or indirectly; cystic fibrosis, arthritis, rheumatism and carcinoma are prime among these diseases. It is obviously of the greatest importance therefore that the chemistry and molecular biology of these complex molecules should be fully understood. We should not, however, allow ourselves to forget that we can learn much of our own biochemistry by reference to molecular events occurring in lower and more simple organisms; our knowledge of intermediary metabolism and molecular genetics does, after all, owe much to the wealth of data which has resulted from studies of bacterial biochemistry. Moreover a great part of the biological chemist's function is the elucidation of the nature and basis of that remarkable phenomenon we call life. We cannot expect to understand how biochemical processes have evolved in complexity (or apparent simplicity) if we confine our attention to only the more highly evolved and hence phylologically younger life forms.

It is primarily in the hope of drawing more biological chemists to the field of invertebrate carbohydrate-protein biochemistry that this book has been written. Yet there are equally enchanting vistas for those who work in other disciplines. For the chemist, the opportunity to study exotic macromolecules of immense complexity surely must hold great attraction. Invertebrates seem capable of almost infinite variety; described in this book are at least two completely new classes of polymeric macromolecule. For the biophysicist there are here many opportunities to accept the challenge of relating complex structure to physical behaviour and function and of examining novel systems of molecular organization. For the zoologist, detailed chemical and physical knowledge of the nature of secretions or tissue components is likely to lead to greater understanding of the manner in which organisms function and interact with their environments. Comparative studies of these macromolecular constituents are likely to provide insights into problems of phylogeny and taxonomy perhaps leading to greater understanding of the mechanisms of evolution.

In this book I have attempted to draw together a diffuse and at present comparatively sparse literature in order to indicate the stimulating field of study that this might be. If the work seems to be more a catalogue of data than an analysis it is largely because in a sense, it is only an interim report, until such time as a sufficient body of information is available to make a critical discussion possible.

Having bemoaned our lack of knowledge, I would like to point out that certain aspects of the field have not been neglected, particularly the study of the chitin-protein complexes and the structure proteins. Moreover, I imply no criticism of those who have assiduously endeavoured to advance our knowledge of protein-carbohydrate interactions in the invertebrates.

In rationalizing the study of invertebrate carbohydrate-protein complexes we must attempt to overcome the attitude that histochemical and cytochemical studies as a means of identifying molecules are sufficient unto themselves. Stain technology never can be sufficiently specific a basis for unequivocal identifications of complex molecular systems. This is not to say that cytochemical-histochemical studies are valueless; far from it, but they must be carried out with the backing of a sound knowledge of the macromolecules which they are being used to detect. Thus for example, while a strong gamma metachromasia with azure A may well indicate a sulphated mucopolysaccharide it most certainly does not identify this material with chondroitin sulphate. In fact one finds this type of faulty over-extrapolation all too often. Fortunately rationalization of the stain technology for acid mucopolysaccharides recently has been considerably advanced by the studies of J. E. Scott using the dye alcian blue (see for example Scott and Willett, 1966). We should also guard against observations of the action of specific enzymes, characterized in mammalian systems and used as histochemical or cytochemical reagents, being uncritically extended to studies of invertebrate tissues as they frequently are. Data obtained in this manner is of course valuable but must at all times be correlated carefully with more detailed chemical and physical studies.

The application of well-founded techniques for the isolation and purification of glycoproteins and mucopolysaccharides must be a primary concern, as must the avoidance of methods likely to cause disruption of delicate carbohydrate-protein interactions. The use of quaternary ammonium salts is likely to be of as much importance for the fractionation of invertebrate acid mucopolysaccharides as it has proved for vertebrate materials. With regard to chromatographic techniques of purification, the newly advanced ion-exchange celluloses are to be preferred infinitely to the older ion-exchange resins, owing to improved separation as well as the reduced possibility of degradation which may result from contact with strongly ionic resin. A word of caution is necessary however; ion-exchange celluloses recently have been implicated as agents responsible for the introduction of xylose contamination into proteins fractionated on them. The range of molecular weights spanned by gel-filtration systems recently has been expanded greatly by the development of new types of agar gel. Density gradient centrifugation of large macromolecules also may prove to be an important preparative method in the future, using perhaps the new high-speed zonal ultracentrifuge heads.

Chemical studies at the micro level are now completely feasible. Szirmai and others have shown that individual tissue sections can be analysed for their mucopolysaccharide contents and the mucopolysaccharides satisfactorily characterized. In this way it has proved possible to build up a picture of the mucopolysaccharide architecture of a tissue from serial sections (Antonopoulos, et al., 1964; Svejcar and Robertson, 1967).

Analytical techniques developed in recent years now permit critical examination of far smaller quantities of materials. It is to be expected that gas-liquid chromatography coupled with mass spectrometry will play an increasingly important role in the study of proteins and carbohydrates. The gas-liquid chromatograph now makes possible the rapid identification and quantitation of complex mixtures of monosaccharides, containing neutral sugars, hexosamines and uronic acids in microgram quantities. Conventional liquid chromatography systems for the analysis of the amino acids in proteins have now been brought to a high degree of sensitivity as well as speed; full analyses are now possible in under two hours. It seems likely that gas-liquid chromatography methods will also prove useful for amino acid analysis in the future although derivatization of all the amino acids simultaneously has proved a stumbling block up to the present time. A recent important advance has been the development of mass spectrometry methods for sequence determination in peptides (Biemann, et al., 1966; Agarwal, et al., 1968). It should be possible to use similar methods for structure determination in complex oligosaccharides.

The elegant methods which have been applied to the elucidation of the carbohydrate-protein linkage regions in vertebrate mucopolysaccharides and glycoproteins should now find ready application in the invertebrate field, though we should be ready to develop new methods, to meet new types of structure, as and when it becomes necessary.

It is likely that physical techniques could be now more generally and most profitably applied to the study of the mucopolysaccharides and glycoproteins of invertebrates as well as to the structure proteins. Infrared spectroscopy has already proved useful and informative in several instances while X-ray diffraction has been extensively applied to the study of chitin and structure proteins. One would hope that the recent promising applications of nuclear magnetic resonance spectroscopy to studies of vertebrate mucopolysaccharides (Inoue and Inoue, 1966; Jaques, et al., 1966) will be equally well applied to invertebrate materials. Valuable information with regard to chain conformation in proteins and some polysaccharides can be gained from the use of optical rotatory dispersion and circular dichroism, and the usefulness of these techniques for acid polysaccharide work can be extended by the use of dye-binding methods.

It is not the purpose of this book to act as a handbook of experimental methods, although wherever possible details of techniques, particularly preparative techniques, have been included. Much practical analytical detail on glycoproteins and mucopolysaccharides can be found in Gottschalk's "Glycoproteins" (1966) and also in "Mucopolysaccharides" by Brimacombe and Webber (1964).

Throughout the book I have retained the now well-established term "glycoprotein", meaning a covalently linked complex of protein with carbohydrate in which the properties of protein are predominant.

For the complexes in which the polysaccharide properties are the dominant feature I have used, as a general descriptive term, "mucopolysaccharide". This is rather unsatisfactory in that it fails to refer directly to the protein part of the complex while the prefix "muco" may be held to imply mucinous, viscous, slimy, properties which are not necessarily characteristic of the material in the natural state; it does however have the advantage of being a term well-established in the literature of vertebrate biochemistry. In this latter context mucopolysaccharide is largely confined to describing linear high molecular weight polysaccharides, comprising repeating sequences of hexosamine alternating with uronic acid or neutral sugar and frequently ester sulphated or N-sulphated; the polysaccharide chain most frequently is linked covalently to protein and also frequently ionically bound to further protein. Examples of this type of molecular system are the chondroitin sulphate-protein complexes of vertebrate connective tissues. In fact, in the vertebrates, the terms mucopolysaccharide and glycosaminoglycan are interchangeable. As a general rule we do not find in the vertebrates neutral high molecular weight polysaccharides with specific protein associations, although the uncommon vertebrate cellulose falls into this class, whereas the situation with regard to vertebrate glycogen and galactan is uncertain, nor is there a simple polymer of uncharged hexosamine comparable with invertebrate chitin. Mucopolysaccharide therefore has been extended from its more generally accepted usage to cover the large charged polysaccharides in all their forms as well as the neutral polysaccharides, such as chitin, cellulose, galactan and glycogen, that are associated specifically with protein in the animal body.

Numerous studies of the covalent complexes of relatively low molecular weight oligosaccharides and protein, found ubiquitously distributed in the tissues, secretions and body fluids of vertebrates, by now have made the concept of a glycoprotein a familiar one. The oligosaccharide moieties of these materials usually contain N-acetylated hexosamine, neutral sugar and sialic acid; our knowledge of the protein part of the molecule in general is more limited. The section devoted to glycoproteins examines the occurrence of such complexes in invertebrate situations. It would have been possible also to have included in this latter section the invertebrate collagens, the silks and a variety of other structure proteins which, because they contain firmly bound oligosaccharide, can be classed as glycoproteins. Since however many of these structure glycoproteins and certain other non-glycoprotein structure proteins have important interactions with structural mucopolysaccharides, it was thought easier to consider the structure proteins as a whole, under a single heading which included the simple structure glycoproteins (e.g. silks, molluscan egg cases), the structure glyco-proteins associated with mucopolysaccharides (e.g. collagens) and the simple structure proteins associated also with mucopolysaccharide (e.g. resilin).

The book is thus divided into three broad sections: mucopolysaccharides, glycoproteins and structure proteins. Because of their great diversity the mucopolysaccharides have been further subdivided into acid mucopoly-saccharides and neutral mucopolysaccharides.

Sir Vincent Wigglesworth, introducing a symposium on insect biochemistry (Wigglesworth, 1965), has said "The virtue of insects as a medium for the study of physiological chemistry lies in their diversity"; the same might be said of invertebrates as a whole. Those who are familiar with the chemistry of carbohydrate-protein complexes in vertebrates will, I hope, be pleasantly surprised at the vast diversity in the invertebrates and feel thus prompted to venture into this delightful unknown.

REFERENCES

Agarwal, K. L., Johnstone, R. A. W., Kenner, G. W., Millington, D. S. and Sheppard, R. C. (1968). *Nature, Lond.*, **219**, 498.

Antonopoulos, C. A., Gardell, S., Szirmai, J. A. and de Tyssonsk, E. R. (1964). *Biochim. biophys. Acta*, **83**, 1.

Biemann, K., Covel Webster, B. R. and Arsenault, G. P. (1966). *J. Am. chem. Soc.* **88**, 5598.

Brimacombe, J. S. and Webber, J. M. (1964). "Mucopolysaccharides." Elsevier, Amsterdam.

Gottschalk, A., ed. (1966). "Glycoproteins." Elsevier, Amsterdam.

Inoue, S. and Inoue, Y. (1966). *Biochem. biophys. Res. Commun.* **23**, 513.

Jaques, L. B., Kavanagh, L. W., Mazurek, M. and Perlin, A. S. (1966). *Biochem. biophys. Res. Commun.* **24**, 447.

Jevons, F. R. (1958). *Adv. Protein Chem.* **13**, 35.

Scott, J. E. and Willett, I. H. (1966). *Nature, Lond.* **209**, 985.

Svejcar, J. and Robertson, W. van B. (1967). *Analyt. Biochem.* **18**, 333.

Wigglesworth, V. B. (1965). *In* "Aspects of Insect Biochemistry" (T. W. Goodwin, ed.), p. xi. Academic Press, London.

MUCOPOLYSACCHARIDES
Acid Mucopolysaccharides

CHAPTER 2

Chondroitin Sulphates, Chondroitin and Keratosulphate

INTRODUCTION

In the vertebrates acid mucopolysaccharides occur most characteristically as constituents of connective tissues, the latter term being used in its broader sense. There are perhaps some six major vertebrate acid mucopolysaccharides; the three chondroitin sulphates A, B and C, keratosulphate, hyaluronic acid and heparin. Of these the chondroitin sulphates A and C together with keratosulphate, although widely distributed in the connective tissues of the vertebrate body, are particularly characteristic of the cartilaginous tissues. Cartilage also occurs in certain invertebrates and we shall consider in this chapter how and to what extent vertebrate and invertebrate cartilaginous tissues relate to one another with regard to the chondroitin sulphates and keratosulphate. We shall also consider the occurrence of chondroitin in invertebrates, since this less-widely distributed vertebrate acid mucopolysaccharide, although not found in cartilage, falls naturally into the same group as the chondroitin sulphates. Hyaluronic acid and heparin will be considered in later chapters. Chondroitin sulphate B has not so far been detected in invertebrates.

CARTILAGE

As much of this chapter is devoted to a consideration of the mucopolysaccharide constituents of a particular tissue type, it is important that we should begin by attempting to obtain some understanding of what actually is implied by the term cartilage for a particular tissue.

The definition of what constitutes a cartilaginous tissue is based largely upon the histological properties of certain tissues occurring in vertebrates and characterized at an early date in the study of the histology of mammalian tissues. The histological character of mammalian cartilage may be summarized as a hyaline matrix, staining metachromatically with certain dyes and being found in such situations as the articular surfaces of the long bones, the junctions between ribs and sternae, in the nasal septae, the larynx and the tracheobranchial rings. This type of cartilage, which is the most abundant, is termed hyaline because of the appearance of its glassy translucent matrix. Also found in vertebrates are fibrous and elastic cartilages containing increased amounts of collagen and other fibrous proteins. Histologists reasonably tend to assume that if a tissue occurs in a situation where cartilage functionally might be expected to be found (in the invertebrates this is likely to be some supporting role) and if the tissue shows similar histological properties to already well-defined cartilaginous tissues, then that tissue may perhaps be called a cartilage. Proceeding upon this basis the presence of what appeared to be cartilaginous tissues in invertebrates was recognized as early as 1818 by Schultze. In a magnificent and monumental review of supporting tissues Schaffer (1930) quotes a considerable literature extending from the early years of the 19th century to the end of the first quarter of the 20th century and describes cartilages from a wide range of invertebrate sources. The large majority of those works quoted by Schaffer however refer only to the histological and developmental aspects of these tissues and have little to say regarding the chemical composition and character of the molecular units of the tissue matrix. Early studies however had begun to realize that vertebrate cartilage tissues had a characteristic and distinctive chemical composition. Even so it was not until quite late in the 20th century that Levene (1941) carried out the first really rigorous studies of the major cartilage constituent chondroitin sulphate. At the time of Schaffer's review it was in some doubt whether or not invertebrate connective tissues contained chondroitin sulphate, chitin or collagen. This particular problem in recent years has been clarified to some extent but if we note the considerable advances which have been made in the field of vertebrate connective tissue chemistry during the last two decades, with the almost complete structural elucidations of the major connective tissue mucopolysaccharides, it is both surprising and disappointing that the corresponding invertebrate tissues have received such scant attention.

VERTEBRATE CHONDROITIN SULPHATES AND KERATOSULPHATE

The structures of the polysaccharide moieties of the three vertebrate chondroitin sulphates A, B and C and of vertebrate keratosulphate are shown in Fig. 1. The work leading to the establishment of these structures has been reviewed comprehensively by Brimacombe and Webber (1964). The basic structure of the three chondroitin sulphates (B is now frequently referred to as dermatan sulphate) is a repeating sequence of a disaccharide comprising a residue of 2-acetamido-2-deoxy-D-galactose linked by a glycosidic bond at position 3 to the anomeric carbon atom of a residue of uronic acid. In chondroitin sulphates A and C the uronic acid is glucuronic acid, in chondroitin sulphate B it is iduronic acid. The hexosamine is

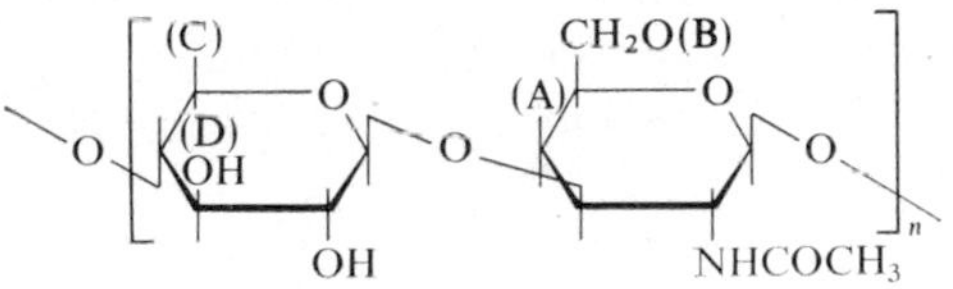

	(A)	(B)	(C)	(D)
Chondroitin sulphate A or chondroitin-4-sulphate	OSO_3H	H	COOH	H
Chondroitin sulphate C or chondroitin-6-sulphate	OH	OSO_3H	COOH	H
Chondroitin sulphate B or dermatan sulphate	OSO_3H	H	H	COOH

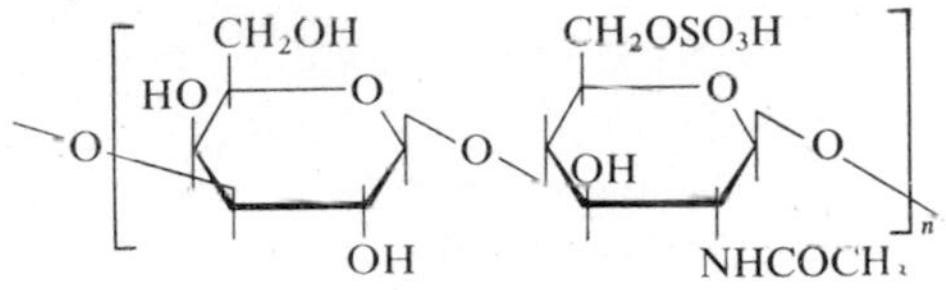

Fig. 1. Keratosulphate.

sulphated at position 4 in chondroitin sulphates A and B and at position 6 in chondroitin sulphate C. The disaccharide sub-unit is joined into the linear repeat sequence by a 1 → 4 glycosidic link between the hexosamine anomeric carbon and the uronic acid position 4 ring hydroxyl. Both types of link are β.

Although the composition is different, keratosulphate resembles the chondroitin sulphates. Galactose replaces uronic acid while the amino sugar is now 2-amino-2-deoxy-D-glucose sulphated at position 6. The mucopolysaccharide contains the characteristic repeat of alternating $\beta1 → 3$, $\beta1 → 4$ glycosidic links but their position is reversed. The disaccharide contains the 1 → 4 link to the 4 position of the amino sugar while the glycosidic link which unites the disaccharides forming the completed linear chain is to position 3 of the galactose.

As already mentioned the distribution in the vertebrate body of these acid mucopolysaccharides is wide; however chondroitin sulphates A and C tend to occur with keratosulphate in cartilage while chondroitin sulphate B is more characteristic of such tissues as skin and tendon. Keratosulphate also occurs in cornea and *nucleosus pulposus*. The amount of keratosulphate in vertebrate cartilage seems to vary as a function of age increasing with advancing age at the expense of chondroitin sulphate.

Chondroitin, the other acid mucopolysaccharide which we shall discuss in this chapter, has the same structure as chondroitin sulphates A and C but lacks the sulphate residues; it has been suggested as the biological precursor of chondroitin sulphates A and C but it has been detected in vertebrates only in bovine cornea.

INVERTEBRATE

Early reports of the chemistry of chondroitin sulphate-like substances in invertebrate cartilage tissues are due to Bradley (1912, 1913) and to Mathews (1923) both of whom reported the presence of reducing sugar and acid-labile sulphate in the enterosterite of *Limulus* ("horseshoe crab", Arthropoda, Arachnida, Xyphosura). Apart from these early references interest in the chemistry of invertebrate cartilage would appear to have lapsed completely until well into the second half of the 20th century.

The comparative biochemistry of invertebrate and vertebrate cartilage in a sense might be said to have started in an unfavourable manner in that the first detailed characterization of an invertebrate cartilage mucopolysaccharide yielded a material completely different in structure to the familiar vertebrate cartilage mucopolysaccharides.

In 1959 Lash, during a study of the odontophore, a chondroid apparatus supporting the rasping organ or radula, of the marine gastropod mollusc *Busycon canaliculatum*, noted two remarkable features of this tissue. While in its gross and histological appearance the tissue closely resembled typical vertebrate cartilage tissues, it contained quantities of myoglobin. Moreover, it failed to yield either uronic acid or hexosamine on acid hydrolysis thereby suggesting the absence of chondroitin sulphates. Preliminary studies suggested that the place of the chondroitin sulphates was filled by a glucan sulphate. This was later substantiated by a more detailed study (Lash and Whitehouse, 1960a) described in Chapter 5. Even so the cartilage polysaccharide, while not a chondroitin sulphate, was similar to its vertebrate counterparts in being a sulphated, linear, probably β linked acid mucopolysaccharide. At this stage therefore it seemed as though invertebrate cartilage mucopolysaccharides might resemble vertebrate ones in broad physical detail but not in more detailed chemical structure. Other studies have shown however that this is not necessarily the case.

SQUID CRANIAL CARTILAGES

In a study of cartilage tissues from two vertebrate sources and one invertebrate source, Lash and Whitehouse (1960b) reported the presence of both chondroitin sulphate and keratosulphate in the cranial cartilage taken from squid (Mollusca, Cephalopoda).

Mucopolysaccharides were extracted from the cartilage tissue by exhaustive tryptic digestion and precipitated from the solubilized material as complexes with hexaminocobaltic cations and with Cetavlon (cetyltrimethyl ammonium bromide). Following removal of the precipitating cations the extracts were found to have strong anticoagulant properties and electrophoretic mobilities similar to those of chondroitin sulphate preparations from bovine nasal septa. Sulphate ions were detected after acid hydrolysis, and analysis by the benzidine precipitation method indicated an ester sulphate content of 20–25%. Hydrolysis of the squid polysaccharide with 1N sulphuric acid for four hours at 105° and paper chromatography or borate buffer paper electrophoresis of the hydrolysates identified 2-amino-2-2-deoxy-D-glucose, 2-amino-2-deoxy-D-galactose, galactose, glucuronic acid and traces of xylose as constituent monosaccharides.

The glucuronic acid and galactose contents of the dried extracts were measured by the carbazole-sulphuric acid and cysteine-sulphuric acid reactions respectively. Squid cartilage from a cranium of 2·0 cm width contained 0·8% galactose and 3·2% glucuronic acid, giving a galactose glucose ratio of 0·25, while cartilage from a 3·5 cm cranium (i.e. an older animal) contained 2·6% galactose and 7·7% glucuronic acid, a ratio of 0·34. The presence of galactose and 2-amino-2-deoxy-D-glucose were considered to be fair presumptive evidence for the existence in the tissue of keratosulphate. The results seemed to indicate that the relative proportions of keratosulphate and chondroitin sulphate varied with the age of the animal, the keratosulphate content rising with increasing age. Such an increase at the expense of the chondroitin sulphate had already been noted for vertebrate tissues by Kaplan and Meyer (1959). The results obtained for squid cartilage were compared with those for embryonic and adult shark cartilage tissues. Here the ratio of galactose to glucuronic acid rose from 0·06 in the embryo to 0·51 in the adult. Embryonic chick tissue culture cartilage showed a ratio of 0·02, while this rose to 0·25 in a 15 day embryo.

Anno, Seno and Kawaguchi (1962) have compared the 2-amino-2-deoxy-D-glucose and 2-amino-2-deoxy-D-galactose contents of cartilages from various sources including the cranium of the squid *Ommastrephes sloani pacificus*. Acetone-dried cartilage powders were hydrolyzed with 3N hydrochloric acid and the amino sugars separated by ion-exchange chromatography on columns of Dowex 50 X8 eluted with 0·3N hydrochloric acid. The squid cartilage used was from 3–4 cm wide crania and contained a total of 5·5% amino sugar. Of this 0·3% was 2-amino-2-deoxy-D-glucose and 5·2% 2-amino-2-deoxy-D-galactose, a 1 : 17·5 ratio. The 2-amino-2-deoxy-D-glucose content of squid cartilage was the lowest for the animals studied; other values ranged from 0·4% in ray foetal cranial cartilage to 2·5% in whale nasal cartilage. The 2-amino-2-deoxy-D-galactose content was also low in comparison to other species although bovine articular cartilage contained only 2·7%. In general the proportion of 2-amino-2-

deoxy-D-glucose and 2-amino-2-deoxy-D-galactose was greater not only in adults as compared with embryos but also in higher evolutionary as compared with lower evolutionary forms. The presence of 2-amino-2-deoxy-D-glucose was considered to suggest identification of keratosulphate and hyaluronic acid as constituents of the cartilage. No other amino sugars were found. These same results have also been reported by Anno and Seno (1962).

In a further study of the polysaccharides of squid connective tissue Anno *et al.* (1964) digested *Ommastrephes* cranial cartilage and skin with the bacterial protease Pronase and precipitated mucopolysaccharides, as their sodium salts, from the extracts with ethanol. The crude precipitates were then further fractionated using cetylpyridinium chloride (CPC). The main skin polysaccharide was found to be non-sulphated containing 2-amino-2-deoxy-D-galactose and glucuronic acid in a 1:1 ratio but no 2-amino-2-deoxy-D-glucose. This material appeared to be chondroitin and we shall have more to say about it later. The cranial mucopolysaccharide was over-sulphated.

TABLE I
Analysis of Squid Cartilage Mucopolysaccharide[a]

	Mixed mucopolysaccharide	MPS-3
Hexosamine (%)	24·6	28·5
Uronic acid (%) (Carbazole)	26·5	29·8
Uronic acid (%) (Orcinol)	16·5	—
Nitrogen (%)	2·0	—
Sulphate (%)	20·0	23·7
$[\alpha]_D$ in water (c, 1)	$-38°$	$-36°$

[a] Kawai, Seno and Anno (1966).

This over-sulphated mucopolysaccharide yielded chondrosine (2-amino-2-deoxy-3-0-(β-D-glucopyranosyluronic acid)-D-galactose) (Fig. 2) on acid hydrolysis. The infrared spectra differed from those of the over-sulphated

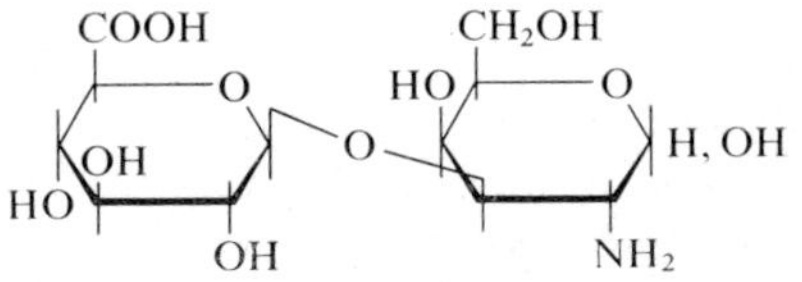

Fig. 2. Chondrosine
2-amino-2-deoxy-3-O-(β-D-glucopyranosyluronic acid)-D-galactose.

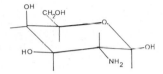

Fig. 3. Preferred conformation of 2-amino-2-deoxy-β-D-galactose.

shark chondroitin sulphates with which it was compared and indicated the presence of both axial and equatorial ester sulphate groups suggesting that both positions 4 (axial hydroxyl in the preferred conformation of 2-amino-2-deoxy-D-galactose, Fig. 3) and 6 (equatorial hydroxyl in the preferred conformation) of the hexosamine might be sulphated. Thus this chondroitin sulphate might be considered to be a hybrid of chondroitin sulphates A and C. It might of course have been possible for the equatorial sulphate to have been sited at positions 2 or 3 of the glucuronic acid residue. This point is clarified by later more detailed studies carried out by the same group.

The method of extraction, fractionation and purification of *Ommastrephes* cartilage mucopolysaccharide was modified and refined by Kawai *et al.* (1966). Homogenized and heated cranial cartilage was extracted with 2% NaOH at 4 to 5° for 24 hours and then treated with Pronase. The digest, with added sodium acetate, was precipitated with ethanol giving mucopolysaccharide sodium salts. The mixed mucopolysaccharides were fractionated on columns of Dowex 1 ion-exchange resin by stepwise elution with solutions of sodium chloride over the range 0–4 M. The major uronic acid-containing fraction, eluted with 3 M NaCl (MPS-3) was dialyzed and freeze-dried. The yield of MPS-3 was approximately 1% of the weight of the wet cartilage or 63% of the mucopolysaccharide extract. Analytical data for the mixed mucopolysaccharide extract and MPS-3 are given in Table I. The molar ratios of hexosamine, uronic acid and sulphate in MPS-3, calculated from the percentage compositions, are 1·00, 0·97 and 1·55 respectively. The mucopolysaccharide thus contains sulphate residues in excess of one per hexosamine residue.

The hexosamine moiety of MPS-3 was identified as 2-amino-2-deoxy-D-galactose both by ion-exchange column chromatography of acid hydrolysates and by paper chromatographic identification of lyxose in ninhydrin degraded acid hydrolysates. No other amino sugar was detected.

Cellulose acetate strip electrophoresis of MPS-3 at pH 3·5 showed it to be electrophoretically homogeneous with a mobility greater than either chondroitin sulphate A or C. An attempt to fractionate MPS-3 further by selective extraction of its CPC complex with sodium chloride solutions resulted in 94% of the total uronic acid being extracted with 2·1 M sodium chloride and only negligible quantities being extracted at molarities of 0·3, 1·2 and 3·0.

The infrared spectrum (Fig. 4) showed characteristic ester sulphate bands at 1250, 928, 850 and 820 cm^{-1}. The latter two band wave numbers suggest

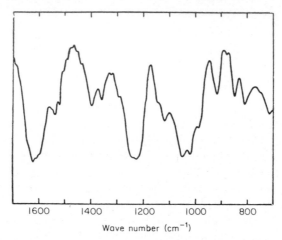

Wave number (cm^{-1})

Fig. 4. Infrared spectrum of *Ommastrephes sloani pacificus* cartilage mucopolysaccharide (from Kawai, Seno and Anno, 1966).

the presence of both axially and equatorially orientated substitutions, as has already been mentioned. Moreover the value for equatorial frequency would suggest substitution upon a primary rather than a secondary equatorial hydroxyl thus eliminating the possibility of substitution at position 2 or 3 of the D-glucuronic acid residue and fixing the group involved as the 6 hydroxyl of the 2-acetamido-2-deoxy-D-galactose residue. The only possible position for the axial sulphate is the 4 hydroxyl of the hexosamine. Such assumptions depend of course upon the molecule having the same basic disaccharide repeat as the chondroitin sulphates. As we shall see, this was in fact the case, as earlier evidence had already suggested. The rate of acid hydrolysis of the ester sulphate groups yielded, on analysis, complex kinetics with two different sets of half-lives, 0·9 and 2·3 hours, suggesting again that there were two classes of ester sulphate group and that in addition the uronic acid residues were probably not sulphated (see Rees, 1963).

The major repeat unit of MPS-3 would appear to be 2-acetamido-2-deoxy-3-0-(β-D-glucopyranosyluronic acid)-D-galactose (chondrosine). Partial acid hydrolysis of the mucopolysaccharide yielded a disaccharide, in 45% yield, identical with authentic chondrosine by the criteria of chemical analysis, paper chromatography, infrared spectroscopy and optical activity.

Testicular hyaluronidase hydrolyzed MPS-3 more slowly than it did vertebrate hyaluronic acid, chondroitin sulphate A or chondroitin sulphate C.

After 48 hours incubation with hyaluronidase the reducing value of MPS-3 was a quarter that of hyaluronate and half that of chondroitin sulphates A and C. More prolonged incubation and further additions of enzyme did however, after 172 hours, bring MPS-3 and the chondroitin sulphates to the same level of digestion. The molar ratio of N-acetylhexosamine (Morgan-Elson) to reducing sugar (as glucose), given by the products of the hyaluronidase digestion of MPS-3, was 0·25. This value was higher than that of chondroitin sulphate A (0·07), in which position 4 of the hexosamine is sulphated, and lower than that of chondroitin sulphate C (1·07), in which the 4 position is free. Thus the combined evidence, obtained from the enzymic digests, from the infrared spectra of MPS-3 and from the sulphate-release kinetics, suggests that about 80% of the 2-acetamido-2-deoxy-D-galactose residues of the mucopolysaccharide are sulphated at position 4 and that there is a high proportion of 4,6-disulphated residues. Shark cartilage over-sulphated chondroitin sulphate in contrast probably carries its excess sulphate at position 2 or 3 of the uronic acid residue while the hexosamine is sulphated at position 6 only (Suzuki, 1960).

TABLE II

Analysis of *Loligo* and *Limulus* cartilage mucopolysaccharides[a]

	Limulus gill	*Loligo* cranial and nuchal
	mole ratios	
Hexosamine	1·0	1·0
Hexuronic acid	1·13	0·99
Nitrogen	1·29	0·99
N-acetyl	0·95	1·09
Sulphur	1·45	1·49

[a] Mathews, Duh and Person (1962).

Over-sulphated chondroitin sulphate was also detected in the cartilage of the squid *Loligo*, as well as in cartilage from the arachnid *Limulus*, by Mathews et al. (1962). Acid mucopolysaccharides were extracted from papain digests of squid cranial and nuchal cartilages and from digests of the cartilage supporting the *Limulus* gills. Analytical data for these acid mucopolysaccharides are given in Table II. The materials from both sources are very similar in composition, closely corresponding to the theoretical values for a vertebrate chondroitin sulphate. The optical activities differ quite markedly however from the values normally expected for vertebrate chondroitin sulphates (chondroitin sulphate A $[\alpha]_D$ $-28°$ to $-32°$, chondroitin sulphate C $[\alpha]_D$ $-16°$ to $-22°$) and from those obtained for over-sulphated chondroitin sulphates (*Ommastrephes* $[\alpha]$ $-36°$, shark $-18°$).

In both cases the sulphate values are approximately 50% in excess over the normal 1 : 1 ratio with hexosamine. The infrared spectra resembled chondroitin sulphate A although the spectrum of the squid material contained an extra band at 800 cm^{-1} while that of the *Limulus* mucopolysaccharide had an extra band at 800 cm^{-1}.

The mucopolysaccharides were completely hydrolyzed by testicular hyaluronidase, as far as a turbidimetric test could show, suggesting that the hexosaminidic linkages are probably similar to those of chondroitin sulphate A or C. As neutral sugars and 2-amino-2-deoxy-D-glucose detected in the crude extracts were absent from the acid mucopolysaccharide fraction (the method of fractionation was not given), it was assumed that keratosulphate probably was not present, in contrast to the results of Lash and Whitehouse (1960b). Analysis of the products of partial acid hydrolysis and enzymic digestion seemed to indicate that the hexuronide linkages were probably to position 3 of the 2-acetamido-2-deoxy-D-galactose and that about three-quarters of the hexosamine residues were sulphated at position 4. The authors suggested that these invertebrate chondroitin sulphates were merely isologues of vertebrate chondroitin sulphates having arisen by biochemical convergence. In their view it seemed also that the two invertebrate materials structurally were more closely related to one another than to the vertebrate chondroitin sulphates and that they at least might have arisen from a common ancestral molecule before the separation of the molluscan and arthropod lines.

Preparations of cartilage from the giant squid *Architeuthis dux* have been found to yield chondroitin sulphates with lower sulphate-hexosamine ratios than other squid over-sulphated chondroitin sulphates; the ratios were in the range 1·00 to 1·30 (Mathews, 1967). These mucopolysaccharides had unusually high molecular weights of up to 107,000. Such molecular weights are well above the range occupied by vertebrate chondroitin sulphates, i.e. 5000 to 50,000. Cartilage from the octopus, *Octopus maculatus*, has also yielded a chondroitin sulphate chemically similar to mammalian chondroitin sulphate A and with only a slight sulphate excess. There were however significant differences in the infrared spectra. The molecular weight was of the order 70,000 (Mathews, 1967). Mathews has commented that there seems to be an inverse relationship between molecular weight and the degree of sulphation in cephalopod chondroitin sulphates.

SQUID CORNEAL STROMA

Anseth (1961) has compared the glycosaminoglycans of the corneal stroma from various vertebrate species and from the squid *Loligo pealii*. The mucopolysaccharides were extracted and then fractionated on columns of Ecteola cellulose. The hexosamine contents of the stroma varied between 1·1% and 1·5% of the dry tissue weight and was significantly lower only in the case of the squid where it was 0·5%. The squid tissue was also exceptional in having a 2-amino-2-deoxy-D-galactose content almost twice that of 2-amino-2-deoxy-D-glucose. The fractions separated on the ion-exchange cellulose were divided into two groups: low molecular weight mucopolysaccharides, which were eluted by 0·02, 0·2 and 0·3 M chloride, and high molecular weight mucopolysaccharides, which were eluted by 2·0 M chloride. Both groups were sub-divided into glucosaminoglycans and galactosaminoglycans. The low molecular weight glucosaminoglycans contained 1·5 μg amino sugar per mg dry material while the high molecular weight ones contained 0.1 μg per mg. Similarly, low molecular weight galactosaminoglycans contained 1·8 μg per mg while high molecular weight ones contained 1·1 μg per mg. In the squid the cornea, in contrast to the other animals studied, is a direct continuation of the skin and might therefore be expected to have a somewhat similar composition to the skin. Anseth and Laurent (1961) consider corneal glucosaminoglycans to be keratosulphates and galactosaminoglycans to be chondroitin sulphates. As we shall see later however the galactosaminoglycans eluted at low ionic strength were more likely to have been chondroitin.

ANNELID CARTILAGE

The presence of a cartilage-like tissue in the marine polychaete, annelid worm *Eudistylia polymorpha* has been reported by Person (1964). Three different types of tissue, all histologically resembling vertebrate cartilage, were found in the head and tentacular region. A chemical analysis of the tissue indicated that hexuronic acid, hexosamine and ester sulphate were present. Later studies (Person and Mathews, 1967) showed that the mucopolysaccharide constituent of the cartilage was in fact closely similar to vertebrate chondroitin sulphate C except for a degree of over-sulphation.

The cartilage yielded 2·5% of its dry weight as chondroitin sulphate; this is roughly within the range one would expect for vertebrate cartilage tissues, i.e. 1·5 to 20%. The composition of the mucopolysaccharide is given in Table III. The electrophoretic mobilities of the mucopolysaccharide on cellulose acetate strip at pH 7 and pH 3 were greater than for mammalian chondroitin sulphate; the relative values at the two pH values were 1·05 and 1·25 respectively. Infrared spectra of this over-sulphated chondroitin sulphate more closely resembled mammalian chondroitin sulphate C than A; in this case it seems difficult to account for the distribution of the excess sulphate. An extra band at 700 cm^{-1} seems an unlikely candidate for an extra sulphate absorption. The number average molecular weight, obtained

TABLE III

Analysis of *Eudistylia polymorpha* cartilage mucopolysaccharide[a]

	Mole ratios to 2-amino-2-deoxy-D-galactose	
	E. polymorpha	Theoretical for mammalian cartilage chondroitin sulphate
2-amino-2-deoxy-D-galactose	1·00	1·00
2-amino-2-deoxy-D-glucose	0·03	0·00
Uronic acid	1·07	1·00
Nitrogen	1·20	1·00
Sulphate	1·88	1·00

[a] Person and Mathews (1967).

by osmometry, was of the order 10,000 while the intrinsic viscosity was 0·32. Testicular hyaluronidase hydrolyzed the mucopolysaccharide more slowly than it did a mammalian chondroitin sulphate.

CHONDROITIN

As we have already noted, chondroitin, the unsulphated form of chondroitin sulphates A and C, is only found in vertebrates as a constituent of the cornea (Davidson and Meyer, 1954); its presence therefore in the skin of an invertebrate is of some interest. In an earlier study, previously mentioned in this chapter, Anno *et al.* (1964) had noted the presence of an unsulphated mucopolysaccharide in the skin of the squid *O. sloani pacificus* and had suggested that it might be chondroitin, as it appeared to be composed of equal amounts of glucuronic acid and 2-amino-2-deoxy-D-galactose. This material was the principal skin mucopolysaccharide and was investigated in greater detail by Anno *et al.* (1964).

The skin was defatted and ground before heating in water at 100° for one hour. Following an exhaustive digestion with Pronase the almost completely liquified material was centrifuged and the supernatant precipitated

TABLE IV

Analysis of Squid Skin Mucopolysaccharide[a]

	(%)	Molar ratio
Hexosamine	29·8	1·00
Uronic acid (Carbazole)	31·4	0·98
Uronic acid (Orcinol)	19·2	—
Nitrogen	2·5	1·06
Sulphate	< 0·7	< 0·04

$[\alpha]_D$ in water (*c*, 1) −23°

[a] Anno, Kawai and Seno (1964).

with ethanol. The precipitate was dissolved in water, the insoluble material removed, and further protein precipitated by first of all the addition of trichloracetic acid and then, after dialysis, by adsorption on a mixture of kaolin and Lloyd's reagent. Finally the mucopolysaccharide was precipitated in the presence of sodium acetate as its sodium salt by the addition of ethanol. The yield of mucopolysaccharide from dry skin was in the region of 2·4%.

The mucopolysaccharide obtained at this stage of the preparation was in fact a mixture of mucopolysaccharides. Further fractionation and purification was achieved by forming the cetylpyridinium chloride complexes and extracting with a range of concentrations of sodium chloride solution. The fraction soluble in 0·3 M sodium chloride was the most abundant, making up rather more than 70% of the total uronic acid of the mixed mucopolysaccharides. This fraction was again subjected to treatment with Lloyd's reagent, dialyzed, concentrated and precipitated with ethanol as before. This fraction (MPS-0·3) represented 63% of the total skin mucopolysaccharides. Analytical data for MPS-0·3 are given in Table IV. MPS-0·3 was homogeneous by cellulose acetate strip electrophoresis at pH 3·5 (1·0 M pyridine acetate), with a mobility corresponding to either hyaluronic acid or bovine corneal chondroitin. Its infrared spectrum lacked the characteristic bands associated with ester sulphate and was in fact almost identical with authentic chondroitin.

The only amino sugar identified in MPS-0·3 was 2-amino-2-deoxy-D-galactose, and when the mucopolysaccharide was subjected to partial hydrolysis a disaccharide was obtained whose infrared spectrum was identical with that of chondrosine (2-deoxy-3-0-(β-D-glucopyranosyluronic acid)-D-galactose) (Fig. 2). Thus the evidence strongly favoured a firm identification of MPS-0·3 as chondroitin. MPS-0·3 contained less than 0·7% ester sulphate; bovine corneal chondroitin has been reported to contain 2% (Davidson and Meyer, 1954).

In commenting upon the occurrence of chondroitin in squid skin it should be noted, as was mentioned earlier, that the squid cornea is a continuation of the skin and that in this respect the squid chondroitin is sited in a situation analogous to that of vertebrate chondroitin. The work of Anseth (1961) with squid cornea does not preclude the presence of chondroitin in the cornea.

LIMPET BUCCAL CARTILAGE

The buccal cartilage of the limpet is a tissue similar to the odontophore cartilage of the whelk Busycon, in which an atypical cartilage mucopolysaccharide was found by Lash and Whitehouse (1960a). The buccal cartilage of the limpet Patella vulgata also contains a mucopolysaccharide or mucopolysaccharides which differ from the chondroitin sulphates, although not perhaps to as great an extent as *Busycon* glucan sulphate (Doyle, 1965).

Acetone-dried powders of the cartilage were analysed and found to contain 11·3% nitrogen and 2·71% hexosamine of which 36·4% was 2-amino-2-deoxy-D-glucose and 63·6% was 2-amino-2-deoxy-D-galactose. A paper chromatographic analysis of hydrolysates revealed the presence of glucuronic acid, mannose, galactose, glucose, xylose and fucose.

Ethanol precipitation of a papain digest yielded a crude mucopolysaccharide preparation which had a nitrogen content of 5·8% and a total hexosamine content of 14·2%. Although values are not given the ratio of the two amino sugars was reported to have changed from that in the whole cartilage. The same monosaccharides were present in this preparation as were present in the cartilage. The crude digest contained 5·7% uronic acid. The author comments that this clearly is not a 1 : 1 ratio with total hexosamine; this is not however the point at issue. It would be more informative if we were to know the ratio of uronic acid to 2-amino-2-deoxy-D-galactose and the ratio of galactose to 2-amino-2-deoxy-D-glucose. This data unfortunately is not available nor is a sulphate content given although the crude mucopolysaccharide was reported to stain metachromatically and have an electrophoretic mobility similar to that of chondroitin sulphate.

COMPLEXES WITH PROTEIN

In vertebrate cartilage the chondroitin sulphates A and C, as well as keratosulphate, are associated with covalently bound protein and also with collagen, the bond to the latter being of weaker character. For many years the whole question of the association of protein with chondroitin sulphate and other connective tissue mucopolysaccharides was in a parlous state and such protein was in fact usually considered to be no more than a troublesome impurity. Recent studies by Rodén and by Karl Meyer together with his co-workers however have considerably clarified this problem.

The linkage of protein to polysaccharide in chondroitin sulphate has long been known to be alkali-labile and this fact has been frequently utilized to remove protein during the preparation of the mucopolysaccharide. Anderson *et al.* (1965) have shown that when both chondroitin sulphates A and C are prepared by proteolytic digestion of cartilage there remains a firmly bound peptide moiety rich in serine, glutamic acid, proline and glycine. Treatment of the complex with alkali ruptures the bond and destroys some or all of the serine. If the cleavage is carried out with 0·5 N sodium hydroxide in the presence of a reducing agent (Adams catalyst and hydrogen) the disappearance of serine is accompanied by the appearance of an equimolar quantity of alanine. It was suggested that the cleavage mechanism was that of a β carbonyl elimination of an alkoxide moiety (the carbohydrate) from the hydroxyl of serine, resulting in the formation of a dehydroalanyl residue which is then reduced giving alanine.

More detailed and conclusive knowledge of the linkage region was obtained by the enzymic degradation of bovine nasal septum chondroitin sulphate A-protein complex to small glycopeptides, first with hyaluronidase and then with papain, leucine aminopeptidase and carboxypeptidase, yielding a glycopeptide fraction which contained serine as the only amino acid (Gregory, Laurent and Rodén, 1964). Examination of the carbohydrate composition of the glycopeptide revealed the expected presence of glucuronic acid and 2-amino-2-deoxy-D-galactose, but unexpectedly showed that it also contained galactose and xylose. An investigation of the products of partial hydrolysis of the glycopeptide revealed the structure of the linkage region as that shown in Fig. 5. The main sequence of the polysaccharide terminates in a uronic acid residue and the chain then continues

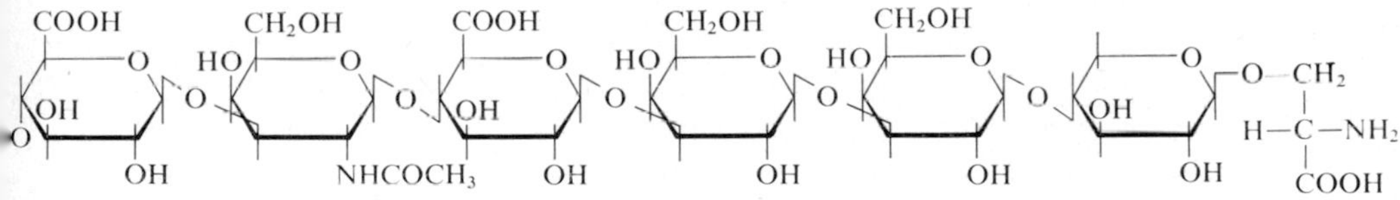

Fig. 5. The polysaccharide-protein linkage region of chondroitin sulphate A.

with a further two residues of galactose, a xylose and then a serine of the protein chain. Thus the terminal sequence is 0-β-D-2-acetamido-2-deoxy-galactopyranosyl-(1 → 4)-0-β-D-glucuronopyranosyl-(1 → 3)-0-β-D-galacto-pyranosyl-(1 → 3)-0-β-D-galactopyranosyl-(1 → 4)-0-β-D-xylopyranosyl-L-serine (Rodén and Smith, 1966).

Much less is known of the chondroitin sulphate C-protein linkage region. However, as already mentioned, the linkage is destroyed by alkali with loss of serine, and the glycopeptides, isolated from chondroitin sulphate C by Schmidt, et al. (1966), have been found to contain xylose and galactose in the molar ratio 1 : 2, as in chondroitin suphate A. These authors have in fact obtained chondroitin sulphates from a wide range of vertebrates including fish, bird and amphibian tissues. In all cases the proportion of 2-amino-2-deoxy-D-galactose to galactose, xylose and serine was 1 to 2 to 1 to 1.

Mathews (1965) has proposed a schematic model for the interaction of collagen with the chondroitin sulphate-protein complex in cartilage. In this model there is a parallel-ordered interaction of collagen fibrils with the chondroitin sulphate side chains of a chondroitin sulphate-protein macro-molecule in which a protein core carries some 60 chondroitin sulphate prosthetic groups each of molecular weight 50,000. The length of the protein core in this giant glycoprotein is about 4000 Å (Mathews, 1966). Evidence obtained by Luscombe and Phelps (1967a, b) from a bovine nasal cartilage chondroitin sulphate-protein complex, of molecular weight $3 \cdot 2$–$5 \cdot 8$ million, suggests that the structure proposed by Mathews might have to be revised by reducing the chondroitin sulphate prosthetic group molecular weights to 25,000 and increasing the number of side chains to 120. It has also been suggested that keratosulphate is linked to the same core protein as the chondroitin sulphate molecules (Rosenblum and Cifonelli, 1967) but the studies of Luscombe and Phelps (1967b) suggest that the association of keratosulphate with the chondroitin sulphate-protein complex takes place at the time of extraction while Fricke and Jeanloz (1966) have been able to isolate the keratosulphate-protein complex free from chondroitin sul-phate. The linkage region in the keratosulphate-protein complex may be different from that of chondroitin sulphate-protein complex; evidence suggests that more than one type of link may be involved and that the types of bond may also vary from source to source (Seno, et al., 1965; Mathews and Cifonelli, 1965; Castellani, 1965). These findings suggest threonine and serine as well as aspartic acid as possible linking amino acids. A glyco-sidic link between 2-acetamido-2-deoxy-D-galactose and serine or threo-nine has been proposed (Bray, et al., 1967).

Preliminary studies suggest that the chondroitin sulphate-protein links in invertebrate cartilage tissues may be similar in character to the vertebrate linkage regions.

Chondroitin sulphate preparations from cartilages of the squids L. pealii and A. dux contain almost equimolar proportions of galactose, xylose and serine; L. polyphemus gill cartilage chondroitin sulphate also has a high xylose content while the chondroitin sulphate from the endoskeletal carti-lage of the polychaete E. polymorpha contains serine and glycine as the principal residual amino acids (Mathews, 1967; Person and Mathews, 1967).

A mucopolysaccharide-protein prepared by Mathews (1967) from the cranial cartilage of the squid L. opalescens contained 10% protein. This material was molecularly polydisperse with the relatively low molecular weight of 300,000. The molecular weight of the polysaccharide chains linked to the protein was of the order 30,000 (compared to the value of 25,000 estimated for the constituent polysaccharide chains of the vertebrate chondroitin sulphate-protein complex) which indicates that the number of polysaccharide units linked to the protein core must be far fewer than in vertebrate cartilage chondroitin sulphate-protein.

The studies of invertebrate cartilage mucopolysaccharides reported in the first half of this chapter are not particularly informative with regard to the possible association of the polysaccharides with protein. In the majority of cases the preparative procedure was specifically designed to remove as much protein as possible from the preparation. Moreover, the workers concerned were not particularly interested at the time in identifying the few (if any) amino acids remaining. The presence of minor quantities of monosaccharides other than those expected was equally likely to pass unnoticed, although Lash and Whitehouse (1960b) had remarked the presence of xylose in hydrolysates of the mucopolysaccharide from squid cranial cartilage and had rejected it as an artefact arising from the decarboxylation of glucuronic acid. This may well be but in the light of recent knowledge the point now requires reappraisal.

With regard to the association of mucopolysaccharide with collagen in invertebrate cartilage the presence of hydroxyprolines has been noted in the cartilage tissues of *Loligo, Limulus* and *Eudistylia*; in the latter case the hydroxyproline content ($3\cdot2\%$) was comparable with that of 18-day embryonic chick articular cartilage ($3\cdot5\%$). Collagen was observed to be present in these same tissues on examination in the electron microscope (Person and Philpott, 1963; Person and Mathews, 1967). An association between collagen and a sulphated acid glycosaminoglycan in the connective tissues of the echinoderm *Thyone* has been suggested by Jeanloz (1967).

FUNCTION

The role of the chondroitin sulphates, keratosulphate and chondroitin in connective tissues cannot probably be tied down to any single specific function. It is instead better to account for the whole complex gamut of physical, chemical and biological properties of the tissues in terms of the complex of constituents from which they are fabricated.

The possibility seems strong that the chondroitin sulphates and keratosulphate act as organizers of the highly-ordered networks of collagen fibres in connective tissues, and the process of calcification in vertebrate cartilage would seem to be intimately involved with collagen and the chondroitin sulphate-protein complexes (Dziewiatkowski, 1966; Mathews, 1965, 1967; Mathews and Decker, 1967). The elastic properties of cartilage owe something to the chondroitin sulphate-protein molecules enmeshed in the network of collagen (Schubert, 1966). The general biological significance of the interactions between acidic macromolecules and proteins or cellular structures has been discussed by Mathews (1967) and by Balazs and Jacobson (1966).

It should not be forgotten that such a highly-charged molecule as chondroitin sulphate closely resembles an ion-exchange resin and that it is likely to show considerable affinity for both large and small charged metabolites. The connective tissue matrix resembles also the now widely-used gel-filtration fractionation systems and with this combination of ion-exchange and molecular sieving properties, coupled with a marked tendency to heavy hydration, it would be surprising if it did not have a significant role to play in the exchange and diffusion of many types of metabolite (Dunstone, 1960; Laurent, 1966).

Attention has been called to the possible role of cartilage mucopolysaccharides in the regulation of aerobic oxidative metabolism in invertebrate cartilage (Person and Philpott, 1963). The presence of active cytochrome oxidase in *Limulus* and *Busycon* cartilage tissues with $Q\,O_2$ values as high as 10 has been noted (Person *et al.*, 1959; Person and Fine, 1959a, b). It was shown however that in *Limulus* the tissues of only very young animals were active while those of older (i.e. larger) animals were completely inactive. The presence in both of these invertebrate cartilage tissues, and also in the cartilage of some higher vertebrates, of substances inhibiting components of the terminal sections of the respiratory chain was demonstrated. In a further study it was found that such mucopolysaccharides as chondroitin sulphate and glucan sulphate would interact with the cytochrome C of the respiratory chain inhibiting its catalytic activity (Person and Fine, 1961). In squid cartilage however it was found to be possible neither to detect cytochrome activity nor to detect substances capable of inhibiting such activities in other circumstances. Person and Philpott (1963) have suggested that chondroitin sulphates and related mucopolysaccharides in addition to their structural function, might act as regulators of enzymic and other metabolic processes. It may perhaps be of interest to note that Ladanyi and Leray (1968) observed that low cytochrome oxidase activity was correlated with high hexosamine content in molluscan muscle although the hexosamine here appeared to be derived from neutral polysaccharide.

BIOSYNTHESIS

The problem of mucopolysaccharide and mucopolysaccharide-protein biosynthesis in vertebrate connective tissue has received considerable attention and the biosynthesis of chondroitin sulphate has been particularly closely studied. Mucopolysaccharide biosynthesis has been reviewed by Böstrom and Rodén (1966).

The pathways leading to the formation of the immediate precursors of the chondroitin sulphates and keratosulphates are shown in Fig. 6. In the case of the monosaccharide constituents the sugars are activated, prior to

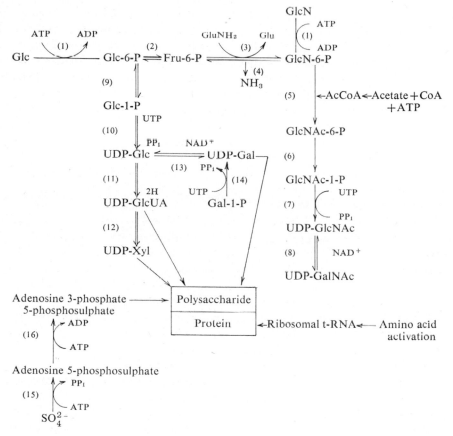

Fig. 6. Pathways leading to the formation of the immediate precursors of the chondroitin sulphates and keratosulphate.

(1) Hexokinase, EC 2.7.1.1.
(2) Glucosephosphate isomerase EC 5.3.1.9.
(3) Glutamine-fructose-6-phosphate aminotransferase EC 2.6.1.16.
(4) Glucosaminephosphate isomerase EC 5.3.1.10.
(5) Phosphoglucosamine transacetylase EC 2.3.1.4.
(6) Acetylglucosamine phosphomutase EC 2.7.5.2.
(7) UDPglucosamine pyrophosphorylase EC 2.7.7.23.
(8) UDPacetylglucosamine epimerase EC 5.1.3.7.
(9) Phosphoglucomutase EC 2.7.5.1.
(10) Glucose-l-phosphate uridyltransferase EC 2.7.7.9.
(11) UDPG dehydrogenase EC 1.1.1.22.
(12) UDPglucuronate decarboxylase EC 4.1.1.35.
(13) UDPglucose epimerase EC 5.1.3.2. Hexose-l-phosphate uridyltransferase EC 2.7.7.12.
(14) Galactosesulphosphate uridyltransferase RC 2.7.7.10.
(15) Sulphate adenyltransferase EC 2.7.7.4.
(16) Adenylsulphate kinase EC 2.7.1.25.

GlcN = glucosamine = 2-amino-2-deoxy-D-glucose
GlcNAc = N-acetylglucosamine = 2-acetamido-2-deoxy-D-glucose
GalNAc = N-acetylgalactosamine = 2-acetamido-2-deoxy-D-galactose

formation of the polysaccharide chain, as nucleoside (di)phosphate sugars. Assembly is mediated by specific transferases which form the glycosidic bonds. Transferase activity would not appear to take place in a haphazard manner with the formation of small units ultimately linked to form the whole chain, but rather appears to be initiated by an acceptor so that growth of the chain takes place in a regular manner from this point. Following commencement of chain synthesis sulphate is transferred to the growing polysaccharide, by specific sulphotransferases, from adenosine 3-phosphate 5-phospho sulphate (PAPS). The process of polysaccharide synthesis appears to be initiated by completion of synthesis of the protein moiety, this being the primary acceptor already referred to. Thus considerable interest has centred around the mode of synthesis of the linkage region. That the synthesis of the whole chondroitin sulphate-protein complex requires prior formation of the protein core was clearly demonstrated by Tesler, et al. (1965). The first reaction in the formation of the carbohydrate chain would then be addition of D-xylose to serine residues on the protein acceptor. This reaction has been demonstrated with UDP-D-xylose as the glycosyl donor (Robinson, Tesler and Dorfman, 1966). It has been suggested that UDP-D-xylose may function as a feedback regulator of the synthesis of the chondroitin sulphate-protein complex (Neufeld and Hall, 1965). UDP-D-xylose, as Fig. 6 indicates, is formed by decarboxylation of UDP-D-glucuronic acid which, in its turn, is derived from UDP-D-glucose. UDP-D-xylose strongly and specifically inhibits the enzyme catalyzing the UDP-D-glucose to UDP-D-glucuronic acid transformation. UDP-D-glucose dehydrogenase. Thus, should the rate of synthesis of the protein core decrease, the resulting build up of unused UDP-D-xylose would inhibit the formation of UDP-D-glucuronic acid and hence the consequent failure to renew the UDP-D-glucuronic acid pool would diminish and halt synthesis of the main polysaccharide chain. A further consequence of course would be a halt in the build up of the UDP-D-xylose pool. Thus the function of the xylose residue as the link to protein, may be suggested by this process.

It is regretable that while so much is known of the synthesis of chondroitin sulphate-protein in vertebrates, little or nothing is known of the same process in invertebrate cartilage. Indirect evidence, such as for example the fact that the glucuronic acid pathway in crustacea is similar to that reported in mammals (Puyear, 1967), that UDP-acetylhexosamines are also found in invertebrates (Lunt and Kent, 1961; Wheat, 1960) and that PAPS is used as the sulphate transfer agent in several invertebrate situations (see Chapters 4 and 5), suggests that the biosynthetic pathway for these mucopolysaccharides may be similar but a cautious approach to such extrapolations obviously is necessary. Establishing the details of biosynthesis is likely to be an important factor in deciding whether the invertebrate and vertebrate chondroitin sulphates and related molecules have arisen completely independently by convergent evolution or whether they originate from some common ancestral gene.

SUMMARY

It would appear from the foregoing that there is fairly conclusive evidence for the presence of acid mucopolysaccharides in the connective tissues of invertebrates, which closely resemble in structure chondroitin sulphates A and C of vertebrate cartilage origin. The evidence accrues from chemical composition, chemical properties, physical properties and susceptibility to enzymes. Similarly the evidence for the occurrence of chondroitin in squid skin seems quite conclusive.

The evidence for the presence of keratosulphate is much weaker, resting mainly upon the presence in the tissue or in unfractionated extracts of galactose (Lash and Whitehouse, 1960) and 2-amino-2-deoxy-D-glucose (Anno, Seno and Kawaguchi, 1962). An age-dependent change in the ratio of galactose to glucuronic acid, while having been shown in vertebrates to be due to an increase of cartilage keratosulphate at the expense of chondroitin sulphate as age advances, cannot be assumed to apply to invertebrates without careful fractionation experiments.

The isolation and identification of a glucan sulphate in *Busycon* odontophore, replacing chondroitin sulphate in what would appear in many respects to be morphologically, histologically and cytologically a typical cartilage tissue (setting aside the presence of myoglobin), raises the question of what can be regarded as normal cartilage tissue. The glucan sulphate and the chondroitin sulphates are after all both linear sulphated polysaccharides and therefore likely to be physicochemically similar. Lash and Whitehouse (1960a) considered however that the similarities mentioned above are of lesser importance than the biochemical differences between glucan sulphate and chondroitin sulphate and suggested that the odontophore could not be considered as normal cartilage tissue. Such an approach might seem to be rather extreme since it is likely that in connective tissue much of the organization and function may depend upon the overall physicochemical properties of the system's constituents rather than upon the type of finer detail which would be required to provide an enzyme with its specificity. It has already been mentioned that the glucan sulphate and chondroitin sulphate are extended polysaccharide chains. Both are polyanionic by virtue, of ester sulphate groups alone in the one case and of ester sulphate alternating with carboxyl in the other; the overall effect probably is rather similar and probably is sufficient to fulfil the requirement for a linear distribution of charge. This type of lack of very strict specificity in mucopolysaccharide structure from species to species, even though the locations and functions may be identical, is common in the invertebrates and also in the marine algae.

REFERENCES

Anderson, B., Hoffman, P. and Meyer, K. (1965). *J. biol. Chem.* **240**, 156.
Anno, K. and Seno, K. (1962). *In* "Biochemistry and Medicine of Mucopoly-saccharides." (F. Egani and Y. Oshima, eds, p. 26. Maruzen, Tokyo.
Anno, K., Seno, N. and Kawaguchi, M. (1962). *Biochem. biophys. Acta.* **58**, 87.
Anno, K., Seno, N. and Kawai, Y. (1964). *Sixth Int. Congr. Biochem. New York.* Abs. VI–6, 502.
Anno, K., Kawai, Y. and Seno, N. (1964). *Biochem. biophys. Acta.* **83**, 348.
Anseth, A. (1961). *Expl. Eye Res.* **1**, 106.
Anseth, A. and Laurent, T. C. (1961). *Expl. Eye Res.* **1**, 25.
Balazs, E. A. and Jacobson, B. (1966). *In* "The Amino Sugars" (E. A. Balazs and R. W. Jeanloz eds). Vol. 2B, p. 361. Academic Press, New York and London.
Böstrom, H. and Roden, L. (1966). *In* "The Amino Sugars" (E. A. Balazs and R. W. Jeanloz, eds). Vol. 2B, p. 45. Academic Press, New York and London.
Bradley, H. C. (1912). *Proc. Am. Soc. biol. Chem.* 40.
Bradley, H. C. (1913). *J. biol. Chem.* **14**, xl.
Bray, B. A., Lieberman, R. and Meyer, K. (1967). *J. biol. Chem.* **242**, 3373.
Brimacombe, J. S. and Webber, J. M. (1964). "Mucopolysaccharides" p. 64. Elsevier, Amserdam.
Castellani, A. A. (1965). *In* "International Symposium on the Chemical Physio-logy of Mucopolysaccharides" (G. Quintarelli and J. E. Scott, eds). Little, Brown, Boston.
Davidson, E. A. and Meyer, K. (1954). *J. biol. Chem.* **211**, 605.
Doyle, J. (1965). *Biochem. J.* **96**, 73P.
Dunstone, J. R. (1960). *Biochem. J.* **77**, 164.
Dziewiatkowski, D. D. (1966). *In* "Structural Organization of the Skeleton," Vol. 2, p. 31. Birth Defects Original Article Series. The National Foundation, New York.
Fricke, R. and Jeanloz, R. W. (1966). *In* "Biochimie et Physiologie du tissu con-jonctif" (P. Compte, ed.), p. 77. Société Ormeco et Imprimerie du Sud-Est a Lyon.
Gregory, J. D., Laurent, T. C. and Rodén, L. (1964). *J. biol. Chem.* **239**, 3312.
Jeanloz, R. W. (1967). Presented at the *Symposium on glycosaminoglycans, glyco-proteins and glycolipids.* Tokyo, Japan.
Kaplan, D. and Meyer, K. (1959). *Nature, Lond.* **183**, 1267.
Kawai, Y., Seno, N. and Anno, K. (1966). *J. Biochem.* **60**, 317.
Ladanyi, P. and Leray, C. (1968). *Marine Biol.* **1**, 210.
Lash, J. W. (1959). *Science, N.Y.* **130**, 334.
Lash, J. W. and Whitehouse, M. W. (1960a). *Biochem. J.* **74**, 351.
Lash, J. W. and Whitehouse, M. W. (1960b). *Archs. Biochem. Biophys.* **90**, 159.
Laurent, T. C. (1966). *Fedn Proc. Fedn Am. Socs exp. Biol.* **25**, 1128.
Levene, P. A. (1941). *J. biol. Chem.* **140**, 267.
Lunt, M. R. and Kent, P. W. (1961). *Biochem. J.* **78**, 128.

Luscombe, M. and Phelps, C. F. (1967a). *Biochem. J.* **102**, 110.
Luscombe, M. and Phelps, C. F. (1967b). *Biochem. J.* **103**, 103.
Mathews, A. P. (1923). *Hoppe-Seyler's Z. physiol. Chem.* **130**, 169.
Mathews, M. B. (1965). *Biochem, J.* **96**, 710.
Mathews, M. B. (1966). *Clin. Orthop.* **48**, 267.
Mathews, M. B. (1967). *Biol. Rev.* **42**, 499.
Mathews, M. B. and Cifonelli, J. A. (1965). *J. biol. Chem.* **240**, 4140.
Mathews, M. B. and Decker, L. (1967). *Seventh Int. Congr. Biochem.* Abs. p. 711.
Mathews, M. B., Duh, J. and Person, P. (1962). *Nature, Lond.* **193**, 378.
Neufeld, E. F. and Hall, C. W. (1965). *Biochem. biophys. Res. Commun.* **19**, 456.
Person, P. (1964). *Biol. Bull. mar. biol. Lab., Woods Hole.* **127**, 357.
Person, P. and Fine, A. (1959a). *Nature, Lond.* **183**, 610.
Person, P. and Fine, A. (1959b). *Archs. Biochem. Biophys.* **84**, 123.
Person, P. and Fine, A. (1961). *Archs. Biochem. Biophys.* **94**, 392.
Person, P. and Mathews, M. B. (1967). *Biol. Bull. mar. biol. Lab., Woods Hole.* **132**, 244.
Person, P. and Philpott, D. E. (1963). *Ann. N.Y. Acad. Sci.* **109**, 113.
Person, P., Lash, J. W. and Fine, A. (1959). *Biol. Bull. mar. biol. Lab., Woods Hole.* **117**, 504.
Puyear, R. L. (1967). *Comp. Biochem. Physiol.* **20**, 499.
Rees, D. A. (1963). *Biochem. J.* **88**, 343.
Robinson, H. C., Tesler, A. and Dorfman, W. A. (1966). *Proc. natn. Acad. Sci. U.S.A.* **56**, 1859.
Rodén, L. and Smith, R. (1966). *J. biol. Chem.* **241**, 5949.
Rosenblum, E. L. and Cifonelli, J. A. (1967). *Fedn Proc. Fedn Am. Socs exp. Biol.* **26**, 282.
Schaffer, J. (1930). *In* "Handbuch der mikroskopische Anatomie des Menschen" (W. von Möllendorf, ed.). Vol. 2, part 1, p. 1. Springer, Berlin.
Schmidt, M., Dmochowski, A. and Wierzbowska, B. (1966). *Biochem. biophys. Acta.* **117**, 258.
Schubert, M. (1966). *Fedn Proc. Fedn Am. Socs exp. Biol.* **25**, 1047.
Schultze, C. A. S. (1818). *Meckel's Deutsch. Arch.* **4**, 334.
Seno, N., Meyer, K., Anderson, B. and Hoffman, P. (1965). *J. biol. Chem.* **240**, 1005.
Suzuki, S. (1960). *J. biol. Chem.* **235**, 3580.
Tesler, A., Robinson, H. C. and Dorfman, A. (1965). *Proc. natn. Acad. Sci. U.S.A.* **54**, 912.
Wheat, R. W. (1960). *Science, N.Y.* **132**, 1310.

Hyaluronic Acid

INTRODUCTION

Hyaluronic acid is an unsulphated acid mucopolysaccharide composed of D-glucuronic acid and 2-acetamido-2-deoxy D glucose in equimolar proportions. Its basic structure is that of a repeating sequence of the disaccharide unit 2-acetamido-2-deoxy-3-0-(β-D-glucopyranosyl uronic acid)-D-glucose (Fig. 1). The anomeric carbon atom of the 2-acetamido-2-

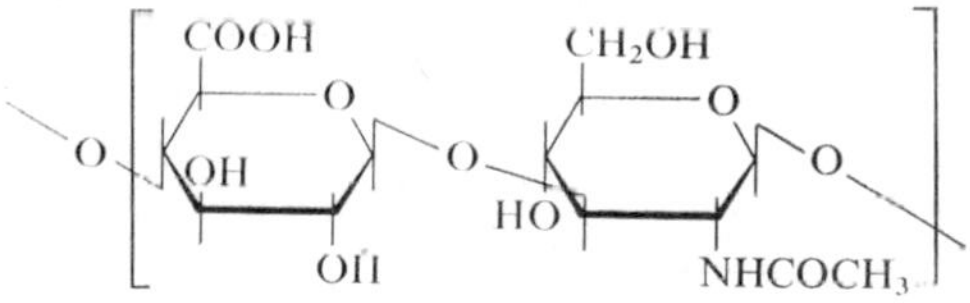

Fig. 1. Disaccharide repeat unit of hyaluronic acid.

deoxy-D-glucose residue is linked β-glycosidically to position 4 on the glucuronic acid residue of the next disaccharide, joining the disaccharides to form the polysaccharide chain. Our knowledge of the chemistry and biology of hyaluronic acid stems largely from studies of material isolated from vertebrate sources where it occurs widely distributed in the tissues. It is found particularly in connective tissue, in synovial fluid and in the vitreous of the eye. As a constituent of the ground substance hyaluronic acid has been ascribed the role of binding water in the interstitial space and of holding the cells together in a jelly-like matrix. It has been suggested that hyaluronic acid, in conjunction with protein and water, resists compression and functions to raise the viscosity of the fluids bathing the joints, thus aiding lubrication and shock-absorption. It has also been suggested that the networks of hyaluronic acid molecules, formed in biological systems, may act as molecular sieves. Since hyaluronic acid occurs in bacteria and certain invertebrates as well as in vertebrates it seems likely that the roles of lubrication and shock-absorption may have arisen in a secondary manner.

INVERTEBRATE HYALURONIC ACID

Hyaluronic acid has been detected, isolated and characterized in certain species of insect. Its presence seems probable in the vitreous of the squid eye and possibly in the palial fluids of certain other molluscs. Hyaluronic acid has also been identified, on the rather precarious basis of histological tests, in the tissues of parasitic worms and annelid worms.

Histochemical Observations: Both the histology and chemistry of hyaluronic acid in invertebrates have been studied principally using insect material. The occurrence of mucoid substances, of acid mucopolysaccharide type, in insects was noted by Day (1949) using histochemical methods. The presence of a polysaccharide with the staining properties of hyaluronic acid was demonstrated in the salivary gland secretions of the larvae of *Chironomus plumosus* by Defretin (1951a). This secretion gave positive reactions to the Hale stain for hyaluronic acid and apparently was broken down by a hyaluronidase preparation. It did not stain strongly metachromatic with toluidine blue; in spite of the fact that the hyaluronidase preparation was not completely specific but also degraded chondroitin sulphate, this was held to suggest the absence of ester sulphate and hence to strengthen the case for identification of hyaluronic acid. The polysaccharide was associated with material staining positively for protein.

The larval salivary cells of *Smittia* and *Bradysia mycorum* have been histochemically demonstrated to synthesize a mixture of neutral and acid mucopolysaccharides which probably includes hyaluronic acid (Kato and Sirlin, 1963).

The dermal glands in the epidermis of *Rhodnius prolixus*, fourth instar larvae, produce a mucilaginous secretion with some of the properties of hyaluronic acid (Baldwin and Salthouse, 1959). The gland cells become distended with alcian blue-positive mucin immediately prior to moulting and this material is discharged, forming an inner lubricatory layer 2 μ thick on the exuvia, during ecdysis. The glands have appeared in the larvae by the eleventh day after feeding; at this time a clear hyaline, alcian blue-positive secretion can be seen in the gland. The secretion is positive also to Myer's mucicarmine and γ-metachromatic with toluidine blue. The contents of the glands lose their alcian blue staining properties when digested with bovine testicular hyaluronidase or with bovine liver, β-glucuronidase, suggesting that the mucin contains hyaluronic acid. Unfortunately these tests cannot be regarded as being unequivocal since mammalian hyaluronidase also attacks chondroitin sulphates A and C (Stacey and Barker, 1962).

The peritrophic membranes of several cocoon-forming insect species have been shown to contain a substance giving positive staining reactions for hyaluronic acid (Kawakita, 1960a, 1961a). *Theophila mandarina* and

Hemerophila artlinera had a chondroitin sulphate-like substance in their membranes but those of *Bombyx mori*, *Attacus ricini* and *Samia cynthia* were positive only for hyaluronic acid.

In a histochemical investigation of the neurones and connective tissues of the meso- and meta-thoracic ganglia in the cockroach *Periplaneta americana* an area surrounding the central mass of nerve fibres was found to be γ-metachromatic to toluidine blue (Ashhurst, 1961). This region separated the neurones and the connective tissue sheath from the main mass of nerve fibres. No comparable region was found in the interganglionic connectives or peripheral nerves and the structure was peculiar to the cockroach. The metachromatic substance was apparently structureless and extra-cellular. Reactions for proteins were negative and the material was not PAS positive. The metachromasia was removed by treatment with hyaluronidase and although this was again a non-specific preparation, chondroitin sulphates A and C, in contrast to hyaluronic acid, might have been expected to give a positive PAS reaction (Stacey and Barker, 1962).

Toluidine blue γ-metachromatic polysaccharides, susceptible to ovine testicular hyaluronidase, have been noted in the neural lamella and dorsal connective tissue masses in the central nervous system of 96-hour onward pupae of the wax moth *Galleria mellonella* L. (Ashhurst and Richards, 1964a) and in the spherules of the spherule blood cells from the fully-grown last instar larvae of the same insect (Ashhurst and Richards, 1964b).

The hepatopancreas of the isopod (crustacea) *Armadillidium vulgare* contains, in addition to other acid mucopolysaccharides, a substance located in the border of the storage and secretory cells of the posterior hepatopancreas and the border of the storage cells of the anterior hepatopancreas which exhibits γ-toluidine blue metachromasia sensitive to hyaluronidase treatment. The metachromasia was strong at pH 2·8 in the outer and inner layers of the cell borders and weak in the middle layers. In the secretory cells of the more cephalic portions of the hepatopancreas the γ-metachromasia of the outer and inner layers gradually diminishes giving way to β-metachromasia in the outer layer only (Steeves, 1965). It seems likely that the more weakly staining material was a polysaccharide resembling hyaluronic acid.

The presence of hyaluronic acid, identified on the basis of staining by Hale's technique, by loss of staining on treatment with bovine testicular hyaluronidase, and by toluidine blue staining, in the glands of a number of annelid polychaete tubiculous worms, has been noted by Defretin (1951b). *Nereis irrorata*, *Hyalinoecia tubicola*, *Leiochene dypeata*, *Petaloproctus terricola*, *Pectinaria* (*Lagis*) *koreni*, *Lanice conchilega*, *Spirographis spallanzanii*, and *Myxicola infundibulum* all possessed a mucopolysaccharide with the staining reactions of hyaluronic acid.

A hyaluronic acid-like polysaccharide is reported to be present throughout the hydatid scolex of cestodes (Chowdhuri, Ray and Bhadhuri, 1956).

A bovine testicular hyaluronidase-sensitive, toluidine blue γ-metachromatic mucopolysaccharide has been reported to be associated with the cellulose fibres in the tunic of the ascidian *Pyura stolonifera* (Endean, 1955). This author however believes the mucopolysaccharide to be sulphated and related to the chondroitin sulphates rather than to hyaluronic acid.

Chemical Studies: Acid hydrolysis of the tubes of the annelid polychaete worms *Onuphis* (*Hyalinoecia*) *tubicola* and *Spirographis spallanzanii* yields glucuronic acid, 2-amino-2-deoxy-D-glucose,2-acetamido-2-deoxy-D-glucose and most of the common amino acids (Defretin, *et al.*, 1949) suggesting the possible presence of hyaluronic acid and lending weight to the histochemical evidence for its presence in certain mucus-secreting glands of these worms (Defretin, 1951b). It is curious however that Defretin and his co-workers did not detect glucose in their hydrolysates of *Onuphis* tubes in view of the presence of a glucose-phosphate polymer (Chapter 7).

The silkworm possesses, in its alimentary canal, a soft thin membrane called the peritrophic membrane. The structure, which is attached to the midgut epidermis, wraps the food taken up by the insect larva and is constantly renewed, the old one being discharged along with the faeces. Physiologically the membrane may protect the epidermis and act to prevent infection by flacherie. On hydrolysis with strong hydrochloric acid the peritrophic membrane of the wild silkworm *Antheraea pernyi* releases a considerable amount of carbon dioxide together with 2-amino-2-deoxy-D-glucose and amino acids (asp., glu., ser., ala., tyr., phe., leu.) (Yamazaki, 1955). It was thus suggested that the membrane might be composed of a uronic acid-containing polysaccharide in addition to chitin and protein. The uronic acid moeity, assuming it to be a hexuronic acid, would comprise some 20% of the dry weight of the membrane (Nishizawa *et al.*, 1956).

The membrane of *B. mori* yields 2-acetamido-2-deoxy-D-glucose and glucuronic acid on acid hydrolysis and it was therefore suggested that the mucopolysaccharide might be hyaluronic acid (Nishizawa *et al.*, 1960; Kawakita, 1960b).

Extraction of silkworm peritrophic membrane with 0·1 M sodium chloride solution saturated with chloroform, precipitation of the extract with ammonium sulphate in the presence of pyridine, precipitation with ethanol, and treatment with a protease gave a 1% yield of acid mucopolysaccharide (Kawakita, 1961b). Acid hydrolysis of this material yielded glucuronic acid and 2-acetamido-2-deoxy-D-glucose (Kawakita, 1961c).

A more detailed study of *B. mori* peritrophic membrane mucopolysaccharide has been carried out by Nishizawa establishing almost unequivocally its identity with hyaluronic acid (Nishizawa *et al.*, 1963). The membranes contained 15% glucuronic acid, estimated by decarboxylation in concentrated acid and determination of the released carbon dioxide. Two extraction procedures were used to obtain two different mucopolysaccharide preparations. Membranes were extracted, first with water for two hours at 0°, with 0.1% sodium hydroxide for a further two hours at 0°, and the latter extract then neutralized and dialysed to give a 13% yield of acid mucopolysaccharide (Extract I). Alternatively the membranes were extracted four times with 0·1% sodium hydroxide and the combined extracts (which contained protein and uronic acid in the ratio 12 : 1) treated with *Streptomyces griseus* protease at pH 6 for 24 hours to give Extract II. After a period of dialysis, Sevag treatment, concentration, and ethanol precipitation, the ratio of protein to uronic acid became 1·7 : 1. Both extracts (I and II) contained 2-amino-2-deoxy-D-glucose and glucuronic acid, identified by paper chromatography in several solvent systems. Only 0·17% ester sulphate could be detected in the whole membranes among a total of 3·4% S. Hydrolysis of Extract I with bovine testicular hyaluronidase yielded three components, F_1, F_2, F_3, separable by paper chromatography. These components were identical with those obtained by degradation of an authentic sample of hyaluronic acid, under the same conditions. F_1, F_2, and F_3 (in order of decreasing Rf) obtained from peritrophic membrane and authentic hyaluronic acid by preparative paper chromatography were examined for reducing power before and after acid hydrolysis. The results were converted to a ratio of reducing power before to reducing power after. Thus F_1 from authentic hyaluronic acid gave a ratio of 1·8 while that from peritrophic membrane extract gave a ratio of 1·5. Similarly the F_2 fractions had ratios of 3·6 and 3·4 respectively while the F_3 fractions had ratios of 5·6 and 7·0 respectively. On this basis it was assumed that the F_1, F_2 and F_3 fractions from Extract I and the authentic hyaluronic acid represented di-, tetra- and hexa-saccharides respectively. F_1 and F_2 fractions were also prepared from Extract II. The molar ratio of glucuronic acid to 2-amino-2-deoxy-D-glucose in F_1 and F_2 was found to be 0·91 and 0·95 respectively, for the whole extract 1·14 and for authentic cockscomb hyaluronic acid 0·99. This suggested that F_1 and F_2 were respectively the same di- and tetra-saccharides as those from authentic hyaluronic acid.

The infrared spectrum (Fig. 2) of the whole membrane, of membrane chitin of Extract II and cockscomb hyaluronic acid were compared. Extract II most closely resembled authentic hyaluronic acid. Since the polysaccharide was prepared by an alkaline extraction procedure it seems likely that it would be in the form of the sodium salt. In view of this it is quite to be expected that the spectrum of Extract II should lack an absorption at 1725 cm^{-1} (or rather 1735 cm^{-1}) in the position of unionized carboxyl rather than it being masked by bound protein as the authors suggested.

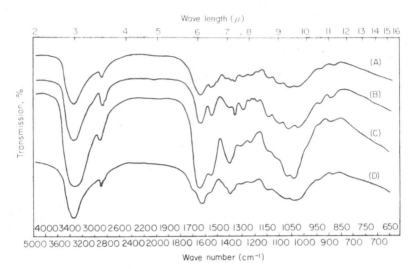

Fig. 2. Infrared spectra of silkworm whole peritrophic membrane (A), crude chitin from the peritrophic membrane (B), hyaluronic acid-like mucopolysaccharide from the membrane (C) and cockscomb hyaluronic acid (D) (from Nishizawa *et al.*, 1963).

A mucopolysaccharide occurring in the midgut (comprising basement membrane, interstitial areas and peritrophic membrane) of the larva of the Greater Wax Moth, *Galleria mellonella* L. has been shown to be similar to, if not identical with, hyaluronic acid (Estes and Faust, 1964). Midgut tissues were homogenized with methanol and methanol-chloroform mixtures to remove lipids followed by treatment with papain and trypsin to solubilize the tissues and remove protein. Following TCA precipitation the soluble material was dialysed and freeze-dried before chromatographic separation on Ecteola cellulose, using the procedure of Ringertz and Reichard (1960). The major portion of the carbozole positive material in the extract (91·3% of total) was eluted at 0·025 M Cl^- as a single peak and in the elution position of authentic hyaluronic acid. There was no evidence to suggest the presence of the more acid mucopolysaccharides, such as heparin or the chondroitin sulphates, which would have been eluted at higher molarities of chloride. The mucopolysaccharide sedimented in the ultracentrifuge as a single component, although the Schlieren patterns indicated a probable polydispersity of degree of polymerization.

The polysaccharide isolated by column chromatography was brown, probably indicating that the coloured material was not in fact extraneous impurity but an integral part of the molecule. Whether this indicates a "browning" reaction or some affect of the reducing power of papain (known to degrade hyaluronic acid) is uncertain. EDTA was also used during the papain degradation and this has been shown to adversely affect hyaluronic acid (Ogston and Sherman, 1959).

Glucuronic acid was identified as the uronic acid component of the mucopolysaccharide on the basis of its colour yields at 510 and 480 mu in the thioglycolic acid-sulphuric acid reaction of Dische (1962), whilst analysis for amino sugar, carried out according to Gardell (1953), indicated the presence of 2-amino-2-deoxy-D-glucose as the only hexosamine constituent.

The laevorotatory properties of the mucopolysaccharide were in general agreement with those reported for hyaluronic acid; exact values were not given. No ester sulphate was present and the molar equivalence ratios among N-acetyl groups, glucuronic acid and hexosamine residues were 1·0 : 0·98 : 0·92, strongly suggesting the close identity of this material with hyaluronic acid.

The meso- and meta-thoracic ganglia of the cockroach P. americana, previously reported on histochemical grounds to contain a hyaluronic acid-like mucopolysaccharide (Ashhurst, 1961; Pipa, 1961), yielded such a muco-polysaccharide when extracted by a modification of the method of Slack (1953) (Ashhurst and Patel, 1963). Trypsinized, homogenized ganglia were treated with TCA to precipitate protein. The supernatant was precipitated, after dialysis, with cetylpyridinium chloride together with borate. The mucopolysaccharide-CPC-borate complex was dissociated with a methanol-saturated sodium chloride-water mixture (1 : 1 : 1) at 37° and the CPC removed by shaking with fuller's earth. The mucopolysaccharide was precipitated with three volumes of ethanol. Electrophoretically the mucopolysaccharide behaved in a similar manner to hyaluronic acid and quite unlike chondroitin sulphate or heparin while chromatograms of acid hydrolysates indicated the presence of 2-amino-2-deoxy-D-glucose.

Whether the ten mucopolysaccharide components noted by Kato, Perkowska and Sirlin (1963) as being present in saponin-extracted fractions of the salivary secretions of Chironomus thummi, and identified on electro-phoretograms by toluidine blue staining, are related to the hyaluronic acid identified histochemically in another species of Chironomus (plumosus) larva (Defretin, 1951a) is uncertain.

The relation between the distribution of protein, acid mucopolysaccharide and PAS positive material in the extra-pallial fluid and calcium carbonate crystals of the shells from a number of molluscs has been studied by electrophoretic techniques (Kobayashi, 1964). The palial fluid of species with an aragonite shell contained much more complex systems of protein, acid mucopolysaccharide and PAS positive material than that of a species with a calcite shell. In all the species studied, however, hyaluronic acid-like substances were commonly found in the extrapallial fluid.

Balazs (1965) in an account of the glycosaminoglycans and glycoproteins in the vitreous humour of various animal species regarded the state of the vitreous of the squid as being liquid and containing 230 μg/ml of glycosaminoglycans estimated as hexuronic acid. Hyaluronic acid is, of course, a principal constituent of the vertebrate vitreous.

Association with Protein: Even in the case of vertebrate hyaluronic acid, the manner in which it associates with protein is still obscure. It seems likely, however, that the mode of combination between the acid polysaccharide and any protein associated with it as a complex, is largely mediated through ionic linkages rather than covalent bonds, although it must be emphasized that the situation is still uncertain. In view of this, it is not surprising that little or nothing is known with regard to invertebrate hyaluronic acid-protein complexes. The identification of protein, histochemically, associated with hyaluronic acid cannot be accepted as conclusive evidence for the presence of a complex. Neither can its absence be deduced from lack of protein staining in tissues containing hyaluronic acid (Ashhurst, 1961) since it is distinctly possible that linkage of acid mucopolysaccharide with the protein would result in blockage of dye-reactive groups. Most of the procedures cited earlier for the isolation of invertebrate hyaluronic acid either were designed specifically to remove and degrade protein or would tend to do so anyway. The identification of several amino acids in hydrolysates of the hyaluronic acid-containing wild silkworm peritrophic membrane by Yamazaki (1955) indicates the presence of protein but goes no way towards proving or disproving its being part of a true complex with the mucopolysaccharide. A similar argument applies to the amino acids associated with mucopolysaccharide in polychaete worm tubes (Defretin *et al.*, 1949). However in the extraction of *Bombyx mori* peritrophic membrane (Nishizawa *et al.*, 1963) treatment with alkali and a powerful proteolytic enzyme still left a considerable amino acid residue. This suggests the possibility of some form of strong associations. Evidence for the association of hyaluronic acid-like substances and proteins in molluscan pallial fluids (Kobayashi, 1964) and in insect salivary secretions (Kato *et al.*, 1963) accrues from electrophoretic data.

Function in Invertebrates: As stated earlier, the primary function of hyaluronic acid is probably to act as the cell cementing ground substance, other functions being likely to have arisen secondarily. Baldwin and Salthouse (1959) believed that the hyaluronic acid-containing mucin secreted by the dermal glands in *Rhodnius prolixus* functioned to lubricate the areas of contact between the inner surface of the old cuticle and the outer waxy layer of the new epicuticle during ecdysis.

Although it has been postulated (Kantor and Schubert, 1957) that acid mucopolysaccharides may act as selective ion barriers, it was thought that this was an unlikely function for the hyaluronic acid present in cockroach ganglia (Ashhurst, 1961; Ashhurst and Patel, 1963) since the neural lamella and sheath cells are known to control the movement of ions into and out of the nervous system, and also since the neurones are not inside this layer containing the hyaluronic acids. Nishizawa *et al.* (1963) suggested that the insect peritrophic membrane hyaluronic acid functioned to give the tissue its soft, flexible, silky character. According to Defretin (1951b) (Defretin *et al.*, 1949), the presence of hyaluronic acid in the secretory glands of certain Annelid polychaete worms and its absence from glands of those not forming tubes, strongly suggests a role in the hardening of the mucoprotein constituents of the tubes. A similar function would account for the presence of hyaluronic acid in the salivary secretions of *Chironomus* larvae (Defretin, 1951a). Hyaluronic acid-like substances present in molluscan pallial fluids and cestode hydatid scolices may be concerned with calcification (Kobayashi, 1964; Chowdhuri *et al.*, 1956).

BIOSYNTHESIS

No experimental evidence exists to the author's knowledge concerning the biosynthesis of hyaluronic acid in invertebrates.

INVERTEBRATE HYALURONIDASES

Enzymes capable of degrading hyaluronic acid have been demonstrated in the venoms of Hymenoptera, particularly in that of the wasp *Vespa* but also in the honey bee and other related insects (Jacques, 1956). Rapid evaporation of bee venom yields a crystalline product containing, amongst enzymic and biological activities, a highly-active hyaluronidase of the β-N-acetyl-glucosaminidase. This enzyme, unlike testicular hyaluronidase, hydrolyses hyaluronic acid to give only a tetra- and hexa-saccharide. It possesses no transglycosidase activity. The hyaluronidase may be purified by passage through columns of Sephadex G200 when it elutes in the first fraction showing absorption at 279 mu (Barker *et al.*, 1963a,b, 1964; Palmer, 1961). It is thought that the enzyme aids the spread of toxin in the tissues of the victim by bringing about breakdown of the ground substance.

The presence of a hyaluronidase has been demonstrated in the salivary glands of *P. americana*, a non-venomous insect (Stevens, 1956). It would appear that the hyaluronidase is taken with the food into the alimentary canal where it possibly increases the permeability of the gut wall, by degrading the peritrophic membrane. Hyaluronidase activity was also detected in the beetle *Passalus cornutus*, but tests of *Aedes aegypti*, *Musca domestica*, and *Apis mellifera* were inconclusive.

Extracts of leech (*Hirudo medicinalis*) degrade vertebrate hyaluronic acid to a saturated tetra-saccharide (A-U-A-U; where A is 2-acetamido-2-deoxy-D-glucose and U is D-glucuronic acid) with hexosamine as the terminal non-reducing unit (Linker *et al.*, 1957; Linker *et al.*, 1960; Weissmann, 1955; Favilli, 1940; Hahn, 1945). Leech hyaluronidase is the most specific enzyme known for the degradation of hyaluronic acid and is the first recorded example of a hyaluronic acid endoglucuronidase. Chondroitin and chondroitin sulphates A and C are not attacked. The enzyme would seem to be specific for both the glucuronic acid and the 2-amino-2-deoxy-D-glucose residues with the emphasis lying on the amino sugar. Presumably the enzyme functions to assist the breakdown and penetration of the victim's cuticle.

An enzyme having hyaluronidase activity is found in oesophageal glands of the nematodes *Strongylus equinus* and *S. edentatus* (Deiane, 1957).

REFERENCES

Ashhurst, D. E. (1961). *Nature, Lond.* **191**, 1224.
Ashhurst, D. E. and Patel, N. G. (1963). *Ann. ent. Soc. Am.* **56**, 182.
Ashhurst, D. E. and Richards, A. G. (1964a). *J. Morph.* **114**, 237.
Ashhurst, D. E. and Richards, A. G. (1964b). *J. Morph.* **114**, 247.
Balazs, E. A. (1965). *In* "The Amino Sugars" (A. Balazs and R. W. Jeanloz, eds) Vol. 2B, p. 401. Academic Press, New York and London.
Baldwin, W. F. and Salthouse, T. N. (1959). *J. Insect Physiol.* **3**, 345.
Barker, S. A., Bayyuk, S. H. I., Brimacombe, J. S. and Palmer, D. J. (1963a). *Nature, Lond.*, **199**, 693.
Barker, S. A., Bayyuk, S. H. I., Brimacombe, J. S., Hawkins, C. F. and Stacey, M. (1963b). *Clin. Chem. Acta.* **8**, 902.
Barker, S. A., Bayyuk, S. H. I·, Brimacombe, J. S., Hawkins, C. F. and Stacey, M. (1964). *Clin. Chem. Acta.* **9**, 339.
Chowdhuri, A. B., Ray, H. N. and Bhadhuri, V. N. (1956). *Bull. Calcutta Sch. trop. Med.* **4**, 160.
Day, M. F. (1949). *Aust. J. scient. Res.* **B2**, 421.
Defretin, R. (1951a). *C.r. hebd. Séanc. Acad. Sci., Paris.* **233**, 103.
Defretin, R. (1951b). *C.r. hebd. Séanc. Acad. Sci., Paris.* **232**, 888.
Defretin, R., Biserte, G. and Montreuil, J. (1949). *C.r. Séanc. Soc. Biol.* **143**, 1208.
Deiane, S. (1957). *Boll. Soc. ital. Biol. sper.* **33**, 104.
Dische, Z. (1962). *In* "Methods in Carbohydrate Chemistry" (R. L. Whistler and M. L. Wolfrom, eds) Vol. 1, p. 8. Academic Press, New York and London.
Endean, R. (1955). *Aust. J. mar. Fresh. wat. res.* **6**, 139.
Estes, Z. E. and Faust, R. M. (1964). *Comp. Biochem. Physiol.* **13**, 443.
Favilli, G. (1940). *Nature, Lond.* **145**, 866.
Gardell, S. (1953). *Acta chem. scand.* **7**, 207.
Hahn, L. (1945). *Ark. kemi. Mineral O. Geol.* **19A**, No. 33.
Jacques, R. (1956). *In* "Venoms" (E. E. Buckley, ed.), p. 291. Am. Assoc. Adv. Sci., Washington, D.C.
Kantor, T. G. and Schubert, M. (1957). *J. Histochem. Cytochem.* **5**, 28.
Kato, K. I. and Sirlin, J. L. (1963). *J. Histochem. Cytochem.* **11**, 163.

Kato, K. I., Perkowska, E. and Sirlin, J. L. (1963). *J. Histochem. Cytochem.* **11**, 485.

Kawakita, T. (1960a). *Medicine Biol.* **55**, 150.

Kawakita, T. (1960b). *Igaku To Seibutsugaku*, **55**, 20. *Chem. Abst.* (1964). **60**, 11100h.

Kawakita, T. (1961a). *Kagaku (Tokyo)*, **31**, 262. *Chem. Abst.* (1961). **55**, 27671g.

Kawakita, T. (1961b). *Nippon Sanshigaku Zasshi*, **30**, 191. *Chem. Abstr.* (1963). **59**, 11941c.

Kawakita, T. (1961c). *Nippon Sanshigaku Zasshi*, **30**, 325. *Chem. Abstr.* (1963). **59**, 11941c.

Kobayashi, S. (1964). *Nippon Suisan Gahkaishi*, **30**, 893. *Chem. Abstr.* (1966). **64**, 2467g.

Linker, A., Hoffman, P. and Meyer, K. (1957). *Nature, Lond.* **180**, 810.

Linker, A., Meyer, K. and Hoffman, P. (1960). *J. biol. Chem.* **235**, 924.

Nishizawa, K., Shikata, S. and Yamazaki, Y. (1956). *J. Japan biochem. Soc.* **28**, 274.

Nishizawa, K., Handa, N. and Yamazaki, Y. (1960). *J. Japan biochem. Soc.* **32**, 652.

Nishizawa, K., Yamaguchi, T., Handa, N., Maeda, M. and Yamazaki, H. (1963). *J. Biochem.* **54**, 419.

Ogston, A. G. and Sherman, T. F. (1959). *Biochem. J.* **72**, 301.

Palmer, D. J. (1961). *Bee Wld.*, 225.

Pipa, R. L. (1961). *J. comp. Neurol.* **116**, 15.

Ringertz, N. R. and Reichard, P. (1960). *Acta chem. scand.* **14**, 303.

Slack, H. G. B. (1953). *Biochem. J.* **69**, 125.

Stacey, M. and Barker, S. A. (1962). "Carbohydrates of Living Tissues". Van Nostrand, London.

Steeves, H. R. (1965). *Ann. N.Y. Acad. Sci.* **118**, 1026.

Stevens, T. M. (1956). *Ann. ent. Soc. Am.* **49**, 617.

Weissmann, B. (1955). *J. biol. Chem.* **216**, 783.

Yamazaki, H. (1955). *Bull. Nagano-Ken Sericult. Exp. Stat.* **10**, 269.

CHAPTER 4

Heparin

INTRODUCTION

Heparin is a highly-sulphated mucopolysaccharide occurring principally in the mast cells of mammalian tissues, but also found in certain species of molluscs in rather different circumstances. In vertebrates it is found in liver, muscle, thymus, spleen and lung tissue firmly bound to protein. Its physiological properties, notably as an anticoagulant and as an antilipaemic agent, have attracted considerable attention to both its chemical structure and biological role, both of which have, however, proved rather difficult to establish. Vertebrate heparin contains D-glucuronic acid and 2-amino-2-deoxy-D-glucose in equimolar ratios. The predominant chain structure is now thought to be that of an alternating sequence of $\alpha 1 \to 4$ linked D-glucuronic acid and 2-sulphoamino-2-deoxy-D-glucose units, upon which are located units of ester sulphate (Wolfrom *et al.*, 1963). Dietrich (1968), using bacterial enzymes, has obtained degradation products of heparin which indicates that the 2-sulphoamino-2-deoxy-D-glucose is also sulphated at the 6 position and that there are probably two basic disaccharide units (Fig. 1) in which the uronic acid is either sulphated at position 3 or unsulphated. The possibility of some L-iduronic acid units, some $1 \to 6$

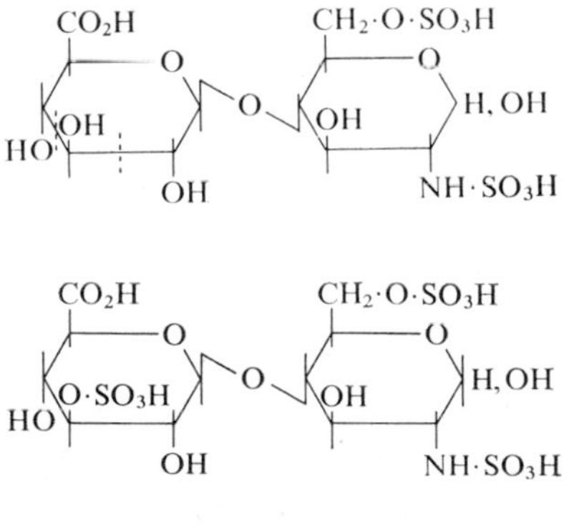

Fig. 1. Fundamental sub-units of the heparin polysaccharide chain according to Dietrich (1968). The trisulphated disaccharide may predominate in the intact molecule.

linkages, and some glucuronidic linkages in the β configuration cannot at present be discounted. As we shall see later Lindahl (1966) has demonstrated the presence of N-acetylated amino sugar in the heparin-protein linkage region.

The importance of heparin as the most potent of naturally occurring blood anticoagulants known has prompted many workers to search for a more readily available source of heparin, other than mammalian tissues where its concentration is relatively low. While a host of synthetic substitutes for heparin have been prepared, only one animal source of what would appear to be a true heparin has been discovered outside the vertebrates.

INVERTEBRATE HEPARIN

Thomas (1951), prompted by reports of metachromatic substances in the surf clam *Spisula solidissima*, examined the tissues of this lamellibranch mollusc after the removal of the shell, viscera and gonads. The remaining tissue, extracted according to the classic method of Charles and Scott (1933), followed by a phenol/water partition, yielded an acid mucopolysaccharide (500 mg/Kg of tissue) containing a trace of protein and having considerable anticoagulant activity. In a further investigation of this material, Frommhagen *et al.* (1953) isolated mucopolysaccharides from the tissues of both *Spisula* (*Mactra*) *solidissima* and also *Arctica islandica* by a process involving autolysis, proteolytic digestion, ethanol precipitation, alkali treatment, heat denaturation of protein and precipitation of mucopolysaccharide as the Na, Ca or Ba salts. Two mucopolysaccharide preparations, one from *S. solidissima* and the other from *A. islandica*, designated "Mactins" A and B respectively were obtained. The yields of anticoagulant activity per kilogram of tissue were twice as high as that normally obtained using beef lung. The purified products had *in vitro* anticoagulant activity of from 70–120 USP units per mg. The *in vivo* intensities of action using *in vitro* assayed doses of 100 units per mg. of Mactins and heparin were more than six times that of heparin and the duration of action was longer. Electrophoretically both Mactins A and B were inhomogeneous showing at least two major components and some minor ones. Both mucopolysaccharides gave strong metachromatic reactions with toluidine blue but had lower sulphate contents than heparin. The infrared spectra of the Mactins and heparin were almost identical but for differences in the positions of bands at wavelengths above 8μ. The optical rotations of the sodium salts were almost identical with that of sodium heparinate. The intrinsic viscosities of the Mactins were higher than heparin, suggesting perhaps higher molecular weights.

Later and more detailed studies of Mactins have shown that these substances must be very closely related to, if not identical with, heparin

(Burson *et al.*, 1956). The preparative procedure used by these workers was a complex one and is summarized as a flow-sheet in Fig. 2.

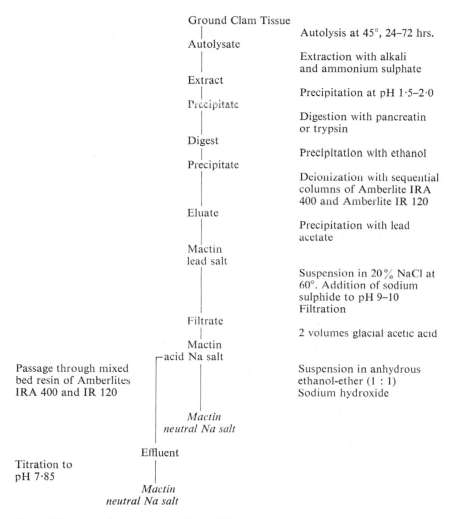

Fig. 2. Flow sheet for the preparation of Mactins according to the method of Burson *et al.* (1956).

In vitro anticoagulant activities of the acid and neutral sodium salts of Mactin A ranged between 130 and 150 USP units per mg, whereas those of the corresponding sodium salts of Mactin B were between 150 and 180 USP units per mg. In contrast to sodium heparinate the Mactin sodium salts could not be converted to crystalline acid barium salts. Analytical data for Mactin

A, Mactin B and heparin neutral sodium salts is given in Table I. The correspondence of the Mactins with heparin appears to be quite close.

TABLE I

Analyses of the Neutral Sodium Salts of Mactins A and B of and Two Preparations of Heparin[a]

	Mactin A	Mactin B	Heparin I	Heparin II
Anticoagulant activity (USP units) per mg	134	170	146	140
C (%)	25·8	26·7	26·0	25·2
H (%)	3·2	2·9	3·4	2·8
S (%)	12·0	11·6	11·6	12·2
N (%)	2·4	2·5	2·6	2·4
Na (%)	12·2	12·3	12·0	12·2
2-amino-2-deoxy-D-glucose (%)	25·2	30·8	28·8	29·8
Glucuronic acid (%)	26·5	34·7	24·8	25·8
N-acetyl (%)	2·0	1·7	1·6	2·0
O-acetyl (%)	2·3	1·5	1·5	2·4
P, Amino-N, N-CH$_3$ and C—CH$_3$	Nil	Nil	Nil	Nil
$[\alpha]_D^{25}$ in water	+71°(c 1·2)	+61°(c 1·5)	+47°(c 1·5)	

All data on an anhydrous basis.

[a] Burson *et al.* (1956).

Thomas (1954), using similar techniques to those described above, had obtained a preparation of Mactin A which after acid hydrolysis gave positive reactions to tests for reducing sugar, hexosamine and ester sulphate. Burson *et al.* (1956) identified the hexosamine in both Mactins A and B as 2-amino-2-deoxy-D-glucose (see Table 1). The presence of glucuronic acid in both Mactins was established by paper chromatography of hydrolysates and was quantitated by estimation of the carbon dioxide released during strong acid hydrolysis (Table 1). These two sugars were the only organic substances detected in the acid hydrolysates other than a low percentage of acetic acid (see Table 1). The low O-acetyl content detected (Table 1) was considered by the authors to be of doubtful validity.

That structural differences exist between the Mactins and heparin, perhaps in the glycosidic bonding, may be implied by the significantly higher specific optical rotations of the Mactins. The infrared spectra obtained by the authors for heparin, Mactin A and Mactin B (Fig. 3) show no very significant differences other than a lesser degree of structural detail in the 12 to 14 μ (850–700 cm^{-1}) region of the Mactin A spectrum; this may again reflect

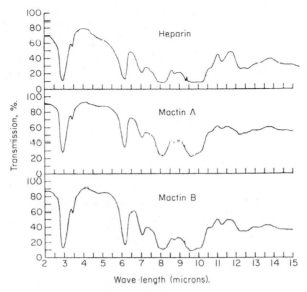

Fig. 3. Infrared spectra of Heparin, Mactin A and Mactin B (from Burson *et al.*, 1956).

TABLE II
Electrophoretic Parameters for Mactins A and B and Heparin[a]

Measurement in phosphate buffer pH 7·5, ionic strength 0·15

	Mactin A	Mactin B	Heparin
Anticoagulant activity in USP units per mg	141	165	146
Mobility ($\times 10^5$)	−16·4	−16·2	−16·1

[a] Burson *et al.* (1956).

TABLE III
Physical Parameters for Mactins A and B and Heparin[a]

	Mactin A	Mactin B	Heparin
$S^\circ_{20,w}$	2·65	2·57	2·28
$D_{20,w}^{0·2\%}$	4·41	3·70	6·63
$M_{w.av.}$	24,800	28,700	14,200
Standard deviation of the diffusion coefficient $\sigma^2 D \times 10^7$	1·5	<0·2	0·8

[a] Burson *et al.* (1956).

differences in glycosidic bonding, No specific absorption was detected in the 210–400 mμ region of the ultraviolet.

Titration curves of Mactins A and B were identical with that of heparin indicating the presence of free carboxyl groups and the absence of free amino groups. The titration data gave an equivalent weight of 193 for the anhydrous neutral sodium salt of Mactin B and a molecular weight of 579 for its disaccharide sub-unit. The corresponding values for Mactin A were 185 and 558.

Moving boundary electrophoresis of Mactins A and B showed them to be essentially homogeneous although there was evidence of a minor component in Mactin A. The electrophoretic mobilities of Mactin A, Mactin B and heparin are given in Table II. Like Mactin A the heparin used for comparison purposes also had a minor component. Sedimentation and diffusion data obtained for both of the Mactins and heparin are given in Table III, together with molecular weights calculated using an average value of 0·41 for the partial specific volume of the three preparations. Calculation of the standard deviation for the diffusion constants from the reduced second and fourth moments indicated that while the Mactin B preparation was of a high degree of homogeneity, Mactin A and the preparation of heparin used were of a lesser degree of purity.

ASSOCIATION WITH PROTEIN

The nitrogen contents of Mactins A and B are in close agreement with the value expected for a polysaccharide containing amino sugar and hexuronic acid in a 1 : 1 ratio. Thus there is little indication of additional nitrogen arising from amino acids (Burson et al., 1956). It should be remembered however that the purification procedures used involved a drastic treatment with alkali. The earlier preparation obtained by Thomas (1951) had been mildly treated and this contained some protein. It has recently been shown (Lindahl, 1966) that vertebrate heparin is covalently linked to protein through an alkali labile bond which involves a glycosidic linkage between xylose and the hydroxyl group of a serine residue in the protein chain. The xylose is in turn linked, through two galactose units, to the terminal hexuronic acid of the mucopolysaccharide chain (Fig. 4). The amino sugar constituents in this region are N-acetylated. If this type of terminal structure is present in the invertebrate heparin this may account for the small percentage of N-acetyl groups detected by Burson et al. (1956). The galactose and xylose residues, if present, might easily have escaped detection, as indeed they did for many years in the vertebrate heparins.

Cifonelli and Mathews (1966) have found varying amounts of 2-amino-2-deoxy-D-galactose and threonine in Mactins A and B. These constituents

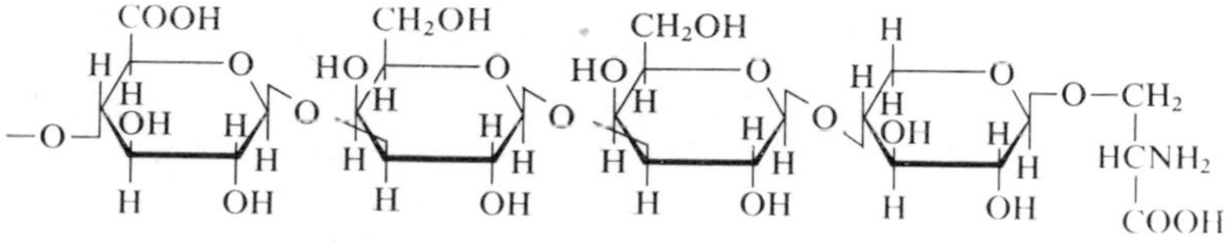

Fig. 4. The heparin-protein linkage region in pig heparin (Lindahl, 1966).

may well be involved in a protein-carbohydrate linkage region. That the 2-amino-2-deoxy-D-galactose and the hydroxyl amino acid were in fact integral with the polysaccharide seems certain since they persisted after fractionation with cetylpyridinium chloride in 1·35 M sodium chloride. Reaction of the Mactins with nitrous acid gave three fractions separable by gel-filtration. The two lower molecular weight fractions corresponded to those which were obtained with heparin but the third high molecular weight fraction contained the 2-amino-2-deoxy-D-galactose and threonine and failed to precipitate when treated with CPC. This fraction, from Mactin B, contained neutral sugar, 2-amino-2-deoxy-D-galactose, threonine and sulphate in the ratio 1 : 0·47 : 0·44 : 1·44 together with smaller amounts of uronic acid and 2-amino-2-deoxy-D-glucose. The high molecular weight fraction, from Mactin A, consisted mainly of sulphated neutral sugars.

LOCATION IN THE TISSUE

It is now generally accepted that heparin originates in the mast cells in the vertebrates. These granular cells are located principally in connective tissue close to the blood capillaries and in the walls of the blood vessels. The cells are characterized by the toluidine blue metachromaticity of the granules.

Mast cells could not be detected in the tissues of *M. solidissima* (Love and Frommhagen, 1953). Evidently the invertebrate heparin (Mactin A) arises in rather different circumstances. It was concluded that the invertebrate heparin probably had its origin in the ground substance of the connective tissues. Thomas (1954) compared the anticoagulant activities of extracts obtained from various organs and tissues of *S. (M.) solidissima* with results obtained from metachromatic staining in tissue sections and also from *in vitro* observations of metachromasia. The most potent preparation obtained had an anticoagulant activity of 130 USP units per mg and was derived from a mucus secretion of the mantle tissue. High activity preparations were also obtained from the gills and palps.

FUNCTION

The role of heparin in the vertebrate body has perhaps occupied an even more obscure position than the question of its structure. It is not thought, however, that its anticoagulant activity is of particular physiological significance, but rather that its principal role is concerned with transport of fat. The role of the invertebrate heparins described here is not understood at all. As this highly-charged macromolecule appears to have its major site in the molluscan mantle folds, a probable function is as a cation-exchanger acting as a vehicle for the transfer of calcium from the mantle to the shell, as a calcium-mucopolysaccharide complex.

Vertebrate heparin has been observed to have bacteriostat properties (Warren and Graham, 1950) at low concentrations; it may therefore be possible that Mactins and other polysaccharide sulphates secreted by the mantle epithelia of molluscs could fulfill a protective role.

BIOSYNTHESIS

With an adequate knowledge of its structure only recently available, the mode of biosynthesis for heparin has not been particularly well established. It would appear that the process of sulphation in mammalian tissue has an absolute requirement for adenosine triphosphate and that incorporation is via the intermediate, adenosine 3-phosphate 5-phosphosulphate (PAPS, Fig. 5) (Korn, 1959) Spolter and Marx, 1959; Ringertz, 1960). Whether the

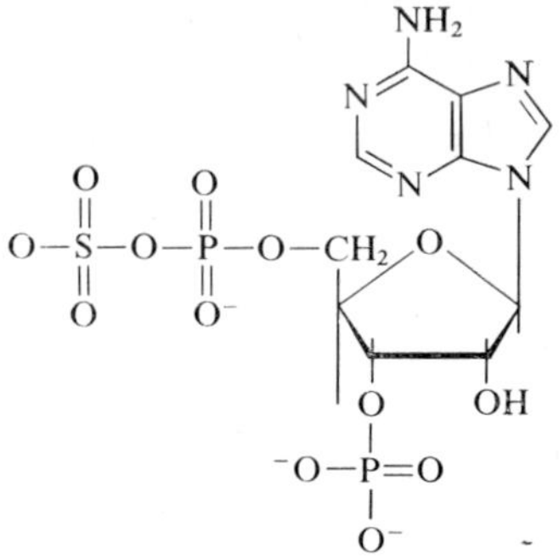

Fig. 5. Structure of adenosine 3-phosphate 5-phosphosulphate (PAPS).

completed polymer or precursors of small molecular weight are the actual receptors of the sulphate is in dispute (Korn, 1959; Spolter et al., 1963). Silbert and Brown (1961) detected uridine 5-(2-amino-2-deoxy-D-glucosyl dihydrogen pyrophosphate) in mouse mast cell tumours during heparin biosynthesis. This nucleotide sugar derivative was also found, together with heparin, when 2-amino-2-deoxy-D-glucose was incubated with tumour slices. Incorporation of uridine 5-(2-acetamido-2-deoxy-D-glucosyl dihydrogen pyrophosphate) and UDP-D-glucuronic acid into an acid polysaccharide by mouse mast cell tumour was detected by Silbert (1962).

In a study of the biosynthesis of Mactin A, DeMeio et al. (1964, 1967) incubated gills or gill preparations in a medium containing [$^{35}SO_4$]. The incubated material was digested with papain, precipitated at 70% ethanol, passed through a column of Sephadex G25 and reprecipitated again with ethanol. Intact gills incorporated sulphate into the mactin. A gill homogenate centrifuged at 600 g for 15 minutes gave similar results when fortified with 0·005 M $MgCl_2$ and adenosine triphosphate. Sulphation was inhibited by 0·005 M molybdate, 0·05 M sucrose and 0·001 M β-napthylamine to the extent of 74(75, 1967), 51 and 76% (90%, 1967) respectively. Incubation with D-glucose $U^{14}C$ led to incorporation of [^{14}C] into the mactin. A particle-free supernatant obtained by centrifugation of gill homogenate at 100,000 g showed no significant activity but 5% of the original activity for sulphation was restored by adding [$^{35}SO_4$] adenosine 3-phosphate 5-phosphosulphate (PAPS[35]). It was therefore concluded that as in the case of vertebrate heparin PAPS participates in the sulphation of mactin, and that there was apparently de novo synthesis of this compound. The inhibition of activity by sucrose was thought to be due to interference with the formation of the substrate for sulphation. The β-napthylamine inhibition was probably not as the earlier paper suggested due to formation of a complex with the polysaccharide amino groups but instead due to inhibition of PAPS formation.

REFERENCES

Burson, S. L., Fahrenbach, M. J., Frommhagen, L. H., Riccardi, B. A., Brown, R. A., Brockman, J. A., Lewry, H. V. and Stokstad, E. L. R. (1956). J. Am. chem. Soc. 78, 5874.

Charles, A. F. and Scott, D. A. (1933). J. biol. Chem. 102, 425.

Cifonelli, J. A. and Mathews, M. B. (1966). Fed. Proc. 25, 742.

De Meio, R. H., Lin, Y. C. and Narasimhulu, S. (1964). Sixth Int. Congr. Biochem., New York. Abs. VI-23, 506.

De Meio, R. H., Lin, Y. C. and Narasimhulu, S. (1967). Comp. Biochem. Psysiol. 20, 581.

Dietrich, C. P. (1968). Biochem. J. 108, 647.

Frommhagen, L. H., Fahrenbach, M. J., Brockman, J. A. and Stokstad, E. L. R. (1953). Proc. Soc. exp. Biol. Med. 82, 280.

Korn, E. D. (1959). J. biol. Chem. 234, 1647.

Lindahl, U. (1966). Biochim. biophys. Acta. 130, 368. Arkiv für Kemi, 26, 101.

Love, R. and Frommhagen, L. H. (1953). Proc. Soc. exp. Biol. Med. 83, 838.

Ringertz, N. R. (1960). Ark. Kemi, 16, 67.

Silbert, J. E. (1962). *Arthritis Rheum.* **5**, 657.
Silbert, J. E. and Brown, D. H. (1961). *Biochim. biophys. Acta.* **54**, 590.
Spolter, L. and Marx, W. (1959). *Biochim. biophys. Acta.* **32**, 291.
Spolter, L., Rice, L. I. and Marx, W. (1963). *Biochim. biophys. Acta.* **74**, 188.
Thomas, L. J. (1951). *Biol. Bull. mar. biol. lab., Woods Hole.* **101**, 230.
Thomas, L. J. (1954). *Biol. Bull. mar. biol. lab., Woods Hole.* **106**, 129.
Warren, J. R. and Graham, F. (1950). *J. Bacteriol.* **60**, 171.
Wolfrom, M. L., Vercellotti, J. R., Tomomatsu, H. and Horton, D. (1963). *Biochem. biophys. Res. Commun.* **12**, 8.

CHAPTER 5

Polysaccharide Sulphates
and Miscellaneous Mucopolysaccharides

INTRODUCTION

Apart from the sulphated polysaccharides of connective tissues and the echinoderm egg jellies, polysaccharide sulphates are found to be widely distributed in the invertebrates, particularly in the secretions of those groups having a marine habitat and notably in molluscs and echinoderms.

MOLLUSCAN POLYSACCHARIDE SULPHATES

GLUCAN SULPHATES

The β-linked glucan sulphate which replaces chondroitin sulphate in the cartilage of the mollusc *Busycon canaliculatum* has been mentioned in Chapter 2. We shall consider this again in more detail later in this chapter. The presence in molluscs, however, of such glucan sulphates as that from *Busycon*, had been recognized at an earlier date by Soda and his collaborators in Japan, working with a secretion derived from the hypobranchial gland of the marine gastropod *Charonia lampas*. This work, spanning more than thirty years of study, has recently been reviewed by Egami and Takahashi (1962).

The principle interest, during the early 1930s, of Soda and his group was the class of enzymes which liberate inorganic sulphate from organic sulphate esters. They found that marine molluscs provided a ready and prolific source of such sulphatases and that the enzymes had specificities towards carbohydrate sulphates (Soda and Hattori, 1933; Soda and Egami, 1938a). It was therefore a logical step to attempt an isolation of the natural substrates for the sulphatases from the organisms which had yielded the enzymes themselves.

The large whelk *C. lampas* (*Tritonalia sauliae*), called "Boshubora" in Japanese, had proved to be an excellent source of a highly active glucosulphatase (Soda and Hatori, 1933) and, in satisfying agreement with prediction, was also shown to contain a polysaccharide sulphate ester as part of the secretion of its hypobranchial gland, capable of acting as a substrate for the enzyme (Soda *et al.*, 1936; Soda and Egami, 1938b,c)

Hypobranchial Glands

Since polysaccharide sulphate secretions from the hypobranchial glands of several molluscan species are to be described, a brief description of this organ would seem to be appropriate.

The body of the prosobranch, gastropod, of which *C. lampas* is an example, is roughly divisable into three distinct external regions: the head, the foot and the visceral hump, the last containing the alimentary, renal and reproductive organs. The body is enclosed by a spiral, conical shell from which the head and foot can be extended but which always encloses and protects the visceral hump. The integument of the visceral hump is developed and extended to produce the characteristic, tent-like, molluscan mantle whose edge is visible around the inner edge of the shell mouth. This mantle or pallium encloses a space, the mantle or pallial cavity, largest on the dorsal and anterior surface of the body, that contains several important organs grouped together as the pallial complex. Into the cavity discharge the anal, renal and genital openings (Fig. 1). Three of the organs of the pallial complex

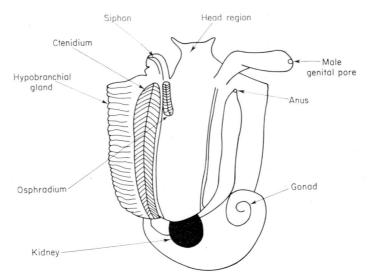

Fig. 1. Semi-diagramatic view of the opened mantle cavity of a gastropod mollusc (*Buccinum undatum*).

are derived from the epithelium of the mantle wall: the ctenidium or gill the osphradium which is a sensory organ acting as a nephelometer and an oxygen detector, and the hypobranchial gland. These three organs lie parallel to each other across the opening into the pallial cavity of the syphon, the tube which carries a constant stream of water (and hence oxygen) into the cavity. Thus the pallial cavity contains organs essential to the life of the animal, namely the respiratory apparatus which comprises and inhalent syphon, sensor and gill and the external openings of the gonads, kidneys and digestive tract.

These molluscs are benthic, living in an environment frequently high in suspended sediment. This sediment is obviously likely to be drawn into the pallial cavity on the respiratory current. The function of the hypobranchial gland would appear to be secretion of a sticky mucin which entraps particles of sediment, together with the animal's own faeces, carrying them away on the exhalent stream, before they can settle upon, and impair the function of, the delicate essential organs present in the pallial cavity. The function, therefore, is largely hygenic. A more detailed account of this organ may be found in Fretter and Graham (1962).

The histology and cytology of the hypobranchial glands of two of the species to be discussed here have been examined in some detail. However, a brief account will serve.

The hypobranchial gland of *B. canaliculatum* contains three types of mucous cell, the relationships of which are not completely clear (Ronkin, 1952a). Two types contain metachromatic material, while one contains non-metachromatic mucin; the latter is probably a precursor gland cell (Fig. 2). As we shall see the mucin is chemically homogeneous and this would suggest that the two types of metachromatic cell must be closely related.

Hunt (1967) found the hypobranchial gland of *Buccinum undatum* to contain two different types of mucous cell, both containing metachromatic material (Fig. 3). As we shall see again, this secretion also contains only one polysaccharide sulphate. Again, there must be a relationship between the two types of cell.

Charonia lampas *Glucan Sulphate*

The polysaccharide sulphate obtained by Soda (1948) from the hypobranchial gland of *C. lampas,* by precipitation as the sodium salt with ethanol, was given the name charoninsulphuric acid. This material contained 15% ester sulphate S, 36% ash, 0–0·2% nitrogen and 10% water. The material was reported at the time of its first isolation to contain 20% uronic acid; this finding later proved to be erroneous. The polysaccharide gave a strong metachromatic reaction with Lison's reagent and liberated sulphate, together with a reducing substance when treated with the enzyme preparations from the mollusc. It had strong anticoagulant activity.

Soda (1948) was to point out many similarities between this material and cellulose sulphate and it was subsequently shown (Soda and Terayama, 1948) that charoninsulphuric acid released glucose as its only monosaccharide constituent on acid hydrolysis. Treatment of charoninsulphuric acid with

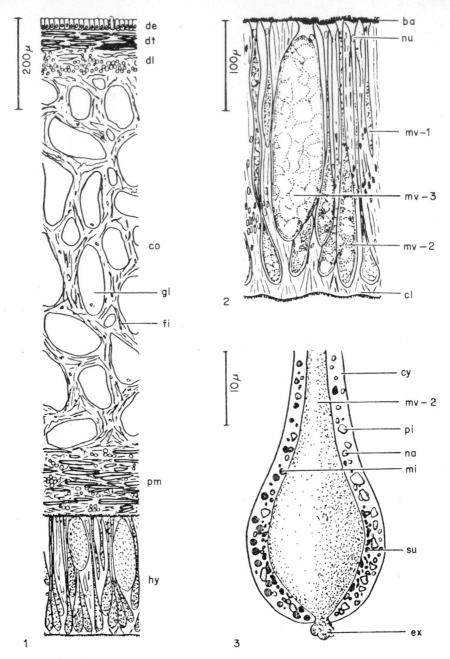

Fig. 2 (1). Transverse section of *Busycon canaliculatum* mantle in the region of the hypobranchial gland: co: connective tissue layer; de: dorsal epithelium; dl: dorsal longitudinal muscle layer; dt: dorsal transverse muscle layer; fi: fibrous tissue; gl: glycogen cell; hy: hypobranchial gland; pm: pallial muscle layer. (2). Cell types in the hypobranchial gland: ba: basement membrane; cl: ciliated cells; mv-1,2,3: mucous vacuoles of cell types 1,2, and 3 respectively; nu: nucleus of a type 2 cell. (3). Semi-diagrammatic drawing of the distal end of a type-2 cell: cy: cytoplasm; ex: extruded mucus; mi: mitochondrion; mv-2: mucous vacuole; na: nadi-positive granule; pi: pigment granule; su: sudanophilic granule (from Ronkin, 1952a).

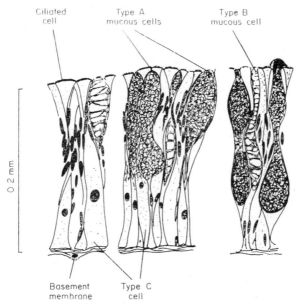

Fig. 3. The glandular epithelium of the hypobranchial gland of *Buccinum undatum* L.; drawing based on sections stained in Ehrlich's haematoxylin, alcian blue 8*GS* and eosin; fixed Bouin and sectioned at 10*μ* (from Hunt, 1967).

methanolic hydrogen chloride, to remove the sulphate residues, yielded a material whose X-ray diffraction pattern was identical with that of cellulose.

Later studies showed that charoninsulphuric acid was in fact a mixture of glucan sulphates (Egami *et al.*, 1955). Mucin was extracted with saturated sodium chloride solution at 80° for three hours, or with 1·7% hydrochloric acid at 40° for 30 minutes. The extract was precipitated with concentrated potassium hydroxide. The precipitate was separated, redissolved in water and re-precipitated. This material (Fraction 3) was washed with ethanol and ether and dried *in vacuo*. On analysis, Fraction 3 was found to contain 10–18% S; it was precipitable by potassium chloride and trypaflavine. It gave a negative colour reaction with iodine-potassium iodide reagent.

The supernatant which remained after the potassium hydroxide precipitation was treated with ethanol yielding a further precipitate; this was dissolved in water and reprecipitated with trypaflavine. The precipitate was extracted with 20% calcium chloride to dissociate the trypaflavine complex and the extract decolourized with charcoal. Ethanol precipitation now yielded a material (Fraction 2) which after washing and drying was found to contain 2–5% S. This also was precipitable by potassium chloride and also gave no colour with I—KI reagent.

The supernatant remaining after trypaflavine precipitation yielded yet another component (Fraction 1) after charcoal decolourization and ethanol precipitation. Fraction 1 contained 1–2% S. It was not precipitable by KCl but gave a red colouration with the I—KI reagent.

More recently, Iida (1963a; Iida *et al.*, 1960) has separated "S-poor" and "S-rich" charoninsulphuric acids by phenol extraction and zone electrophoresis. Glands were homogenized and extracted with 47·5% phenol and the aqueous phase precipitated with ethanol. Redissolving the precipitate in water and adding KCl to a concentration of 2% yielded an "S-rich" precipitate and an "S-poor" supernatant. The precipitate was dissolved in water and digested with Pronase to remove protein. The clear viscous solution remaining was precipitated with 20% KCl and the precipitate dissolved in 1·7% hydrochloric acid. Precipitated protein was removed by centrifugation and the supernatant precipitated with ethanol. This precipitate of "S-rich" charoninsulphuric acid contained 10–14% S and corresponds to Fraction 3 of Egami *et al.* (1955). It contains less than 1% protein.

The supernatant remaining after the initial potassium chloride precipitation was precipitated with two volumes of ethanol and the precipitate extracted with 1 M NaCl, leaving behind protein-rich material. The sodium chloride extract was re-precipitated with ethanol and the resulting material subjected to zone electrophoresis on columns of Hyflo Super-cel in 0·05 M veronal buffer, pH 8·7. Two fractions separated, the major one of which contained only 0·7% S and less than 1% protein. This is the material called S-poor charoninsulphuric acid.

The sulphated glucans of *C. lampas* mucin are of a complex character and the determination of their structure has occupied a protracted period of study. Moreover, recent work by Iida and his colleagues has made necessary considerable reassessment of earlier aspects thought to be well established.

Desulphation of S-rich charoninsulphuric acid (Fraction 3, hereafter referred to as F3) with methanolic-hydrogen chloride yields a product which has been called charonin (Egami *et al.*, 1955). This material is soluble in Schweizer's cuprammonium reagent (a cellulose solvent). Neutralization of the solution with acetic acid precipitates only part of the charonin, giving fraction AF3. A further precipitate is formed in the supernatant on adding methanol, fraction BF3. This latter fraction, which contains 1–2% S, seems to be chemically identical to fraction 1(F1).

The staining reactions of AF3 and BF3, with a variety of iodine reagents, when compared with cellulose, amylose and starch, were not particularly informative, although fraction BF3 was remarkable for giving a red colour with these reagents. Fraction AF3 had an equivalent reducing power of 0.02 mg glucose per mg of material while fraction BF3 had a value of 0·03 mg per mg material.

Acid hydrolysis and paper chromatography of AF3 yielded D-glucose only while acetolysis gave octa-acetylcellobiose. Treatment of this fraction with Irpex-cellulase resulted in hydrolysis and the release of glucose and cellobiose. AF3 was not susceptible to degradation by either α-amylase or β-amylase. In contrast, the water-soluble BF3, which also yielded glucose only, after acid hydrolysis, was hydrolysed by both α and β-amylases with the release of maltose and glucose. The increases in reducing power obtained after treatment of BF3 with the two enzymes were comparable with the increases obtained for glycogen. BF3 was also hydrolysed by the cellulase although to a lesser extent than AF3. Cellobiose and glucose were detected in the hydrolysis products. The increases in reducing power, as mg cellobiose per mg substrate on cellulase treatment, were 0·33 and 0·04 for AF3 and BF3 respectively. The interpretation of the degradation of BF3 by cellulase requires some caution, in view of the possibility of contamination by other glycosidases and the presence in the enzyme preparation of transglycosidases capable of forming cellobiose from glucose. The authors believed, however, that BF3 was a glucan which contained both $\alpha 1 \rightarrow 4$ and $\beta 1 \rightarrow 4$ glycosidic linkages. BF3 consumed periodate, quantitatively, as expected for an $\alpha 1 \rightarrow 4$ structure; the periodate consumption for AF3 however was only 60% of the expected value for a cellulose-like molecule.

The infrared spectra of AF3 and F1 closely resembled those of cellulose and amylose, respectively (Nakanishi et al., 1956), while BF3 also resembles amylose in its spectrum. The positions of the C—O—S ester sulphate absorptions in F1 and BF3, at 803 and 802 cm^{-1} respectively, and the corresponding absorption in unfractionated charonin, at 812 cm^{-1}, appear to lie at slightly abnormal positions. If the glucose residues are in the preferred equatorial conformation, then one might have expected the ester sulphate C—O—S vibrations to have been in the region of 820 cm^{-1}.

Iida's S-rich charoninsulphuric acid, which was homogeneous both electrophoretically and in the ultracentrifuge, where it had a sedimentation constant of 15S at infinite dilution, had an optical rotation of $[\alpha]_D^{15} - 60°$ (Iida et al., 1960; Iida, 1963a). Compare this with a value of $-67°$ for cellulose-polysulphate. Iida's preparation was not hydrolysed by bacterial of fungal α-amylases, by bean β-amylase or by Rhizopus delemar amylase. Nor did it inhibit the action of these enzymes on glycogen or S-poor charoninsulphuric acid. As with earlier preparations the desulphated S-rich material was insoluble in water and soluble in cuprammonium reagent. In this case, however, no water-soluble fraction (BF3) was produced during desulphation.

The highly-sulphated fraction of charoninsulphuric acid would appear to be based upon a $\beta 1 \rightarrow 4$ linked glucan structure, similar to cellulose. The anamolous periodate consumption of charonin may possibly be accounted for by partial degradation of a proportion of the glucose residues during the desulphation process. The positions of the sulphate residues have not so far been established.

So far as the various fractions of charoninsulphuric acid have been analysed, there would seem to be a parallelism between sulphate content and cellulose structure (Egami *et al.*, 1955). Fractionation of F3 and F2 with cuprammonium reagent yielded AF and BF fractions, in both cases, the approximate ratios being 4 : 1 and 1 : 20 for F3 and F2 respectively. As I have already mentioned, F1 and BF3 appear to be equivalent.

The structure of the sulphate-poor fraction, BF3, seems to be still in some doubt. The results obtained by Egami *et al.* (1955) and by Nakanishi *et al.* (1956) suggested a glucan structure containing both $\alpha 1 \to 4$ and $\beta 1 \to 4$ glucosidic bonds. Iida's S-poor material (Iida *et al.*, 1960; Iida, 1963a) yielded, with this probably purer preparation, rather different results.

S-poor charoninsulphuric acid is soluble in water, contains less than 1% protein and has a sulphur content (0·7%) lower than that obtained previously. It is homogeneous in the ultracentrifuge and does not exhibit the marked concentration dependence of the S-rich material. The sedimentation constant has a remarkably high value, 80S, and assuming a similar fractional ratio to glycogen, a molecular weight of the order one million was suggested. Determinations of reducing power suggested, however, a molecular weight of ten thousand; the discrepancy is perhaps due to branching in the polysaccharide chain. The optical rotation ($[\alpha]_D^{15} + 180°$) is similar in magnitude to values obtained by glycogen and amylose.

S-poor charoninsulphuric acid can be partially degraded by amylases without prior desulphation. α- and β-amylases from various sources degraded the S-poor material to the extent of 42% and 25% respectively; contrast this with values of 73% and 48% for soluble starch. The degradation of the S-poor material by these enzymes gave rise to non-dialysable resistant cores. These limit glucans had lengths of 19 and 62 glucose units after treatment with α- and β-amylases respectively. *R. delemar* amylase, which in addition to hydrolysing $\alpha 1 \to 4$ glucosidic bonds also hydrolyses $\alpha 1 \to 6$ bonds, degraded S-poor charoninsulphuric acid almost completely (94%), leaving a non-dialysable residue with a mean chain length of only two glucose residues. Treatment of the S-poor polysaccharide with iso-amylase (amylo-1,6-α-glucosidase from brewers yeast), which has a specificity for the $\alpha 1 \to 6$ glucosidic linkages of glycogen and amylopectin, brought about an 11% degradation in comparison with 6% for glycogen. Iso-amylase in conjunction with β-amylase completely degraded both S-poor charoninsulphuric acid and glycogen. The rate curves for the degradation of these materials were closely similar. This, in contrast to earlier results, suggests a structure closer to glycogen than amylose.

Matsuoka *et al.* (1956, 1962) have noted that S-rich charoninsulphuric acid has strong anticoagulant activity and that its effect in mammals, *in vivo*, may be to inhibit the plasma factor concerned with thromboplastin generation.

Inoue and Egami (1963) have reported the presence of a sulphated glycoprotein in the hypobranchial gland of *C. lampas* but further details have not yet been presented.

Buccinum undatum *Glucan Sulphate*

B. undatum is a large prosobranch marine gastropod closely related to *C. lampas* and found widely distributed in the waters of northern Europe.

The excised hypobranchial glands can be stimulated to produce a copious flow of mucin when they are agitated in 0·6 M saline at 4° (Hunt and Jevons, 1963). The mucin can be drawn from the glands in thick, ropey strands. This material, which is presumably in a similar form to the *in vivo* secretion, has rather marked rheological properties, exhibiting syneretic and viscoelastic behaviour. Masses of the mucin stirred with a rotating rod exhibit the Weisman effect and show strong elastic recoil when the stirring is stopped. Strands of the mucin, under tension, have considerable strength and exhibit a tendency to expel solvent as beads distributed along the length of the strand. Fresh mucin, collected from several glands and pooled, compacts into elastic masses. If a strand of mucin is drawn from one of these masses, and placed in a second container, complete transference will readily take place. In fact, if a small strand becomes a bridge between laboratory bench and container, elastic transfer *en masse* from beaker to bench and thence to floor will occur. Such properties, while amusing, are of course of considerable practical advantage to the living animal, for the entrapment and removal of material from the mantle cavity.

The mucin disperses readily in 0·6 M saline on stirring (simulation of wave action?). The dispersate lacks the viscoelastic and sineretic properties of the fresh mucin. Dispersion is probably a progressive breakdown of aggregates and is probably not taken to completion merely by stirring since dispersates subjected to repeated shearing in a capillary viscometer show a progressive and irreversible decrease in viscosity approaching a constant value. The viscous behaviour of these completely sheared solutions is typically macromolecular, being non-Newtonian at low shear rates and Newtonian only at higher rates of shear. The non-linear reduced viscosity plots obtained with dispersed solutions are typical of highly charged macromolecules.

The mucin proved to be composed of two major components, a glyco-protein and a single highly sulphated glucan. The infrared spectrum of the whole mucin clearly shows characteristics of both components (Fig. 4). The glycoprotein component is best removed by phenol treatment leaving the glucan sulphate in a pure state. Both components have been characterized (Hunt and Jevons, 1963, 1965a,b, 1966; Hunt and Rudall, 1967). The glycoprotein is discussed in Chapter 12.

The glucan sulphate and the glycoprotein appear to be united by weak forces only. They were readily separated by mild procedures and also separated from one another in the analytical ultracentrifuge and on electro-phoresis (Figs. 5 and 6). The aqueous portion of a 45% phenol extract contains the glucan sulphate which can be precipitated at about 4° (but not at room temperatures) by the addition of potassium chloride to 0·2 M concentration. The physical basis of this latter process is still not clear.

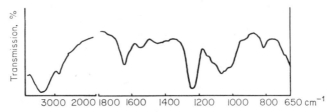

Fig. 4. Infrared spectrum of *Buccinum undatum* hypobranchial mucin. Freeze-dried film (from Hunt and Jevons, 1965a).

The physical parameters of the glucan sulphate have been determined and are given in Table I.

TABLE I

Physical Parameters of *Buccinum undatum* Hypobranchial Gland Glucan Sulphate[a]

$S_{20}{}^{1\%}$ in phosphate buffer pH 7·9 I 0·15 with respect to NaCl	$5·25 \times 10^{-13}$
$S_{20}{}^{0}$	$9·75 \times 10^{-13}$
$\bar{v}_{20}$	$0·6$
$D_{20}{}^{0·5\%}$	$3·28 \times 10^{-7}$
$\eta sp./C$ in water at 25°	$21,320$ ml/g
$[\eta]$ in 0·5 M NaCl at 25°	500 ml/g
Mol.wt. (from $S_{20}{}^{0}$ and $D_{20}{}^{0·5\%}$)	$1·72 \times 10^{5}$
Mol.wt. (from $S_{20}{}^{0}$ and $[\eta]$)	$1·00 \times 10^{7}$
Mol.wt. (from "Approach to Sedimentation Equilibrium") at meniscus	$1·70 \times 10^{5}$
Mol.wt. (from "Approach to Sedimentation Equilibrium") at base	$1·10 \times 10^{5}$
Electrophoretic mobility in phosphate buffer (as above)	$-17·8 \times 10^{-5}$

[a] Hunt and Jevons (1966).

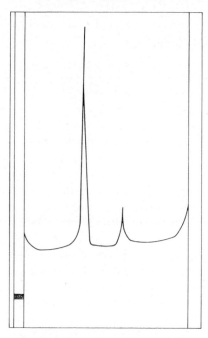

Fig. 5. Sedimentation pattern of *Buccinum undatum* hypobranchial mucin at 59780 rev/min (260000g), 20° in 0·02 M-sodium phosphate–0·15 M-NaCl buffer, pH 8·0; concentration of material, 1 % (w/v) boundary at 120 minutes (from Hunt and Jevons, 1967).

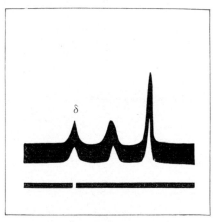

Fig. 6. Moving-boundary electrophoresis diagram of a 1 % (w/v) solution of *Buccinum undatum* hypobranchial mucin in 0·02 M-sodium phosphate–0·15M-NaCl buffer, pH 8·0, at 4° with a potential of 100 V at 23mA. The arrow indicates the direction of migration and δ is the stationary buffer-effect boundary in the ascending limb (from Hunt and Jevons, 1965a).

c

The value for the molecular weight, determined from the Flory-Mandelkern relationship, Equation 1,

$$M = \left[\frac{S[\eta]^{1/3}\eta_0 N_0}{\phi^{1/3}p^{-1}(1-v\rho_0)} \right]^{3/2} \tag{1}$$

appears vastly in excess of that which might be expected for a molecule having a sedimentation coefficient of 9·75 S. This probably stems from the abnormal properties of very highly-charged polyelectrolytes in solution and suggests marked departure from random coil behaviour even in the presence of a high gegenion environment. The molecular weight values, which are of the order of 10^5, are more in keeping with the S value, but even so it is doubtful whether the diffusion and equilibrium sedimentation properties of these highly electroviscous solutions can be relied upon to be truly representative.

The presence of glucose, and glucose only, in hydrolysates of the glucan sulphate was demonstrated by paper chromatography and paper electrophoresis. The glucose was identified as the D form by its susceptibility to oxidation, by glucose oxidase (E.C.1.1.3.4.). The polysaccharide contains 54% glucose (orcinol-sulphuric acid) and 30% sulphate as SO_4^{--}.

Infrared spectra of the polysaccharide sulphate and of material desulphated according to Kantor and Schubert (1956) (Figs. 7 and 8)

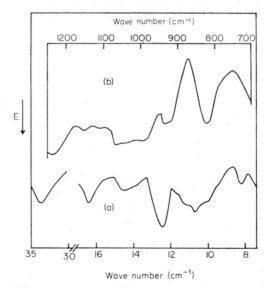

Fig. 7. Infrared spectra of the glucan sulphate from *Buccinum undatum* hypobranchial mucin: (a) as a freeze-dried film; (b) (inset) as a Nujol mull with scale expansion X2 of the region 700–1250 cm^{-1} (from Hunt and Jevons, 1967).

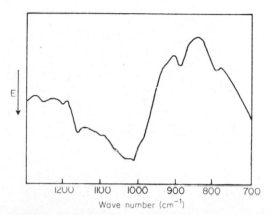

E

1200   1100   1000   900   800   700
Wave number (cm^{-1})

Fig. 8. Infrared spectrum of the desulphated form of the glucan sulphate from *Buccinum undatum* hypobranchial mucin (Hunt and Jevons, 1966).

suggested a $\beta 1 \rightarrow 4$ linked structure, similar to that of cellulose. The water-insoluble desulphated glucan gave no staining reaction with iodine-potassium iodide but stained comparably to cellulose with Schultze's chlor-zinc-iodine reagent.

The desulphated material was readily degraded by a cellulase from *Myrothecium verucaria* with the release of 66% of the available reducing groups. Glucose and cellobiose were identified in the enzymic hydrolysates. Partial acid hydrolysis of the glucan also yielded glucose and cellobiose. A cellulose-like structure was thus indicated and X-ray diffraction studies on the desulphated material (Hunt and Rudall, 1967) have confirmed this view. The diffraction pattern (Fig. 9) clearly demonstrates a structure of a hydrate or regenerated cellulose type. The presence of a low proportion of branch points has not been eliminated; a peak at 790 cm^{-1} in the infrared spectrum coupled with the fact that the sulphate-free glucan is insoluble in cuprammonium solutions supports this possibility.

The positions of the sulphate residues on the glucan chain have not been completely established. The ratio of sulphate to glucose suggests that each monosaccharide unit carries one sulphate group. The substitution is undoubtedly as ester, since desulphation removes characteristic ester sulphate, S$=$O, and C—O—S stretching vibrations, at 1250 and 800–840 cm^{-1}, from the infrared spectrum of the glucan sulphate (Fig. 8). Absorption bands at 940, 920 and 1000 cm^{-1} also disappear on desulphation; their exact origin in the sulphate grouping is not known but a number of other workers have reported the presence of bands in the region between 920 and 955 cm^{-1} for polysaccharide sulphate spectra, without positively identifying them with sulphate groups. For example, absorptions at 928 cm^{-1}

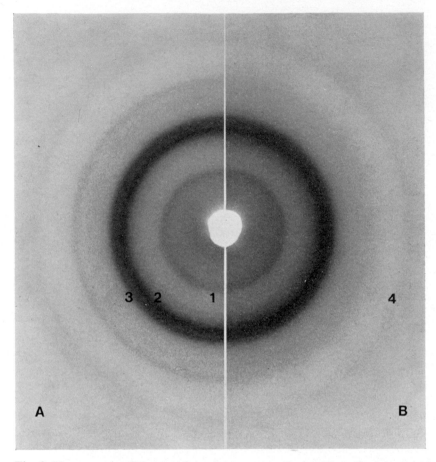

Fig. 9. X-ray powder diagrams of A, viscose rayon; B, desulphated polysaccharide from hypobranchial mucin of *Buccinum undatum* (from Hunt and Rudall, 1967).

in the spectrum of chondroitin sulphate A (Mathews, 1958) at 955 cm^{-1} in the spectrum of a new chondroitin sulphate (Bettelheim and Philpott, 1960) and at 925 cm^{-1} in the spectra of λ and κ-carrageenins (Bayley, 1955) have been noted, but their disappearance on desulphation has not been recorded. Mathews (1958), however, reported a 1000 cm^{-1} absorption in the spectrum of chondroitin sulphate A and the loss of this band was reported by Hoffman *et al.*, (1958).

The hydroxyl groups of β-D-glucose, in its preferred conformation (Reeves, 1951), should all be equatorial and hence, wherever the ester sulphate residues are placed on the *Buccinum* glucan sulphate chain, the characteristic C—O—S band should be found between 840 and 820 cm^{-1} with no indication of an 855 cm^{-1} axially substituted absorption. The position of the peak maximum (Fig. 7), slightly below 820 cm^{-1}, suggested primary equatorial substitution, i.e. at position 6 on the glucose ring. But the absorption band also shows a pronounced shoulder in the 830–840 cm^{-1} region and this suggested a proportion of secondary equatorial sulphate groups, i.e. at positions 2 or 3 (position 4 being occupied by the glycosidic bond). The degree of secondary equatorial sulphation cannot be predicted with any confidence from this result as infrared spectra are difficult to interpret quantitatively. In fact, determination of the rate of hydrolysis of sulphate by dilute acid, according to the method of Rees (1963), gives that half-life for the reaction of about 0.41 hours, a rate comparable with that expected for β-D-glucose-3-sulphate. The linearity of the rate curve obtained suggested that there was only one type of ester substitution. The infrared data and the evidence from reaction kinetics are thus in conflict. Attempts to determine the distribution of the sulphate groups, by model building, on the basis of results obtained from dye-binding experiments and potassium chloride precipitation were unhelpful. All the structures tried seemed almost equally probable. The problem of potassium chloride precipitation is an interesting one, as we have already seen the β 1 → 4 linked glucan sulphate in *C. lampas* mucin is precipitated by potassium ions as is the algal galactan sulphate κ carrageenin and synthetic cellulose sulphate (Schweiger, 1966). The metachromatic complex which the glucan sulphate forms with toluidine blue is an extremely stable one and the dye shows no tendency to unstack in the presence of large excesses of anionic sites. This would suggest that the distribution of the sites on the polymer is even and close, but even so the 2 and 3 positions, and to a slightly lesser extent the 6 position, would fit the requirement. The failure to unstack in the presence of excess polymer would also suggest that there is little or no secondary structure to the molecule.

The glucan sulphate contains some 2·22% protein or peptide bound firmly enough to be insusceptible to removal by physical procedures, i.e. denaturation, repeated potassium chloride precipitation, extensive dialysis. The composition of this peptide is shown in Table II. No N-terminal residues could be detected but this may be due to a particularly efficient masking by the highly charged and bulky polysaccharide chain. Extensive proteolytic degradation by Pronase and bacterial protease (Novo) lowered the peptide content to 0·20%, i.e. 11% of the original peptide. Glutamic acid, serine and glycine are the predominant amino acids in this degraded mucopolysaccharide. The link between peptide and polysaccharide seemed to be alkali-stable (0·5 N NaOH at 2° for 24 hours) and it seems likely therefore that the linkage region cannot contain the type of 0-glycosidic bond to serine found in many of the vertebrate mucopolysaccharide-protein complexes. The intact polysaccharide-peptide complex contains an unidentified

TABLE II

Amino Acid Composition of the Glucan Sulphate-Peptide Complex and of Fragments from this Complex, from *Buccinum undatum*[a]

Amino acids expressed as (a) g of amino acid per 100 g substance and (b) as μmoles amino acid per 100 μmoles of total amino acids. 1, pure polysaccharide sulphate–peptide complex; 2, proteolytically degraded complex; 3, peptide fragment containing component X.

Components	1		2		3
	a	b	a	b	b
Aspartic acid	0·23	8·15	0·02	10·00	5·03
Threonine	0·17	6·77	0·01	6·52	2·71
Serine	0·26	11·58	0·04	20·28	5·03
Glutamic acid	0·16	16·16	0·07	23·60	13·56
Proline	0·08	3·14	—	—	—
Glycine	0·22	13·26	0·02	17·28	2.71
Alanine	0·25	12·75	0·01	8·00	5·03
Valine	0·17	6·73	0·01	5·86	—
Methionine	0·12	3·79	—	—	—
Isoleucine	0·04	1·61	—	—	—
Leucine	0·08	2·84	0·02	8·46	—
Tyrosine	0·03	0·86	—	—	—
Phenylalanine	0·05	1·42	—	—	—
Ornithine	0·07	2·46	—	—	—
Lysine	0·10	3·23	—	—	—
Histidine	0·07	3·07	—	—	—
Arginine	0·12	3·18	—	—	—
% of total weight	2·22	—	0·20	—	—
Component X	Present, not determined		Absent		65·89[b]

[a] Hunt and Jevons (1965).
[b] Assumed to give same absorbance at 570 mμ with ninhydrin as cysteic acid.

component which is released as part of a small peptide during the proteolytic degradation. The amino acid composition of this latter peptide, isolated and purified by dialysis of the digest followed by ion-exchange chromatography and preparative electrophoresis of the dialysable fraction, is given in Table II. The unidentified component of this peptide was isolated from acid hydrolysates by preparative paper electrophoresis (Hunt and Jevons, 1965b). It has a strong positive charge, and reacts strongly with ninhydrin and the Elson-Morgan reagent; these are properties suggesting both an amino acid and an amino sugar character. A weak reaction on paper with aniline hydrogen phthalate was obtained. The infrared spectrum (Fig. 10) is reminiscent of amino acid hydrochlorides. On the Technicon amino acid

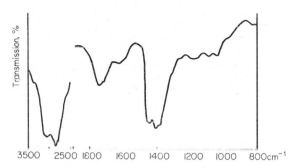

Fig. 10. Infrared spectrum of the unidentified component of the polysaccharide-peptide complex of *Buccinum undatum* hypobranchial mucin. Nujol mull (from Hunt and Jevons, 1965a).

analyser column this unknown compound emerges 12 ml, after cysteic acid and well before the methionine sulphoxides.

Thus, apart from minor differences, the glucan sulphate of *Buccinum* mucin and the S-rich glucan of *Charonia* mucin are probably closely similar in structure. That there are minor differences is suggested by the fact that the *C. lampas* glucan sulphate precipitates at room temperatures with potassium ions while the *B. undatum* material requires a lower temperature. It is almost certain that the S-rich *C. lampas* glucan sulphate also contains covalently bound peptide since it seems to be necessary to resort to pro-teolytic digestion to obtain material containing less than 1% protein (Iida, 1963).

The high viscosity of *Buccinim* hypobranchial mucin appears to be de-rived from the extended chains of the highly-charged polysaccharide sul-phate molecules. Treatment of the whole mucin, with Pronase, digests 70% of the associated protein without any accompanying fall in viscosity (Fig. 11 Hunt and Jevons, 1966). The structural properties of the fresh mucin, i.e. viscoelasticity principally, however may be conferred by the protein or the network association of protein and polysaccharide sulphate chains. The situation may be comparable to that found in vertebrate synovial fluid where hyaluronic acid and protein are associated. Here also 65% degrada-tion of the protein by trypsin and chymotrypsin failed to affect the high viscosity of the material (Ogston and Sherman, 1959). Comparison of the viscosities and chain lengths of the glucan sulphate and hyaluronic acid serves to demonstrate well the effect which chain length has on the viscous behaviour of highly-charged polymers in solution. In contrast with a value of 5500 ml/g for the intrinsic viscosity of native hyaluronic acid, with a chain length of 25,000 Å (Blumberg and Ogston, 1957; Schubert, 1964) dissolved in 0.2 M sodium chloride, the viscosity of the glucan sulphate, with a chain length of 3200 to 3500 Å, under similar conditions is only 500 ml/g. The

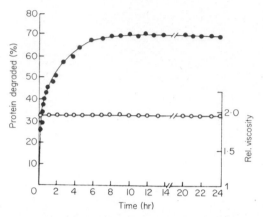

Fig. 11. Time course of proteolytic digestion of whole hypobranchial mucin from *Buccinum undatum* in aqueous solution. ●, Proteolysis measured by ninhydrin reaction and expressed as a percentage of the total degradation obtained by hydrolysis of an equal quantity of mucin, in 5·7 N-hydrochloric acid at 100° for 16 hours. ○, Viscosity of the mucin relative to water at the same temperature (37°) (Hunt and Jevons, 1966).

difference would seem to be due to the disparity in molecular domains which the polysaccharides occupy in solution. Long chains will fill larger domains than short ones and hence increase the possibilities of interaction even at low concentrations.

While the rheological properties of whole mucin suggest a fibrous microstructure, electron microscopic examination of the same material shows, at 36,000 × magnification, the presence only of large amorphous masses of spherical or ellipsoidal particles with diameters in the range 250 to 300 Å.

Pecten maximus *Glucan Sulphate*

Doyle (1967a) has recently demonstrated the presence of a glucan sulphate in the mantle mucin of the lamellibranch mollusc *Pecten maximus*. The secretion is analogous to that of the gastropod hypobranchial gland since both originate in gland cells of the mantle epithelium.

Papain-digested preparations of *Pecten* mantle tissue were subjected to ion-exchange chromatography on DEAE-Sephadex A50, elution being stepwise by sodium chloride solutions. The fraction eluted by 4·0 M sodium chloride contained a polysaccharide sulphate which on hydrolysis yielded only glucose and a trace of pentose. As with the *B. undatum* hypobranchial mucin this glucan sulphate was associated with a glycoprotein component (Chapter 12).

Busycon canaliculatum *Glucan Sulphate*

The final glucan sulphate to be described does not originate in the gland cells of the molluscan epithelium. This material, a constituent of the cartilage tissue in the odontophore of *Busycon canaliculatum*, has already been referred to in Chapter 2.

The presence of what appeared to be a glucan sulphate, replacing chondroitin sulphate in the odontophore cartilage of *Busycon*, was first noted by Lash (1959) and this view was confirmed by the more detailed study of Lash and Whitehouse (1960). The glucan sulphate was extracted from the tissue by tryptic digestion. Degraded protein was removed from the digest by dialysis and undigested protein precipitated with trichloracetic acid. The material which remained in solution was metachromatic to thionin and had an electrophoretic mobility, on paper, between chondroitin sulphate and heparin.

Hydrolysis of the polysaccharide yielded glucose as the only monosaccharide component and this seemed to be the D isomer since it was attacked by glucose oxidase. The sulphate content, as SO_4^{--}, was 35% while the reducing equivalent after hydrolysis, as glucose ,was 57%. The monosaccharide to ester sulphate ratio is therefore approximately 1 to 1, as is the case for *Buccinum* glucan sulphate. The unhydrolysed reducing equivalent was 0·5% suggesting a degree of polymerization of at least 100.

The intact mucopolysaccharide had no colour reaction with iodine nor did a partially desulphated preparation. This would suggest that an α-linked structure is unlikely and this is also suggested by the low optical rotation which was obtained. Thus a β structure seems likely bringing this material into the same category as the *Buccinum* and *Charonia* cellulose sulphates although definite evidence for the position of the glycosidic links was lacking.

Whether this glucan sulphate has any specific association with protein is not clear since it was of course isolated by methods designed to remove as much protein as possible. The α amino content of the purified polysaccharide was 0·08%. There must therefore be some residual peptide moderately firmly attached.

Biosynthesis of Glucan Sulphates

The biosynthesis of the charoninsulphuric acids has been the subject of a number of detailed studies by Suzuki and his co-workers and several aspects of the problem have been well elucidated.

In any study of polysaccharide sulphate biosynthesis one of the most important details which the investigator wishes to establish is the point along the biosynthetic path at which actual sulphation of the carbohydrate takes place. Does sulphation proceed concomitantly with the formation of glycosidic bonds or are the sulphate groups added later in the synthesis of the polysaccharide, perhaps even after the chain is complete? Alternatively, are monosaccharide residues sulphated first and then built up to give the polymer? Each of these possibilities may be biochemically

c*

feasible and the tissue may, *in vitro*, be capable of performing the synthesis by one or more routes to a greater or lesser extent. The investigation of the sulphation of mucopolysaccharides in the oviduct of hen has in fact yielded confusing and contradictory results of this type. For example, in one series of investigations sulphated monosaccharide precursors, activated as nucleotide derivatives, have been isolated while from another study of the same system evidence has accrued to suggest the sulphation of oligosaccharides.

In early studies of *C. lampas* hypobranchial gland Suzuki demonstrated, by analysis of gland extracts and by autoradiography of gland sections, that inorganic ^{35}S-labelled sulphate, injected into the muscle tissues of the living animal, was incorporated into charoninsulphuric acids located around the nuclei of the gland cells. *In vitro* incorporation, using gland tissue slices, was also demonstrated (Suzuki, 1956; Suzuki and Ogi, 1956). Takahashi (1959) was able to show that *in vivo* labelled sulphate incorporation involved both the S-poor and S-rich glucan sulphates and that the incorporation was not due to mere exchange of ^{35}S with ^{32}S. The rate of incorporation into S-poor glucan sulphate was greater than into the S-rich material.

No incorporation into charoninsulphuric acids of inorganic ^{35}S-sulphate could be demonstrated by Suzuki *et al.* (1957) when they used an acetone dried powder of the hypobranchial gland as the enzyme source. Sulphate, however, was incorporated when the inorganic source was replaced with p-nitrophenyl ^{35}S-sulphate as the sulphate donor. Cell-free extracts also could be demonstrated to have the ability to utilize p-nitrophenyl sulphate as an ester sulphate source (Suzuki *et al.*, 1959).

While these crude gland extracts could transfer sulphate to polysaccharide from aryl sulphate, two aryl sulphatases with different pH optima at pHs 5 and 6, separated from the extracts and purified by zone electrophoresis, were not able to bring about the transfer. An aqueous extract of the liver of *Charonia*, however, did show transfer activity and a purified enzyme preparation from this extract, having both aryl sulphatase and glucosulphatase activities, was able to transfer sulphate from p-nitrophenyl sulphate to a non-dialysed acceptor preparation of charoninsulphuric acid. A dialysed preparation was inactive as an acceptor (Suzuki *et al.*, 1959). Egami and Takahashi (1962) have reported that there appears to be a substance with the properties of an aryl sulphatase present in dialysates of *Charonia* hypobranchial gland, and that the transfer reaction appears not to be affected by addition of ATP. The sulphation scheme shown in Fig. 12 was proposed to account for the observed phenomena (Suzuki *et al.*, 1959).

The presence of aryl sulphatases and the transfer of sulphate from aryl sulphate to polysaccharide, however, is most probably a "red herring". In higher animals sulphation of carbohydrate has been shown to proceed directly from inorganic sulphate via a nucleotide intermediate, adenosine

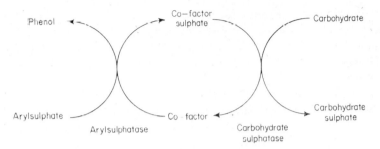

Fig. 12. Scheme for the enzymic sulphation of carbohydrate according to Suzuki *et al.* (1959).

3-phosphate 5-phosphosulphate (see Fig. 6, Chapter 2). Not only has this pathway of sulphation been demonstrated for a variety of vertebrate sulphated mucopolysaccharides but also for seaweed polysaccharide sulphates. We have therefore good reason to expect its occurrence in lower animals.

In this light therefore Yoshida and Egami (1965) re-examined the sulphation of charoninsulphuric acids by gland extracts and found that such extracts, freed from polysaccharide sulphate, could form adenosine 3-phosphate 5-phosphosulphate (PAPS) from inorganic sulphate and ATP. PAPS, labelled with ^{35}S prepared in this way, could be utilized by the gland to form ^{35}S-labelled charoninsulphuric acids. Only a small proportion of the labelled sulphate was actually transferred, but this was thought to be probably due to the presence in the gland and gland extracts of highly active PAPS sulphatase. The reason for the presence of this enzyme is uncertain since it would seem to be unfavourable for the formation of polysaccharide sulphate; perhaps its function is a regulatory one. In contrast to the earlier *in vivo* experiments, no significant incorporation into the S-poor glucan sulphate could be detected, over 95 % of the incorporated sulphate being found in the S-rich fraction. No other natural acceptor (other than water) for the sulphate was found in the glands. It must therefore be concluded that the phenolic sulphate intermediate, postulated in earlier reports, does not exist, or at least cannot derive sulphate from PAPS. The lack of an acceptor other than polysaccharide may also suggest that the polymerization of sulphated monosaccharides does not take place. What one would think we must accept however, is that sulphation of the cellulose sulphate (S-rich, glucan sulphate) takes place at the oligosaccharide level in part at least. If this were not so, sulphation of an insoluble molecule within the gland cell would have to be carried out. The suggestion by Egami and Takahashi (1962) that sulphation might be accompanied by metabolic transformation of the glycogen-like (S-poor) structure to a cellulose-like (S-rich) structure seems rather unlikely and without precedent. Hunt (1967) failed to detect free cellulose in the cellulose sulphate-secreting hypobranchial gland of *B. undatum*. To avoid the problem of insolubility some measure of sulphation would have to be initiated at least before the octasaccharide stage of biosynthesis.

The incorporation of ^{14}C-D-glucose into charoninsulphuric acids by hypobranchial gland tissue slices from *C. lampas* was observed by Iida (1961). A 0·2 to 0·3% utilization of the glucose for polysaccharide sulphate synthesis was obtained and of this a slightly greater incorporation into S-poor glucan sulphate than S-rich polysaccharide was noted. Glucose incorporation was inhibited by the uncoupler 2,4 dinitrophenol.

Later studies (Iida, 1963b) demonstrated that glucose incorporation into the glucan sulphates by cell-free extracts of gland could also be achieved. This incorporation had an absolute requirement for ATP and magnesium. Almost all of the ATP was converted to AMP during the incorporation reaction and it was shown that UTP could neither replace ATP in the reaction nor enhance incorporation in the presence of ATP. Almost all of the glucose incorporating activity of the extract remained in the supernatant after centrifugation at 6000 r.p.m. The sediment, which contained most of the S-rich glucan sulphate and which was inactive enzymatically, had the ability to stimulate the incorporation of glucose into S-rich glucan sulphate when recombined with the enzymically active supernatant. Incorporation of glucose actually proceeded more rapidly into S-poor glucan sulphate, in agreement with the earlier observation, but this rate was diminished by the presence of the S-rich material. It is worth noting that both sulphate and glucose incorporation rates are higher for S-poor charoninsulphuric acid, suggesting a higher overall rate of synthesis of this molecule over the S-rich material. In spite of this the S-rich glucan sulphate seems to be present in higher concentrations *in vivo*. The cell-free system also was demonstrated to form glucose-6-phosphate from the labelled glucose precursor while the 6000 r.p.m. supernatant was observed to produce inorganic phosphate from glucose-1-phosphate. This process was stimulated by AMP, S-poor charoninsulphuric acid or glycogen and resulted in the synthesis of a product giving a red-brown colour with iodine. Whether this represents the action of a glycogen phosphorylase having a role in the synthesis of the polysaccharide core of S-poor charoninsulphuric acid is not clear.

The failure to involve UTP in the incorporation reactions seems to require further clarification in view of the now well-established role of UDP-glucose in the biosynthesis of glucose-containing polysaccharides; it has been implicated in both glycogen and cellulose synthesis (Hassid, 1962). The synthetic role of phosphorylase and glucose-1-phosphate together would seem to have been largely refuted as a pathway for *in vivo* glycogen synthesis. In

Acetobacter cellulose synthesis by a cell-free system has an absolute requirement for ATP although UDP-glucose is believed to be involved in the synthesis. Starch synthetase from bean uses ADP-D-glucose while cellulose synthetase in higher plants seems to require GDP-D-glucose (Siegel, 1968).

Hunt (unpublished) has observed the incorporation of ^{14}C-labelled glucose into *B. undatum* glucan sulphate following injection into the pedal arteries of living animals. Autoradiography of gland sections showed incorporation to have taken place principally around the nuclei of the mucous cells at the extreme tips of the glandular leaflets. The implication would seem to be that synthesis has halted in the lower regions of the leaflets where the gland cells are storing mucin. This is born out by microscopic examination which indicates that cells at the leaflet tips are smaller (younger?) and contain less mucin than their large distended counterparts lower down the leaflet.

GLYCIDAMINAN SULPHATE

The hypobranchial secretions of at least two marine gastropod molluscs thus contain glucan sulphates, both of which include a sulphated form of cellulose. Not all hypobranchial mucins fall into this pattern, however.

An unusual and apparently unique acid polysaccharide has been isolated from the hypobranchial mucin of the gastropod *B. canaliculatum*. This is the same organism which contains a glucan sulphate in its cartilage tissue. Since the function of this mucin presumably is largely similar to that of the mucins from the animals already described it is hardly surprising that many of the physical properties should be similar. Thus it exhibits properties of syneresis, thixotropy and anomalous viscous behaviour which closely parallel those of *B. undatum* mucin (Ronkin, 1955).

Sedimentation of the mucin at about 30,000 g yields two components: one a sediment, which upon examination in the electron microscope appears to consist of agregates up to 4000 Å diameter, the other a supernatant containing particles with diameters between 250 and 1000 Å. The ropy-flow characteristics of the mucin seem to be associated with the 4000 Å material.

The fresh mucin has the property of lowering the surface tension of sea water by up to 30%. Ronkin (1955) considered this property to be of some biological significance since it would impart a detergent action to the mucin in its role as a mantle cavity cleaning agent. Application of the Gibb's adsorption isotherm to the data revealed the proportion of dissolved mucin adsorbed in the surface layer at various concentrations and it was shown that the most efficient wetting action would take place at about ten-fold dilution. Dilution, at the surface of the mucin sheet, would of course take place *in vivo*. It has also been suggested that the surface activity of such mucins would aid detritus feeding, wetted particles being more readily susceptible to enzyme action.

The intact mucin seems to show more ultrastructure than *Buccinum* material. Chromium shadowed smears of mucin, examined in the electron microscope, contained large elliptical bodies, length $1·06 \pm 0·07 \mu$ and width $0·28 \pm 0·025\mu$, with fine transverse striae spaced at $0·022 \pm 0·0015\mu$ intervals and an abundance of sculpturing. Apart from these particles there were also present masses of oriented fibres about 0.028μ diameter and of indeterminate length (Ronkin, 1952b). Strands pulled from masses of the mucin contained similar bodies as well as broad ribbons, also of indeterminate length, of $1·41 \pm 0·27 \mu$ diameter. The margins of these ribbons had double edges appearing as two opaque lines, $0·037 \pm 0·0038 \mu$ apart, separated by a clear space.

The mucin contains both protein and polysaccharide. Bacila and Ronkin (1952) separated the polysaccharide from the protein by stirring the fresh mucin with chloroform and precipitating the aqueous phase with ethanol. Acid hydrolysis of the polysaccharide precipitate yielded products which when treated with phenyl hydrasine gave glucosazone, galactosazone and lactosazone. These derivatives appeared from other evidence to have been derived from 2-amino-2-deoxy-D-glucose, 2-amino-2-deoxy-D-galactose and the amino sugar equivalent of lactose. This would suggest a structure for the polysaccharide based upon an alternation of $\beta 1 \rightarrow 4$ linked 2-amino-2-deoxy-D-glucose and 2-amino-2-deoxy-D-galactose residues.

A more detailed study of the physical characteristics and chemical structure of this mucin has been made (Kwart and Shashoua, 1957, 1958; Shashoua and Kwart, 1959).

Optimum conditions for the extraction of mucin from the excised glands were determined and the effects of a range of ions on the secretion process examined. The largest quantities of mucin were obtained when the glands were placed in solutions at pH 7·2. Monovalent cations seemed to stimulate secretion while divalent cations had an inhibitory effect. The gland appeared to have a specific affinity for calcium and barium ions.

The freshly extracted mucin, which contains as little as 0·3% solids, had a viscosity of 60 to 100 centipoises in 0·5 M sodium chloride solution. Plots of reduced viscosity against concentration for mucin solutions, in distilled water or 0.5 M sodium chloride, were non-linear. This is a type of behaviour typical of polyelectrolytes in solution. In the presence of calcium ions however these curves became linear and an extrapolation to infinite dilution could be made yielding a value of $1·73 \times 10^2$ ml/g for the intrinsic viscosity. These results would suggest that the charge dependent forces in the molecules are too great to be overcome by the gegenion environment provided by 0·5 M sodium chloride, but that calcium ions exert a stabilizing effect on concentration dependent molecular expansion, possibly by forming intra- and intermolecular cross links. The viscosity was also strongly pH dependent being at a maximum in the region between pH 7 and 8. Charge on the molecule seemed to be at a minimum in this pH range since viscosity determination carried out in a range of phosphate buffers indicated that this was the only region in which reduced viscosity showed a linear dependence on $\sqrt{C}$ (Fuoss-Strauss equation, $\eta_{sp}/C = A/1 + B\sqrt{C}$, where A and B are constants and C is concentration).

Surprisingly, the light scattering studies of Kwart and Shashoua (1958) carried out on mucin solutions showed only a low dependence of angular dissymmetry on concentration. Evidently the mucin particles maintain a roughly spherical shape on dilution. Particle sizes, calculated from dissymmetry measurements, ranged from diameters of 1760 Å in 0.5 M sodium chloride through 1890 Å in water to 2380 Å in 7% sodium chloride solution. The weight average molecular weight of the mucin particles, calculated from 90° turbidity measurements, was of the order 2·7 to $3·5 \times 10^7$, while a determination carried out on the polysaccharide alone (extracted according to Bacila and Ronkin, 1952) gave a value of $2·5 \times 10^6$. Calculation of the diameter of the anhydrous mucin particle from the molecular weight gave a value of 280 Å, a value close to that obtained by direct measurement in the electron microscope (200 to 300 Å). Comparison of these values with the light scattering results suggests an 800 to 1000-fold increase in volume on swelling in solvent.

Analytical data for the mucin, obtained by Shashoua and Kwart (1959), are given in Table III. Sodium and calcium ions seem to be integral parts of the molecular system.

TABLE III

Analysis of Fractions from *Busycon canaliculatum* Hypobranchial Mucin[a]

Mole Ratio Na/Ca in whole mucin after dialysis for		
3 days	5·5	
	Polysaccharide fraction	Protein fraction
C (%)	23·4	47·9
H (%)	4·6	6·6
N (%)	4·86	11·3
S (%)	7·2	—
Ash (%)	12·2	0

[a] Shashoua and Kwart (1959).

The protein and polysaccharide components were separated by stirring with chloroform, as already described. This procedure seems to work by denaturation of the protein at the solvent-water interface and, according to Shashoua and Kwart, requires in this particular case the presence of electrolyte in the solution. No fractionation could be achieved when dialysed mucin solutions were stirred with chloroform.

The polysaccharide had an isoionic point of 5·9. Treatment with barium chloride of mucin solutions acidified with hydrochloric acid gave an immediate precipitate of barium sulphate. Some form of readily released sulphate thus appears to be present.

Acid hydrolysates of the polysaccharide were shown, by paper and ion-exchange chromatography, to contain 2-amino-2-deoxy-D-glucose and 2-amino-2-deoxy-D-galactose as the only monosaccharide constituents. The polysaccharide thus appears to be a sulphated polymer of the two hexosamines with probably the $\beta 1 \rightarrow 4$ linked structure already suggested. The mode of sulphation suggested by Shashoua and Kwart is an unusual one and seems to be open to some measure of criticism. In essence they suggested that the amino groups of the hexosamines, rather than being acetylated or even sulphamated as in heparin, carried the sulphate residues in labile form as bisulphate groups with either sodium or calcium as the counter ion. This hypothesis was based largely on the observation already mentioned that acidified polysaccharide gave a precipitate of barium sulphate with barium chloride and on the observation of certain features of the infrared spectrum which was unfortunately not published in full. The infrared spectral data reported are given in Table IV. It was also reported that bands due to car-

TABLE IV

Infrared Spectral Characteristics for *Busycon canaliculatum*
Hypobranchial Mucin[a]

Wave length μ[b]	2·9	6·15		8·15	9·4
Reported correlation	—OH	—NH₂,	RNH₃⁺	—SO—	—SO₂—

[a] Shashoua and Kwart (1959).
[b] Where relevant wave length has been converted to wave number in the text.

boxyl, carboxylate anion and —CO—NH— were absent and we can therefore assume that there were no strong bands in the 1500 to 1600 cm⁻¹ and 1700 to 1800 cm⁻¹ ranges. The band at 1626 cm⁻¹ was therefore interpreted as being due to RNH_3^+. This band however requires more careful interpretation. It should be noted that while —NH_3^+ deformations do indeed occur in this region, the infrared spectra of hexosamine hydrochlorides show bands at 1581 and 1538 cm⁻¹ as well as at 1613 cm⁻¹ and one might reasonably expect some similar characteristics for the sulphate salts of amino sugars.

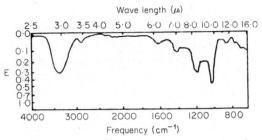

Fig. 13. Infrared spectrum of 2-sulphamido-2-deoxy-D-glucose (from Dietrich, 1968)

Moreover the sulphate absorptions are more in keeping with covalent complexes than inorganic sulphate, which usually manifests itself as a strong absorption between 1000 and 1150 cm^{-1} rather than the more typically covalent band at 1200 to 1250 cm^{-1}. In fact the details of the spectra given closely resemble the spectrum of 2-deoxy-3-sulphamino-D-glucose recently published by Dietrich (1968) (Fig. 13). Sulphation of the same type found in heparin seems a far more likely interpretation of the data available.

The protein moiety of the mucin contains a high percentage of acidic amino acids (Table V), the mole ratio to basic amino acids being approximately 3 to 1. As with the protein obtained from *Buccinun* mucin (Chapter

TABLE V

Amino Acid Composition of the Protein Fraction of *Busycon canaliculatum* Hypobranchial Mucin[a]

Amino acid	μ moles/100 μ moles total amino acids
Cysteic acid	1·8
Aspartic acid	10·0
Threonine	5·9
Serine	5·0
Glutamic acid	12·0
Proline	2·70
Glycine	7·12
Alanine	7·23
Valine	6·00
Isoleucine	4·39
Leucine	8·13
Tyrosine	4·04
Phenylalanine	1·05
Histidine	2·28
Ornithine	0·30
Lysine	3·41
Arginine	1·42

[a] Shashoua and Kwart (1959).

12) ornithine is present and in fact the amino acid analyses of the two proteins are rather similar. Shashoua and Kwart do not say whether or not they detected any protein or peptide covalently bound to the polysaccharide.

It was suggested that the protein component of the mucin was linked to the polysaccharide sulphate via calcium bridges joining secondary carboxyl groups of acidic amino acids to sulphate groups on the polysaccharide. What we have already said about the form of the sulphate group need not necessarily affect the overall picture of this structure since sulphamino groups or ester sulphate groups would equally carry the negative charge required for the formation of a bridge. In fact, accepting a covalently bound sulphate residue structure would make the calcium bridge structure more credible. Certainly removal of calcium ions from the mucin, by dialysis against EDTA at pH 4·2 and then against water, caused the protein to separate from the polysaccharide as a precipitate. Since an amide content for the protein was not published we do not know what proportion of the aspartic and glutamic acid residues had acidic side chains available for this type of cross-linking. It was suggested that the sodium ions which persisted in the polysaccharide-protein complex after dialysis were bound to those sulphate groups not involved in the divalent calcium bridges.

Setting aside details of structure, the overall organizational pattern of this mucin is similar to that of *Buccinum*, where a highly-charged polysaccharide is associated with a protein molecule.

Neptunea antiqua Hypobranchial Mucin

A fourth hypobranchial mucin acid mucopolysaccharide has been described which seems to fall between those of *C. lampas* and *B. undatum* on the one hand and *B. canaliculatum* on the other; this is from the gastropod *Neptunea antiqua* (Doyle, 1964).

Papain digestion of the mucin and fractionation of the digest on columns of Sephadex G75 yielded a high molecular weight metachromatic polysaccharide. The metachromasia was stable at pH2 and therefore appeared to to be due to ester sulphate. Analysis of hydrolysates showed that hexosamine was present in a 1 to 4 ratio with glucose. The hexosamine was made up of 2-amino-2-deoxy-D-glucose and 2-amino-2-deoxy-D-galactose in a 39 to 61 ratio. Fucose was also present but no uronic acid could be detected. Paper electrophoresis of the papain-digested polysaccharide yielded a single component with an electrophoretic mobility approximately half that of mammalian chondroitin sulphate. The evidence available is probably insufficient for one to decide whether this is a homogeneous heteroglycan or whether it is in fact a mixture containing glucan sulphates and glycidaminan sulphate.

LIMACOITIN SULPHATE

Both the mucus mucin and the foot mucin of a Japanese snail (species unspecified) were found by Suzuki (1941) to contain mucopolysaccharide. Hydrolysis gave 2-amino-2-deoxy-D-glucose, identified as the penta-benzoate derivative. The amino sugar contents of the mucus mucin and foot mucin were 11·6 and 4·3% respectively, while the acetyl contents were 2·83 and 1·4%. Both mucins contained galactose (5.4 and 4·07% for the mucus and foot mucins respectively) but only the mucus mucin contained hexuronic acid (6·1% as glucuronic acid) and hydrolysable sulphur (1·2%).

The mucus mucin of the snail was further characterized in a series of studies by Masamune and his co-workers (Masamune and Osaki 1943, Masamune et al., 1947a,b; Masamune and Yoshizawa, 1950, 1956). A polysaccharide sulphate was isolated from the mucin after treatment with formaldehyde. This material was given the name limacoitin sulphate and was shown to contain ester sulphate, L-fucose, galactose, mannose, galacturonic acid and 2-acetamido-2-D-glucose. Galacturonic acid occurs only rarely in animal polysaccharides although it is found widely in plants; the only known sources in vertebrates are a polysaccharide component of defatted human brain (Stary et al., 1964) and an acid mucopolysaccharide secretion of cartilaginous fish (Doyle, 1967b).

The slime secreted by the snail Otella lactea also contains a sulphated mucopolysaccharide in which galacturonic acid is a major component (Pancake and Karnovsky, 1967). The slime contains about 5 mg protein per ml and up to 1 mg uronic per ml. The mucopolysaccharide was isolated by cetylpyridinium chloride precipitation followed by chromatography on DEAE cellulose, or by treatment with Pronase. The polysaccharide obtained by either method contains about 5% protein. The mucopolysaccharide was homogeneous by paper electrophoresis and contained N-acetyl hexosamine, uronic acid and ester sulphate in equimolar ratios. The amino sugar was identified as 2-amino-2-deoxy-D-glucose and the uronic acid as galacturonic acid. Glucose, a trace of fucose and an unidentified sugar were also present.

The glandular region of the common duct in the reproductive system of the slug Ariolimax columbianis contains a highly sulphated acid mucopolysaccharide (Meenakshi and Scheer, 1968). Hydrolysis of this material yielded galactose, fucose and 2-amino-2-deoxy-D-glucose.

HORATIN SULPHURIC ACID

A new type of sulphated polysaccharide has been isolated from the liver (digestive gland) of the whelk C. lampas by Inoue (1965; Inoue and Egami, 1963). This heteropolysaccharide has been given the name horatin sulphuric acid. The livers were homogenized in water, boiled for 5 minutes, centrifuged (4000 r.p.m. for 15 minutes), the supernatants precipitated with one volume of ethanol, the precipitate collected by centrifugation and redissolved in water. After reprecipitation with two volumes of ethanol and dissolving once again in water the crude mucopolysaccharides were precipitated as their trypaflavine complexes. The complex with trypaflavine was dissolved in 5 M sodium chloride and horatin sulphuric acid precipitated from the solution with ethanol. A further resolution in water and reprecipitation with ethanol followed. The yield of mucopolysaccharide from twelve animals (600 g) was 900 mg at this stage of the preparation. Approximately 30 % of this crude preparation was ribonucleic acid. Chromatography of the crude mucopolysaccharide on DEAE Sephadex A50 by stepwise elution with sodium chloride yielded six sulphate-containing fractions which were rechromatographed on continuous gradient eluted columns.

Horatin sulphuric acid is in fact a generic name given to a group of acid mucopolysaccharides found in the liver of *Charonia*. The compositions of the six mucopolysaccharides separated are given in Table VI. All of the polysaccharides contained sialic acid, hexose and hexosamine in addition to sulphate. The composition is thus rather typically that of a glycoprotein prosthetic group setting aside the presence of sulphate. The protein contents are however low and the molecular weights, which were estimated roughly by dialysis or gel-filtration, of the order 5000.

After the second chromatographic purification several sulphated polysaccharides, apparently homogeneous by paper electrophoresis, were obtained. Two of these fractions were investigated in further detail. The compositions of these two fractions are given in Table VI. The sialic acid present in both fractions was identified as the N-glycolyl derivative (see Chapter 13). The infrared spectra of the mucopolysaccharides (Fig. 14)

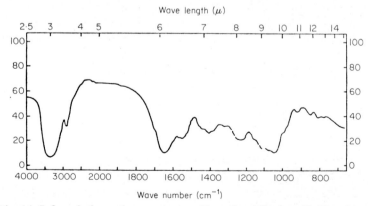

Fig. 14. Infrared absorption spectrum of horatinsulphuric acid, Fraction 4.

clearly show that the sulphate is present in the ester form and in the case of one of the fractions probably situated in an axial situation.

TABLE VI

Analytical Data for Horatinsulphuric Acids Obtained by Chromatography on DEAE Sephadex[a]

First column Peak no. and eluant	N (%)	SO₄ (%)	Hexose (%)	Hexos- amine (%)	Sialic Acid (%)
1. 0·01 M Tris-HCl pH 8·0	6·4	31	23	5·5	1·5
2. 0·3 M NaCl	6·3	11	15	6·1	3·2
3. 0·4 M NaCl	6·5	10	16	6·4	3·8
4. 0·5 M NaCl	5·4	12	18	7·2	2·8
5. 0·7 M NaCl	5·5	16	18	11·8	2·6
6. 1·0 M NaCl	5·7	24	26	11·8	2·1

Second column. Rechromatography of peaks 3 and 4 separately
Carbohydrate composition of the major sialic acid-containing fractions 3′ and 4′

Fraction no.	Molar ratio					
	Fucose	Mannose	Glucose	Galactose	2-amino- 2-deoxy- D-glucose	2-amino- 2-deoxy- D-galac- tose
3′	2	1	1	2	1	1
4′	2	1	1	2	2	2

[a] Inoue (1965).

The possibility that the horatin sulphuric acids had an exogenous origin was rejected. Dialysis effusates from the crude ethanol-precipitate were shown to contain sulphated oligosaccharides. *In vivo* incorporation of injected ^{35}S-labelled inorganic sulphate into these oligosaccharides could be demonstrated, although not into the horatin sulphuric acids themselves.

The nitrogen contents of the various horatin sulphuric acids do not tally exactly with the amino sugar nitrogen, suggesting that a small proportion of peptide may be present.

Lloyd and Lloyd (1963) have also demonstrated the presence of acid mucopolysaccharides in the viscera of the limpet *Patella vulgata*. This material, which was prepared by potassium chloride extraction, tryptic digestion, Sevag treatment, dialysis, alcohol precipitation and cetylpyridinium chloride precipitation, contained fucose and/or xylose, galactose and sulphate. The authors believed at the time of publication that the poly-saccharide was of exogenous origin, probably from seaweed. The identifi-cation of similar substances in *C. lampas* may lay this interpretation open to reassessment.

SULPHATED MUCOPOLYSACCHARIDES IN MOLLUSCAN SHELLS

Numerous writers have suggested that the formation of the crystals of calcium carbonate in the calcified tissues of molluscan shells may depend upon induction by an organic matrix (Wilbur and Watabe, 1963; Wilbur and Simkiss, 1968). While many workers consider this matrix to be a collagen-like protein (Chapter 15) in analogy to the formation of calcified tissues in vertebrates, it has been suggested, largely on the basis of histochemical tests, that sulphated acid mucopolysaccharides secreted by the mantle may also be involved. Precedents exist here also in the field of vertebrate tissue calcification (Bevelander and Benzer, 1948; Ransom, 1952; Tsuji, 1955, 1960; Durning, 1957; Trueman, 1957; Abolins-Krogis, 1963). Mucopoly-saccharides detected by paper electrophoresis have been identified in the extra-pallial fluids of several molluscan species (Kobayashi, 1964).

A sulphated acid mucopolysaccharide has been isolated from the organic matrix of the shell of the oyster *Crassostrea virginica* where it occurs in conjunction with protein (Simkiss, 1965) (see also Chapter 15). Shells were decalcified by dialysis against 20% EDTA at pH 8·0. The matrix obtained was dialysed against water and freeze-dried. Extraction of this material with a solution of potassium chloride (30% w/v) and potassium carbonate (1% w/v) yielded a solution which, when dialysed and made 80% with respect to ethanol and 2% with respect to acetic acid, gave a precipitate of acid mucopolysaccharide. The infrared spectrum of this material (Fig. 15) contained bands characteristic of carbohydrate, ester sulphate, amide and carboxyl groups. The data available seems inadequate to justify the hypothesis that the material was similar to chondroitin sulphate. The muco-polysaccharide gave an X-ray diffraction pattern consisting of two diffuse rings corresponding to spacings of 5·3 and 9·4 Å.

Horiguchi (1956, 1959) has separated sulphated polysaccharides from the mantle tissues of the molluscs *Pteria martensii* and *Hyriopsis schlegelii* and has postulated their involvement in the calcification process. The tissues were homogenized in water, incubated with pepsin for three days and then with trypsin for one day. The digest was filtered and the filtrate made 0·25 N with respect to acetic acid and 2·5% with respect to calcium acetate. Following the addition of 1 to 2 volumes of ethanol to the solution a precipitate formed and this was recovered and dissolved in 5% sodium acetate-0·5 N acetic acid solution. This solution was deproteinized by Sevag treatment and the aqueous phase precipitated with ethanol. The precipitate

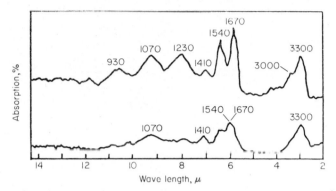

Fig. 15. Infrared spectra of the intact organic matrix from the shell of *Crassostrea virginica* (bottom) and the polysaccharide sulphate from the same source (top). (From Simkiss, 1965.)

was redissolved and precipitated again with ethanol. The content of crude sulphated mucopolysaccharide in the mantle of *P. martensii* was 37% of the dry weight and in the mantle of *H. schlegelii* 13·6%. The compositions of the two mucopolysaccharide preparations are given in Table VII. The adductor muscle, mantle residue, gill and other organs of *P. martensii* also contained sulphated mucopolysaccharides with rather similar compositions as did the viscera, foot and gill tissues of *H. schlegelii*. The glucan sulphate of *P. maximus* mantle tissue has already been mentioned.

The significance of sulphated polysaccharides in the calcification process is in some dispute. Their presence in the shell matrix might be interpreted simply as a consequence of the accidental trapping of mantle mucin. Alternatively it is possible that they may play a role as ion-exchange agents binding calcium ions and thereby initiating or inhibiting the deposition of these ions as inorganic calcium carbonate crystals. Simkiss (1964, 1965) has reported that calcite does not form readily in sea water, the inhibition apparently being due to the presence of a variety of interfering cations and anions. The presence of natural ion-exchange polymers may therefore be necessary to remove or act as a barrier to these ions in order that the insoluble calcite, rather than the less stable and more soluble form of calcium carbonate aragonite, may be deposited.

MOLLUSCAN ENZYMES WHICH DEGRADE POLYSACCHARIDE SULPHATES

The discovery of the charoninsulphuric acid group of polysaccharide sulphates originated in a search for the natural substrates of certain carbohydrate sulphatases found to occur in relatively large quantities in molluscs. The study of enzymes, derived from invertebrates, capable of releasing

TABLE VII

Compositions of Mucopolysaccharide Preparations from the Tissues of *Pteria martensii* and *Hyriopsis schlegelii*[a]

	mg/g wet tissue	mg/g dry tissue	N (%)	Protein (%)	Hexos-amine	Reducing Sugar (%)	S (SO$_4$) (%)	Ca (%)
P. martensii								
Mantle mucilage	19·30	371·15	6·89	43·08	4·35	39·0	1·76	2·3
Mantle residue	4·04	31·36	5·21	31·99	2·91	34·40	0·48	1·90
Gill	1·50	21·90	2·40	15·01	3·24	37·60	0·91	2·20
Adductor muscle	17·40	87·00	0·48	3·00	tr.	55·5	0·48	6·0
Other organs	3·07	14·90	1·98	12·40	2·70	28·80	2·70	3·75
H. schlegelii								
Mantle	6·80	136·00	0·43	2·61	tr.	59·90	1·12	1·10
Viscera and foot	18·55	125·34	0·64	3·92	tr.	73·70	1·28	1·10
Gill	0·05	0·42	—	—	—	—	—	—

[a] Horiguchi (1956).

ester sulphate from polysaccharide and monosaccharide sulphates and of splitting the glycosidic linkages of polysaccharide sulphates, has in fact been almost if not completely confined to the gastropod molluscs.

A glucosulphatase which would release the ester sulphate residue, as inorganic sulphate, from glucose-6-sulphate was first noted by Soda and Hattori (1931) in the snail *Eulota* and later in the liver (digestive gland) of *C. lampas* (Soda and Hattori, 1933). *C. lampas* glucosulphatase has its greatest enzymic activity towards glucose-6-sulphate, but it will also hydro-lyse the sulphate ester groups of other mono- and disaccharides (Soda, 1936; Soda and Egami, 1934; Soda and Kawamatsu, 1936; Soda *et al.*, 1936).

In the case of *Charonia* liver the greater part of the sulphatase activity occurs in the 10,000 g supernatant which remains after centrifugation of a tissue homogenate (Takahashi, 1960 a, b, c). Fractionation of this super-natant on CM-cellulose separated a variety of enzymic components which included, in addition to glucosulphatase: cellulosesulphatase, chondrosul-phatase, arylsulphatase, steroid sulphatase and myrosulphatase. The column chromatography resulted in a 160-fold purification of the gluco-sulphatase and a six-fold purification of the cellulose sulphatase.

The glucosulphatase had a pH optimum at pH 5·8 and was inhibited by fluoride and iodide. Phosphate inhibited competitively while cyanide, 8-hydroxyquinoline and EDTA hardly affected its activity. The enzyme has been reported to hydrolyse adenosine monosulphate (Yamashina and Egami, 1953) but not inositol hexasulphate (Takahashi and Egami, 1959).

Crude extracts of the liver of *C. lampas* have been found to degrade fully sulphated celluloses to glucose and it has been suggested that the effect is due to the sequential action of three enzymes. Cellulose polysulphatase, it was thought, might act first, removing some but not all of the sulphate residues (those at the 6 position being more resistant to attack), the chain then being broken down to sulphated monosaccharides by a polysaccharase when the remaining sulphate is removed by glucosulphatase (Egami and Takahashi, 1962; Takahashi and Egami, 1960). Treatment of cellulose polysulphate with *Charonia* liver enzyme preparations, which had been either freed from associated polysaccharase activities or treated with fluoride, resulted in the formation of S-poor cellulose sulphates. Sulphatase-free polysaccharases were, on the other hand, far less active towards cellulose polysulphate than to S-poor cellulose sulphates, suggesting that the hypo-thesis of sulphate release preceding the breaking of glycosidic bonds is a correct one. Degradation of S-poor cellulose sulphates with purified *Charonia* liver polysaccharase resulted in the release of glucose-6-sulphate and glucose. Treatment of cellulose polysulphate, in dialysis bags, with crude liver extracts yielded dialysable intermediates consisting of glucose-2- and/or-3-monosulphates together with small amounts of glucose tri-sulphate.

The effects of partially purified preparations of cellulose polysulphatase on a variety of polysaccharide sulphate substrates were examined by Takahashi and Egami (1961). S-rich charoninsulphuric acid was rapidly desulphated while S-poor charoninsulphuric acid and dextran polysulphate were desulphated at much lower and similar rates. Chrondroitin sulphate A and shark cartilage chondroitin sulphate were desulphated at only very low rates and it seemed probable that the activity which was detected was due not to the cellulose sulphatase activity but rather to contaminating chrondrosulphatase. Heparin, amylose polysulphate, glycogen sulphate, sea urchin (*Hemicentrotus pulcherimus*) egg jelly polysaccharide sulphate (Chapter 6) and seaweed (*Chondrus* sp.) polysaccharide sulphate were not desulphated to any significant extent. Rather surprisingly the glucan sulphate from *B. caniliculatum* cartilage tissue was not desulphated by this enzyme preparation.

The association of a cellulose sulphate (S-rich charoninsulphuric acid) and a cellulose sulphatase in the same organism need not necessarily imply that the former is the natural substrate of the latter. The mucin, we should remember, is usually released into the sea as part of a cleaning mechanism and thus would not appear to require the presence of degradative enzyme; nor is the enzyme present in the secretory gland itself where turnover of the polysaccharide sulphate might be brought about. Moreover, the accompanying glycogen sulphate seems to lack a complementary degradative enzyme.

Apart from glucan sulphatase and glucosulphatase the *Charonia* liver extracts also contain chondrosulphatase and chondroitinase (Soda and Egami, 1938a). Chondrosulphatase has been separated from glucosulphatase (Soda and Egami, 1938a; Soda *et al.*, 1940) and shown to have a pH optimum at pH5. Phosphate inhibits the action of chondrosulphatase and permits the action of chondroitinase to proceed independently with the resulting release of small ester sulphated units.

Glycosulphatase activity has been demonstrated in extracts of the digestive organs of certain marine molluscs from British waters, particularly the common limpet *Patella vulgata* and the large periwinkle *Littorina littorea* (Dodgson and Spencer, 1953, 1954; Lloyd and Lloyd, 1961). The digestive gland of *B. undatum* was shown to be inactive towards glucose-6-sulphate. Dodgson (1961) was able to bring about a 20-fold purification of the glycosulphatase of *L. littorea* digestive organs, eliminating largely the accompanying chondroitinase, β-glucuronidase and β-N-acetylglucosaminidase activities. The enzyme showed its greatest activity with glucose-6-0-sulphate but would also degrade at a low rate glucose-3-0-sulphate and galactose-6-0-sulphate. The pH optima for these activities were in the range pH 5·2 to 5·9. Glycosulphatase activity in *Littorina* appears to be at its highest in July and falls to a minimum during the winter months. Preparations of *Helix pomatia* digestive juice showed little or no glucose-6-0-sulphatase activity.

Lloyd and Lloyd (1961) have found carbohydrate sulphatases in the digestive organs of *P. vulgata* with specificities for glucose-6- and -3-0-sulphates, galactose-6-0-sulphate, fucose sulphate, chondroitin sulphate and fucoidin. As already mentioned, these workers have also noted the presence of fucose and galactose-containing polysaccharide sulphates in the viscera of *P. vulgata* (Lloyd and Lloyd, 1963); it seems likely therefore that in this case the natural substrate has been identified.

A fucoidinase capable of splitting fucoidin to fucose-4-0-sulphate has been identified in the hepatopancreaii of *Abalone* sp., *Haliotus refescens* and *Abalone corrugata*; sulphatase activities were also present (Thanassi and Nakada, 1967).

ECHINODERM POLYSACCHARIDE SULPHATES

Aqueous extracts of ethanol-dried tissues from the trepang (sea-cucumber) *Stichopus japonicus* have yielded, on the addition of benzidine, a precipitate containing a sulphated polymer of L-fucose complexed with protein. The polysaccharide contained 60·1% fucose, 10·9% hydrolysable sulphur and 1·2% N. In addition the viscera were shown to contain a polysaccharide composed of 2-amino-2-deoxy-D-glucose, glucuronic acid and ester sulphate (Maki, 1956a; Maki and Hiyama, 1956). This material was considered to be similar to the elusive vertebrate mucoitin sulphate, once thought to be a constituent of gastric mucin and now believed to not actually exist (Levene and López-Suàrez, 1916; Brimacombe and Webber, 1964).

The tube feet or podia of echinoderms contain gland cells secreting substances with the histochemical properties of highly sulphated mucopolysaccharides and aminopolysaccharides. These materials have been demonstrated histochemically in *Astropecten aurantiacus*, *A. irregularis*, *Asterias rubens*, *Marthasterias glacialis*, *Echinid* (Defretin, 1952), *Antedon bifida*, *Echinocyanus pusillus*, *Echinocardium cordatum* (Nichols, 1959a, b, 1960) and *Amphipholis squamata* (Fontaine, 1955).

The integumentary mucous secretions of the ophiuroid *Ophiocoma nigra* have been investigated by Fontaine (1955, 1964). This animal when it is alarmed secretes a copious mucin from large multicellular, anastomosing glands located deep in the calcareous integument. The mucin probably fulfills a defensive role. Histologically the mucin has the properties of a highly sulphated polysaccharide.

The freshly collected mucin has a high viscosity and dialysed solutions show an intense gamma metachromasia with toluidine blue which is stable to incubation with testicular hyaluronidase. The mucin contains protein (biuret and Millon tests) which can be removed by tryptic digestion. The intact mucin gave a weak positive response to the Elson-Morgan reaction and it was suggested that there might be some terminal hexosamine residues. Acid hydrolysates of the mucin gave a strong colouration with the Elson-Morgan reaction, a positive reaction for sulphate (barium precipitate) and a weak reaction for phosphate (molybdate precipitate). The mucin exhibited marked anti-coagulant properties comparable with heparin.

The glands of the tube feet contain a secretion histochemically similar to the integumentary mucin while the integument itself also contains superficial unicellular glands whose secretion is an acid mucopolysaccharide associated with the feeding process.

Histological studies have suggested the presence of sulphated mucopolysaccharides and glycoproteins in the gut glands of the echinoids *Strongylocentrotus intermedius* (Fuji, 1961) and *Echinus esculentus* (Stott, 1955).

INSECT POLYSACCHARIDE SULPHATES

Evidence for the occurrence of sulphated acid mucopolysaccharides in the tissues and secretions of insects is sparse.

Marshall (1966a) has demonstrated by histochemical tests the presence of very highly charged acid polysaccharides in the spittle, proximal segments of the Malpighian tubules, hindgut contents and midgut intima of cercopid larvae (*Homoptera*) as well as in the granule zones of the Malpighian tubules from machaerotid larvae. The acid mucopolysaccharides are probably ester sulphated since they stained strongly with alcian blue and toluidine blue.

Paper chromatography of hydrolysed spittle from *Aenolamia varia saccharina*, *Cosmocarta abdominalis* and *Neophilaenus campestris* indicated the presence in each case of glucuronic acid, 2-amino-2-deoxy-D-glucose, glucose and rhamnose in addition to one or two unidentified sugars. Hydrolysates of the proximal segments of the Malpighian tubules from *C. abdominalis* also contained the same monosaccharides. No sugars could be detected in distal segment hydrolysates. The acid polysaccharide in the spittle appeared to be associated with protein. The evidence suggested that the spittle was a secretion of the Malpighian tubules. The mucopolysaccharide present in the Malpighian tubules is sensitive to testicular hyaluronidase (Marshall, 1966a, b).

Marshall (1966a) has suggested that the highly charged water-soluble mucopolysaccharide in the spittle functions as a surface active agent responsible for the maintenance of bubble stability and, in conjunction with protein, increases the cohesion and surface viscosity properties of the bubble film.

Marshall's results were not completely in agreement with those of Wilson and Dorsay (1957) who found pentoses, glucose and amino acids in both unhydrolysed and hydrolysed cercopid spittle, nor with those of Ziegler and Ziegler (1958) who found no sugars at all after hydrolysis.

Marshall (1968) has also investigated the chemical nature of the dwelling tubes of machaerotid larvae. Hydrolysates were found to contain glucuronic acid, 2-amino-2-deoxy-D-glucose and glucose.

MISCELLANEOUS MUCOPOLYSACCHARIDES

Unsulphated mucopolysaccharides are surprisingly rare in the invertebrates apart from such molecules as chitin or hyaluronic acid. Some of the materials cited here are very probably unsulphated or even uncharged: some of them may be sulphated but the evidence, usually histochemical, is inadequate to decide whether the charge on the molecule is due to sulphate or not. In any event they fall into no very definite category and they have therefore been brought together under the heading miscellaneous.

In an attempt to confirm Maki's observation of a polyribose phosphate in the liver of the squid *Ommastrephes* (Maki, 1965b) Rosenberg and Zamenhof (1962), while failing to detect this substance, succeeded in isolating a new polysaccharide from the same source. The squid livers were homogenized in 1·5 volumes of a mixture of 1 volume of 0·5 N NaOH and 0·12 volumes of saturated ammonium sulphate at pH 9·0. After incubation for 30 minutes at 50° the extract was centrifuged (3000 g) and the precipitate and oily upper layer rejected. The supernatant remaining was divided into three parts which were precipitated with 2, 3 and 4 volumes of 95% ethanol respectively. Each yielded a precipitate homogeneous on electrophoresis in borate buffer. Polysaccharide prepared in this manner was deproteinized by Sevag and phenol treatment and further purified by reprecipitation with ethanol and a final precipitation with lead acetate. The polysaccharide obtained, in a yield of 0·58% (of the original liver wet weight), contained 10·2% pentose and less than 0·15% phosphate. Acid hydrolysates of the polysaccharide contained xylose, arabinose, probably galactose, 2-amino-2-deoxy-D-glucose, 2-amino-2-deoxy-D-galactose, probably glucuronic acid, probably rhamnose and no ribose.

The polysaccharides of the shore crab *Hemigraspus nudus* have been examined by two groups of workers. Extraction of frozen, powdered crab with 2% perchloric acid solution yielded a solution which, on precipitation with ethanol and after removal of oligosaccharides and nitrogenous impurities on Norit A, yielded a polysaccharide consisting chiefly of glucose but also containing galactose, fucose, an unknown sugar and trace amounts of amino acids (Hu, 1958). Administration of uniformly labelled ^{14}C-glucose into intermolt crabs resulted in labelling of respiratory carbon dioxide, the polysaccharide but not chitin.

Meenakshi and Scheer (1959) extracted *Hemigraspus* digestive gland and cuticle with alkali and precipitated the extracts with ethanol. The cuticle contained an average of 937 mg of toluidine blue-positive acid mucopolysaccharide per 100 g of dry tissue. The digestive gland also yielded a polysaccharide which contained glucose, galactose and fucose, glucose being the most abundant sugar. The cuticle and digestive gland polysaccharides appeared to be identical. The toluidine blue metachromasia and the reported electrophoretic mobility suggests that this material is a sulphated polysaccharide.

Travis (1957, 1960a, b, 1963) has demonstrated histochemically the presence of acid mucopolysaccharides in the cuticle and digestive organs of Decapod crustacea. The presence of these acid mucopolysaccharides may well be closely linked to the processes of calcification.

Sulphated polysaccharides have been identified histochemically in the nematocysts (stinging capsules) of the coelenterate *Ablyopsis tetragona* (Hamon, 1955), in the mucus from the pharyngeal gland of the earthworm *Eisenia foetida* (Semal-Van Gansen, 1957) and in the gizzard-cuticle of the earthworm *Lumbricus terrestris* (Izard and Broussy, 1957) where the sulphated polysaccharide proved insensitive to testicular and bacterial hyaluronidase. Substances with the histochemical properties of acid mucopolysaccharides have been shown to be present in the mucous secretions involved in smoothing the burrow walls of abareicolid lugworms (Healy, 1963) and in the digestive tract of the polychaete *Hermodice carunculata* (Rattenbury-Marsden, 1963).

Acid mucopolysaccharides which contained both sulphate and carboxyl groups were identified by Pedersen (1965) in the mucous secretions from the gland cells of the acoel turbellarian *Convoluta convoluta* by histochemical methods. Protein was not considered to be an important constituent of this mucin. It seemed that the rhabdites might be composed of sulphated acid mucopolysaccharide also. Acid mucopolysaccharides were found in the slime-secreting cells of the planarians *Planaria vitta* and *Dugesia tigrina* (Pedersen, 1963).

Histochemically the urochordate tunic appears to contain acid mucopolysaccharide material associated with protein and cellulose (Bierbauer and Vagas, 1962; Endean, 1955; Pérès, 1948a, b).

The proboscis epidermis of Enteropneustan hemichordates secretes a mucin which is utilized for the capture and injestion of food as well as for cementing the walls of their burrows (Barrington, 1965). The contents of the mucous cells in *Sacoglossus ruber* stain readily with alcian blue and probably therefore contain sulphated mucopolysaccharide (Knight-Jones, 1953). In appearance the glandular region of the proboscis and the mucous cells themselves closely resemble the hypobranchial glands of molluscs. The mucin contains protein as well as mucopolysaccharide. Norrevang (1965) found there to be several types of cell in the mucin secreting region of the proboscis of *Harrimania kupfferi*. Acid mucous cells secreted a mucin staining metachromatically with toluidine blue at pH 2·2; the secretion did not stain with the periodic acid-Schiff reaction or ninhydrin-Schiff and appeared to be entirely acid mucopolysaccharide which was probably sulphated. The so-called goblet cells found by Knight-Jones (1953) to contain protein in *S. horstii* also gave positive staining reactions for protein in *H. kupfferii*, although there appeared to be neutral polysaccharide present. The association of protein with acid mucopolysaccharide in the entire secretion is therefore probably of secondary origin. Knight-Jones (1953) has commented upon the fibrous nature of the tube lining secreted by *S. horstii*, suggesting that the protein moiety of the acid mucopoly-saccharide-protein complex is an important factor in determining this particular property. The tube lining has a high tensile strength. Apart from the mucin secreted by the proboscis epidermis large quantities are also secreted by the collar epidermis which is differentiated into several zones one behind the other. While there is a general extensive production of acid mucopolysaccharide the ultimate zone is almost entirely free of this material but rich in protein; it is probable therefore that the strengthening of the secretion takes place at this point. Burden-Jones (1951) has noted that in *S. horstii* both eggs and sperm are produced enveloped in mucous.

Chaet and Philpott in an extensive series of cytological studies claim to have identified a new class of subcellular particle concerned with the secretion of acid mucopolysaccharide. The mucopolysaccharides in question act as media for the substrate attachment of the coelenterate *Hydra pirardi* and the starfish *Asterias forbesi*. The secretory regions of the walking surface on the tube feet of the echinoderm and the basal disc of the coelen-terate contained quantities of PAS and alcian blue-positive granules; it is therefore likely that the secretions contained polysaccharide sulphate. The granules have a complex ultrastructure. In the case of the echinoderm these were ellipsoidal in shape, 0·5 to 0·75 micron in size and containing approx-imately 50 hollow fibrils coalescing at both ends of the packet. These particles when released at the foot surface disintegrate as they contact the external medium. Similar particles made up of two concentric spheres and containing ribbons of material within the inner sphere are found in the gland cells of the coelenterate basal discs (Chaet, 1965; Chaet and Philpott, 1960, 1961, 1964; Philpott and Chaet, 1962; Philpott *et al.*, 1966).

The egg jellies of three species of mollusc—*Succinea putris, Limnaea stagnalis* and *Planorbis corneus*—were found by Jura and George (1958) to contain acid mucopolysaccharide (identified by their metachromasia) in association with protein and lipid. Smith (1965), also using histochemical methods, found that the haustoria or side pockets of the female duct of the slug *Arion ater* secreted a strongly acid mucopolysaccharide with irregularly distributed sulphate residues. This secretion is involved in the formation of the inner egg shell membrane.

The granular cell complex in the integument of the slug *Lehmania poirieri* secretes an acid mucopolysaccharide (Arcadi, 1963, 1965). The mucus cells of the mantle and foot of *Helix pomatia* secrete acid mucopolysaccharides which are alcian blue-positive and metachromatic to azure A at pH 3 and 4·5. The mantle mucopolysaccharide is also metachromatic at pH 1·5, strongly suggesting the presence of ester sulphate. Both secretions are hyaluronidase resistant (De Calventi, 1965). The cuticle of the alimentary canal wall of *Octopus vulgaris* has been shown histochemically to contain mucopolysaccharide (Zaccheo and Graziadei, 1957).

Reviewing the structure and composition of the helminth cuticle Lee (1966) has noted that the epidermis of the digenetic trematodes is covered by a layer of acid mucopolysaccharide and mucoprotein. The outer region of the cuticle contains mucopolysaccharide or mucoprotein, probably present in the invaginations and vesicles of this part of the epidermis. The mucopolysaccharides are apparently secreted to form an outer layer of mucin which may function to inhibit or resist the host's digestive enzymes. In the cestodes the main epidermis, supported on a collagenous basal membrane, contains non-glycogen polysaccharide, acid mucopolysaccharide and protein, while the microvillus border, which contains glycoprotein, also contains acid mucopolysaccharide between the microvilli (Monné, 1959).

The epicuticle of the *Acanthocephala* also contains acid mucopolysaccharide (Monné, 1959; Crampton, 1963; Crampton and Lee, 1963; Nicholas and Mercer, 1965). Mucopolysaccharides have been identified in the basement membrane of the embryophore and the embryo proper of the cestode *Taenia saginata* (Chowdhuri *et al.*, 1956), in the scolex and first proglottids of the same organism (Marzullo *et al.*, 1957), and in the secretions from the ventral cells of the cercaria of *Paragonimus kellicotti* (Kruidenier, 1948, 1953). Extracellular mucopolysaccharides have been detected in hookworm (nematode) eggs by Chowdhuri *et al.* (1957).

Acid mucopolysaccharide secretions have been noted in a wide range of invertebrates by Ewer and Hanson (1945) using histochemical methods. Mucopolysaccharides were observed in the foot ectoderm of *Hydra vulgaris*, the epidermis and mesentaries of *Anemonia sulcata*, in the gland cells, parapodia, epidermis, filaments and pinnules of the worms *Nereis*, *Chaetopterus*, *Arenicola*, *Sabella*, *Myxicola* and *Pomatoceros*, in the mucus cells of the filaments lining the gills of the mollusc *Anodonta cygnea*, in the gland cells of the foot of *Aplysia, Helix aspersa* and *Arion ater* and in the gland cells of the collar of *Helix*.

REFERENCES

Abolins-Krogis, A. (1963). *Ark. Zool.* **15**, 461.

Arcadi, J. A. (1963). *Ann. N.Y. Acad. Sci.* **106**, 451.

Arcadi, J. A. (1965). *Ann. N.Y. Acad. Sci.* **118**, 987.

Bacila, M. and Ronkin, R. R. (1957). *Biol. Bull. mar. biol. lab., Woods Hole.* **103**, 296.

Barrington, E. J. W. (1965) "The Biology of Hemichordates and Protochordates". Oliver and Boyd, Edinburgh and London.

Bayley, S. T. (1955). *Biochim. biophys. Acta.* **17**, 194.

Bettelheim, F. A. and Philpott, D. E. (1960). *Nature, Lond.,* **188**, 564.

Bevelander, G. and Benzer, P. (1948). *Biol. Bull mar. biol. lab., Woods Hole.* **94**, 176.

Bierbauer, J. and Vagas, E. (1962). *Acta biol. hung.* **13**, 139.

Blumberg, B. S. and Ogston, A. G. (1957). *Biochem. J.* **66**, 342.

Brimacombe, J. S. and Webber, J. M. (1964). "Mucopolysaccharides". Elsevier, Amsterdam.

Burden-Jones, C. (1951). *J. mar. biol. Ass. U.K.* **29**, 625.

Chaet, A. B. (1965). *Ann. N.Y. Acad. Sci.* **118**, 921.

Chaet, A. B. and Philpott, D. E. (1960). *Biol. Bull.* **119**, 308.

Chaet, A. B. and Philpott, D. E. (1961). *Biol. Bull.* **121**, 373.

Chaet, A. B. and Philpott, D. E. (1964). *J. Ultrastruct. Res.* **11**, 354.

Chowdhuri, A. B., Das Gupta, B., Ray, H. N. and Bhadhuri, V. N. (1956). *J. Ind. med. Ass.* **26**, 295.

Chowdhuri, A. B., Ray, H. N. and Bhadhuri, V. N. (1957). *Bull. Calcutta Sch. trop. Med.* **5**, 24.

Crampton, D. W. T. (1963). *Parasitology.* **53**, 663.

Crampton, D. W. T. and Lee, D. L. (1963). *Parasitology.* **53**, 3.

De Calventi, I. B. (1965). *Ann. N.Y. Acad. Sci.* **118**, 1015.

Defretin, R. (1952). *C.r. hebd. Séanc Acad. Sci., Paris.* **234**, 1806.

Dietrich, C. P. (1968). *Biochem. J.* **108**, 647.

Dodgson, K. S. (1961). *Biochem. J.* **78**, 325.

Dodgson, K. S. and Spencer, B. (1953). *Biochem. J.* **53**, iv.

Dodgson, K. S. and Spencer, B. (1954). *Biochem. J.* **57**, 310.

Doyle, J. (1964). *Biochem. J.* **91**, 6P.

Doyle, J. (1967a). *Biochem. J.* **103**, 41P.

Doyle, J. (1967b). *Biochem. J.* **103**, 325.

Durning, W. C. (1957). *J. Bone Jt. Surg.* **39A**, 377.

Egami, F. and Takahashi, N. (1962). *In* "Biochemistry and Medicine of Mucopolysaccharides" (F. Egami and Y. Oshima, eds), p. 1. University of Tokyo, Tokyo, Japan.

Egami, F., Asahi, T., Takahashi, N., Suzuki, S., Shikata, S. and Nishizawa, K. (1955). *Bull. chem. Soc. Japan.* **28**, 685.

Endean, R. (1955). *Aust. J. mar. fresh wat. Res.* **6**, 139.

Ewer, D. W. and Hanson, J. (1945). *J. Roy. micr. Soc.* **65**, 40.

Fontaine, A. R. (1955). *Nature, Lond.,* **175**, 606.

Fontaine, A. R. (1964). *J. mar. biol. Ass. U.K.* **44**, 145.
Fretter, V. and Graham, A. (1962). "British Prosobranch Molluscs", p. 125. Ray Society.
Fuji, A. (1961). *Bull. Fac. Fish. Hohkaido Univ.* **11**, 195.
Hamon, H. (1955). *Nature, Lond.* **176**, 357
Hassid, W. Z. (1962). *In* "The Structure and Biosynthesis of Macromolecules" (D. J. Bell and J. K. Grant, eds). p. 63. Cambridge University Press (for the Biochemical Society), Cambridge.
Healy, E. A. (1963). *Ann. N.Y. Acad. Sci.* **106**, 444.
Hoffman, P., Linker, A. and Meyer, K. (1958). *Biochim. biophys. Acta.* **30**, 184.
Horiguchi, Y. (1956). *Bull. Japan Soc. scient. Fish.* **22**, 463.
Horiguchi, Y. (1959). *Rept. Fac. Fish. prefect. Univ. Mie.* **3**, 399.
Hu, A. S. L. (1958). *Archs Biochem. Biophys.* **75**, 387.
Hunt, S. (1967). *Nature, Lond.*, **214**, 395.
Hunt, S. and Jevons, F. R. (1963). *Biochim. biophys. Acta.* **78**, 376.
Hunt, S. and Jevons, F. R. (1965a). *Biochem. J.* **97**, 701.
Hunt, S. and Jevons, F. R. (1965b). *Biochim. biophys. Acta.* **101**, 214.
Hunt, S. and Jevons, F. R. (1966). *Biochem. J.* **98**, 522.
Hunt, S. and Rudall, K. M. (1967). *Carbohydr. Res.* **4**, 259.
Iida, K. (1961). *J. Biochem.* **49**, 173.
Iida, K. (1963a). *J. Biochem.* **54**, 181.
Iida, K. (1963b). *J. Biochem.* **53**, 37.
Iida, K., Watanabe, S. and Egami, F. (1960). *J. biochem. Soc. Japan.* **32**, 284.
Inoue, S. (1965). *Biochem. biophys. Acta.* **101**, 16.
Inoue, S. and Egami, F. (1963). *J. Biochem.* **54**, 557.
Izard, J. and Broussy, J. (1964). *Nature, Lond.*, **201**, 1338.
Jura, C. and George, J. C. (1958). *Proc. ned. Akad. Wet.* **C61**, 590.
Kantor, T. G. and Schubert, M. (1956). *J. Am. chem. Soc.* **79**, 152.
Knight-Jones, E. W. (1953). *Proc. zool. Soc. Lond.* **123**, 637.
Kobayashi, S. (1964). Quoted by Wilbur, K. M. *in* "Physiology of Mollusca" (K. M. Wilbur and C. M. Yonge, eds). Vol. 1, p. 243. Academic Press, New York and London.
Kruidenier, F. J. (1948). *J. Parasit.* **34**, Suppl. 22.
Kruidenier, F. J. (1953). *J. Morphol.* **92**, 531.
Kwart, H. and Shashoua, V. E. (1957). *Trans. N.Y. Acad. Sci.* II, **19**, 595.
Kwart, H. and Shashoua, V. E. (1958). *J. Am. chem. Soc.* **80**, 2230.
Lash, J. W. (1959). *Science, N.Y.* **130**, 334.
Lash, J. W. and Whitehouse, M. W. (1960). *Biochem. J.* **74**, 351.
Lee, D. L. (1966). *Adv. Parasitology.* **4**, 187.
Levene, P. A. and López-Suàrez, J. (1916). *J. biol. Chem.* **25**, 511.
Lloyd, P. F. and Lloyd, K. O. (1961). *Biochem. J.* **80**, 4P.
Lloyd, P. F. and Lloyd, K. O. (1963). *Nature, Lond.*, **199**, 287.
Maki, M. (1956a). *Hirosaki med. J.* **7**, 150.
Maki, M. (1956b). *Hirosaki med. J.* **7**, 211.
Maki, M. and Hiyama, N. (1956). *Hirosaki med. J.* **7**, 142.
Marshall, A. T. (1966a). *J. Insect Physiol.* **12**, 635.
Marshall, A. T. (1966b). *J. Insect Physiol.* **12**, 925.
Marshall, A. T. (1968). *J. Insect Physiol.* **14**, 1435.
Marzullo, F., Squadrini, F. and Taparelli, F. (1957). *Boll. Soc. med.-chir. Modena.* **57**, 327.

Masamune, H. and Osaki, S. (1943). *Tohoku. J. exp. Med.* **45**, 121.
Masamune, H. and Yoshizawa, Z. (1950). *Tohoku J. exp. Med.* **53**, 155.
Masamune, H. and Yoshizawa, Z. (1956). *Tohoku J. exp. Med.* **65**, 57.
Masamune, H., Yasuoka, T., Takhashi, M. and Asagi, Y. (1947a). *Tohoku J. exp. Med.* **49**, 117.
Masamune, H., Yasuoka, T., Takahashi, M. and Asagi, Y. (1947b). *Tohoku. J. exp. Med.* **49**, 177.
Mathews, M. B. (1958) *Nature, Lond.* **181**, 421.
Matsuoka, M., Oguchi, G. and Murayama, S. (1956). *J. Jap. Soc. int. Med.* **45**, 1334.
Matsuoka, M., Oguchi, G. and Murayama, S (1962). *In* Biochemistry and Medicine of Mucopolysaccharides" (F. Egami and Y. Oshima, eds), p. 241. University of Tokyo, Tokyo, Japan.
Meenakshi, V. R. and Scheer, B. T. (1959). *Science, N.Y.*, **130**, 1189.
Meenakshi, V. R. and Scheer, B. T. (1968). *Comp. Biochem. Physiol.* **26**, 1091.
Monné, L. (1959). *Ark. Zool.* **12**, 343.
Nakanishi, K., Takahashi, N. and Egami, F. (1956). *Bull. chem. Soc. Japan.* **29**, 434.
Nichols, D. (1959a). *Q. Jl. microsc. Sci.* **100**, 73.
Nichols, D. (1959b). *Q. Jl. microsc. Sci.* **100**, 539.
Nichols, D. (1960). *Q. Jl. microsc. Sci.* **101**, 105.
Nicholas, W. L. and Mercer, E. H. (1965). *Q. Jl. microsc. Sci.* **106**, 137.
Norrevang, A. (1965). *Ann. N.Y. Acad. Sci.* **118**, 1052.
Ogston, A. G. and Sherman, T. F. (1959). *Biochem. J.* **72**, 301.
Pancake, S. J. and Karnovsky, M. L. (1967). *Fedn. Proc. Fedn. Am. Socs. exp. Biol.* **26**, 282.
Pedersen, K. J. (1963). *Ann. N.Y. Acad. Sci.* **106**, 424.
Pedersen, K. J. (1965). *Ann. N.Y. Acad. Sci.* **118**, 930.
Pérès, J. M. (1948a). *Arch. Anat. Micr. Morp. Exp.* **37**, 230.
Pérès, J. M. (1948b). *C.r. hebd. Séanc Acad. Sci., Paris.* **226**, 1220.
Philpott, D. E. and Chaet, A. B. (1962). *Biol. Bull. mar. biol. lab., Woods Hole.* **123**, 509.
Philpott, D. E., Chaet, A. B. and Burnett, A. L. (1966). *J. Ultrastruct. Res.* **14**, 74.
Ransom, G. (1952). *C.r. hebd. Séanc. Acad. Sci., Paris.* **234**, 1485.
Rattenbury-Marsden, J. (1963). *Can. J. Zool.* **41**, 165.
Rees, D. A. (1963). *Biochem. J.* **88**, 343.
Reeves, R. A. (1951). *Adv. Carbohyd. Chem.* **6**, 107.
Ronkin, R. R. (1952a). *Biol. Bull. mar. biol. lab., Woods Hole.* **102**, 252.
Ronkin, R. R. (1952b). *Biol. Bull mar. biol. lab., Woods Hole.* **103**, 306.
Ronkin, R. R. (1955). *Archs Biochem. Biophys.* **56**, 76.
Rosenberg, E. and Zamenhof, S. (1962). *J. Biochem.* **51**, 274.
Schubert, M. (1964). *Biophys. J.* **2**, 119.
Schweiger, R. G. (1966). *Chemy. Ind.* **22**, 900.
Shashoua, V. E. and Kwart, H. (1959). *J. Am. chem. Soc.* **81**, 2899.
Siegel, S. M. (1968). *In* "Comprehensive Biochemistry" (M. Florkin and E. H. Stotz, eds). Vol. 26A, p. 1. Elsevier, Amsterdam.
Simkiss, K. (1964). *Biol. Rev.* **39**, 487.
Simkiss, K. (1965). *Comp. Biochem. Physiol.* **16**, 427.
Smith, B. J. (1965). *Ann. N.Y. Acad. Sci.* **118**, 997.
Soda, T. (1936). *J. Fac. Sci. Tokyo.* **3**, 149.

Soda, T. (1948). *Chem. Res. Tokyo.* **1**, Biochem., 51.
Soda, T. and Egami, F. (1938a). *J. chem. Soc. Japan.* **59**, 1202.
Soda, T. and Egami, F. (1938b). *Bull. chem. Soc. Japan.* **13**, 652.
Soda, T. and Egami, F. (1938c). *J. chem. Soc. Japan.* **59**, 1417.
Soda, T. and Hattori, C. (1931). *Bull. chem. Soc. Japan.* **6**, 258.
Soda, T. and Hattori, C. (1933). *Bull. chem. Soc. Japan.* **8**, 65.
Soda, T. and Kawamatsu, S. (1936). *J. chem. Soc. Japan.* **57**, 604.
Soda, T. and Terayama, H. (1948). *J. chem. Soc. Japan.* **69**, 659.
Soda, T., Hattori, C. and Terasaki, H. (1936). *J. chem. Soc. Japan,* **57**. 981.
Soda, T., Egami, F., Koyama, S. and Horigome, T. (1939). *J. chem. Soc. Japan.* **60**, 1225.
Soda, T., Katsura, T. and Yoda, D. (1940). *J. chem. Soc. Japan.* **61**, 1227.
Stary, Z., Wardi, A. and Turner, D. (1964). *Biochim. biophys. Acta.* **83**, 242.
Stott, E. C. (1955). *Proc. zool. Soc. Lond.* **125**, 63.
Suzuki, H. (1941). *J. Biochem.* **33**, 377.
Suzuki, S. (1956). *J. Biochem.* **43**, 691.
Suzuki, S. and Ogi, K. (1956). *J. Biochem.* **43**, 697.
Suzuki, S., Takahashi, N. and Egami, F. (1957). *Biochim. biophys. Acta.* **24**, 444.
Suzuki, S., Takahashi, N. and Egami, F. (1959). *J. Biochem.* **46**, 1.
Takahashi, N. (1959). *J. Biochem.* **46**, 1167.
Takahashi, N. (1960a). *J. Biochem.* **47**, 230.
Takahashi, N. (1960b). *J. Biochem.* **48**, 508.
Takahashi, N. (1960c). *J. Biochem.* **48**, 691.
Takahashi, N. and Egami, F. (1959). *J. chem. Soc. Japan.* **80**, 1364.
Takahashi, N. and Egami, F. (1960). *Biochim. biophys. Acta.* **38**, 375.
Takahashi, N. and Egami, F. (1961). *Biochem. J.* **80**, 384.
Thanassi, N. M. and Nakada, H. I. (1967). *Archs Biochem. Biophys.* **118**, 172.
Travis, D. F. (1957). *Biol. Bull. mar. biol. lab., Woods Hole.* **113**, 451.
Travis, D. F. (1960a). *Biol. Bull. mar. biol. lab., Woods Hole.* **118**, 137.
Travis, D. F. (1960b). *Fedn. Proc. Fedn. Am. Socs. exp. Biol.* **1**, 1.
Travis, D. F. (1963). *Ann. N.Y. Acad. Sci.* **109**, 177.
Trueman, E. R. (1957). *Nature, Lond.,* **180**, 1492.
Tsuji, T. (1955). *Bull. biogeogr. Soc. Japan.* **16**, 88.
Tsuji, T. (1960). *J. Fac. Fish. pref. Univ. Mie Tsu.* **5**, 270.
Van Gansen-Semal, P. (1957). *Bull. biol. Fr. Belg.* **91**, 225.
Wilbur, K. M. and Simkiss, K. (1968). *In* "Comprehensive Biochemistry" (M. Florkin and E. H. Stotz, eds). Vol. 26A, p. 229. Elsevier, Amsterdam.
Wilbur, K. M. and Watabe, N. (1963). *Ann. N.Y. Acad. Sci.* **109**, 82.
Wilson, H. A. and Dorsay, C. K. (1957). *Ann. ent. Soc. Am.* **50**, 399.
Yamashina, I. and Egami, F. (1953). *J. biochim. Soc. Japan.* **25**, 281.
Yoshida, H. and Egami, F. (1965). *J. Biochem.* **57**, 215.
Zaccheo, D. and Graziadei, P. (1957). *Boll. Soc. ita. Biol. sper.* **33**, 1068.
Ziegler, H. and Ziegler, I. (1958). *Z. vergl. Physiol.* **40**, 549.

CHAPTER 6

Egg Jelly Coats

INTRODUCTION

Animal eggs are usually encased in a capsule which, in the case of marine invertebrates, more often than not takes the form of a coating of acid mucopolysaccharide. The nature and function of this so-called "jelly coat" not unnaturally has excited considerable interest in view of its probable role in the process of fertilization.

The suggestion that the egg coat might fulfill some specific fertilization function rather than simply acting as a protective agent was probably first suggested by F. R. Lillie in the early years of this century (Lillie, 1913, 1914). In his fertilization theory Lillie postulated that the properties of sperm agglutination and motility stimulation, exhibited by sea water washings of unfertilized sea urchin eggs, might represent the manifestation of an active substance, "fertilizin", diffusing out from the egg and saturating the jelly coat.

The fertilizin substance provided, according to Lillie, the basis for the reaction preventing polyspermy. Fertilizin was envisaged as being a dipolar molecule, orientated perpendicularly to, and lying in, the surface jelly coat of the egg. The two poles of the dipolar molecular unit would be the spermophile and ovophile groups respectively. At fertilization, receptors at the sperm surface would react with the fertilizin spermophile groups. The effect of only one such interaction would be an activation of the fertilizin molecules all over the cell surface such that their spermophile groups would react with a second type of molecule also postulated as being present in the egg, "anti-fertilizin", blocking the reaction with the sperm receptors. Concomitant with this process, preventing polyspermy, the egg receptors would then react with the ovophile group on the activated fertilizin molecule and thus bring about activation of the egg.

Fertilizin is now thought to be the only, or at least the major, component of the jelly coat and to be a secretion directly from the gonads rather than from the egg itself. Fertilizins lie within the realm of this work since the jelly coats of the invertebrate eggs studied have all proved to have acid mucopolysaccharide character.

The physiological basis of the fertilization process has recently been well reviewed by Monroy (1965) and by Runnström (1964). A comprehensive work dealing with all aspects of fertilization is that edited by Metz and Monroy (1967). Earlier reviews include those of Runnström (1949, 1951, 1952), Vasseur (1949a, 1954) and Dorfman (1963).

ISOLATION

Echinoderm fertilizin substances figure prominently among the few more easily extracted and purified animal mucopolysaccharides. The jelly coats readily dissolve away from the eggs at acid pH in sea water. The degree of acidity required for optimum solubilization varies somewhat from species to species but even so is not particularly critical.

Tyler (1949) extracted the eggs of *Strongylocentrotus purpuratus* with pH 3·5 sea water and precipitated the sperm agglutinin by making the solution 0·004 N with respect to sodium hydroxide. The precipitate was redissolved in 3·3% sodium chloride solution and dialysed before precipitation with 1·25 volumes of 95% ethanol. The material obtained was further purified, either by reprecipitation with ethanol, or by ammonium sulphate precipitation.

Previously, Vasseur (1948a) had found the optimum pH values for sea water extraction of *Echinus esculentus*, *Strongylocentrotus droebachiensis*, *Echinocardium cordatum*, and *Psammechinus miliaris* eggs to be 5·8, 5·6, 5·1 and 5·3 respectively. In an earlier paper Vasseur (1947) extracted the eggs of *S. droebachiensis* with pH 4·0 sea water. Almost saturated ammonium sulphate was required to precipitate the very viscous jelly coat material from the *E. esculeatus* extract, while 70% ammonium sulphate precipitated the *S. droebachiensis* jelly as a firm fibrous precipitate (Vasseur 1948a).

An extract of *Paracentrotus lividus* egg jelly was precipitated by Vasseur, with 2 volumes of 95% ethenol. This material gave a viscous solution on dialysis.

Ando (1958) extracted the eggs of *Hemicentrotus pulcherrimus* and *Psendocentrotus depressus* with sea water at pH 4·8 to obtain the egg jelly coats.

In studies even earlier than these, Tyler (1940) had extracted fertilizin substances from the eggs of *S. purpuratus* and the keyhole limpet *Megathura crenulata* either with a 1% solution of chymotrypsin in sea water or with sea water acidified to pH values in the range of 3·5 to 4·5. Unfortunately for the cause of comparative biochemistry the material extracted from the molluscan eggs was never further characterized apart from its nitrogen content, which was of the order 4·5%. At this time Tyler and Fox (1940) believed the fertilizin substances to be proteinaceous or at least closely associated with protein. The extracts were non-dialysable, precipitable by ammonium sulphate and gave positive reactions to tests for proteins. More-over they could be shown to contain nitrogen. The nitrogen content of *S. purpuratus* fertilizin was 5·2%.

COMPOSITION AND STRUCTURE

In their belief that the fertilizin substances were protein, Tyler and Fox were only partly correct. Wallenfels (1940) however had suggested that the jelly coat of *Arbacia* eggs might consist of "mucin". In the ensuing decade Vasseur and Tyler, working in independent laboratories, were able to demonstrate that the sperm agglutinins or fertilizin substances were in fact highly-charged polysaccharide molecules associated with protein. In many cases the protein moiety constituted a high percentage of the total mucopolysaccharide complex, thus accounting for Tyler's earlier con-clusions.

In 1942 Runnström, Tiselius and Vasseur (1942) were able to demonstrate definitely the presence of a carbohydrate component in the egg jelly of *E. cordatum*. This material, extracted at acid pH with sea water, precipi-tated with ammonium sulphate, and dialysed, gave a strong colour re-action for carbohydrate. On being subjected to moving boundary electro-phoresis it behaved as a homogeneous polyanion of high charge and hence high mobility (Table I). The high mobility coupled with the presence of

TABLE 1

Electrophoretic Properties of *Echinocardium cordatum* Egg Jelly[a]

Buffer ($\mu = 0\cdot1$)	pH	Mobility cm sec^{-1} volt^{-1}
Acetate	3·76	$-13\cdot2 \times 10^{-5}$
Acetate	4·67	$-14\cdot2 \times 10^{-5}$
Phosphate	6·02	$-16\cdot2 \times 10^{-5}$

[a] Runnström, Tiselius and Vasseur (1942).

carbohydrate suggested that the jelly coat was composed of an "acid-carbohydrate glycoprotein". Ultracentrifugation of a solution, which con-tained 0·264 mg N/ml, in 0·2 N sodium chloride gave a well-defined main component with a sedimentation constant equal to 2·90s. A more rapidly sedimenting minor component was also present. The sedimentation con-stant of the major component was strongly concentration dependent, a property typically exhibited by polyanionic, highly asymmetric, extended, linear polysaccharide molecules. The material presented a strong flow birefringence.

Further early evidence for the polyanionic character of fertilizin was obtained when Vasseur (1947) found a sulphate content of 25·5–27·6% in a freeze-dried preparation of *S. droebachiensis* egg jelly. This material had been prepared by pH 4 sea water extraction, precipitation with ethanol and with ether, and dialysis. The sulphate was released by acid hydrolysis under oxidizing conditions and must therefore have been a firmly bound, integral part of the jelly coat substance.

That the fertilizin substance is indeed polysaccharide sulphate was clearly shown by Vasseur (1948a) in analyses of the egg jellies of *S. droebachiensis*, *E. esculentus*, *E. cordartum*, *P. lividus*, *Brissopsis lyrifera*, *Arbacia Lixula* and *P. miliaris*. The jelly materials, prepared as already described, consisted of methyl pentosan in the case of *S. droebachiensis*, *E. cordatum*, and *P. lividus* and hexosan in the case of *E. esculentus*. The carbohydrates were identified spectroscopically by virtue of their colour reaction with orcinol-sulphuric acid reagent (Vasseur, 1948b). The methyl-pentose was thought to be fucose and the hexose, galactose. A quantitative determination of the carbohydrate content of *S. droebachiensis* egg jelly, by CrO_3 oxidation, indicated a value of 45% methyl pentose. No hexosamine or hexuronic acid could be detected in the *E. cordatum* jelly nor was there any detectable hexuronic acid in the *S. droebachiensis* material.

Elementary analyses of the *E. cordatum* and *S. droebachiensis* jellies yielded high sulphate ash contents; an average of the order of 40%. Acid hydrolysable sulphate contents were as high as 32% (as SO_4) in some of the samples examined. Values obtained for total sulphur and acid hydrolysable sulphur supported the view that not all or most of the sulphur was present as ester sulphate.

The values obtained by Vasseur for total N coupled with amino acid analyses indicated the presence in the egg jellies of 20–25% of a peptide or protein component. Amino acid compositions were determined for *P. lividus*, *E. esculentus*, *S. droebachiensis* and *E. cordatum* jellies (Table II). The low basic amino acid contents of these materials suggest that the protein-carbohydrate link may be other than ionic.

The ester sulphate residues of these materials appeared to be unusually labile, starting to hydrolyze even at pH 3–4; this observation may explain the rapid loss of agglutinin activity at pH 2, noted by Tyler and Fox (1939, 1940), for the fertilizin substance of *S. purpuratus*. *B. lyrifera* egg jelly had a low sulphur content and this may correlate with observations that the egg water and egg jelly solutions of this genus do not agglutinate its own spermatozoa (Vasseur and Hogström, 1946; Vasseur and Carlsen, 1948).

Tyler (1948a, b, 1949) found galactose to be the principal monosaccharide component of *S. purpuratus* egg jelly. This material contained 25% galactose and possibly 1·6%–5% hexosamine. *Lytechinus anamesus* jelly had a

TABLE II

Microbiological Amino Acid Analyses of Echinoderm Egg Jellies[a, b]

Amino Acid	Paracentrotus lividus	Echinus esculentus	Strongylocentrotus droebachiensis	Echinocardium cordatum
Lysine	1·6	1·7	1·5	0·3
Arginine	2·0	1·8	1·9	0
Histidine	—	0·2	—	—
Valine	6·5	8·7	8·3	3·2
Isoleucine	—	1·8	1·0	—
Threonine	2·0	1·6	2·1	2·0
Phenylalanine	1·0	0·7	0·9	0·6
Tyrosine	0·7	0·5	0·7	0·3
Proline	0·7	0·6	0·9	0·3
N (%)			4·02	

Aspartic and Glutamic Acids by paper chromatography present in 1·0 2·0% concentration.

[a] Vasseur (1948a).
[b] Percentages by weight.

similar composition. No glucuronic acid or pentose could be detected. Paper chromatography suggested the presence in the *S. purpuratus* jelly of a protein containing aspartic acid, glutamic acid, serine, and α-amino adipic acid, glycine, threonine, cystine, lysine, alanine, histidine, norleucine, leucine, arginine, tryptophane, methionine, phenylalanine, and isoleucine. These identifications however were based on the interpretation of only seven spots detected on the chromatograms. The total amino acid content was about 20%. This same material had a sedimentation coefficient of 6·3 S at 20°. Assuming a spherical shape this yields a molecular weight of 82,000. The nitrogen content of *S. purpuratus* jelly was 5·6–5·8% and the content of sulphate about 23%. The electrophoretic properties of this egg jelly are compared with those of *Lytechinus pictus* (Tyler, 1949), *Arbacia punctulata* and *Echinarachnus parma* (Tyler *et al.*, 1954) in Table III.

A. punctulata fertilizin has hydrodynamic properties, indicative of a high degree of asymmetry (Tyler, 1956) the axial ratio being about 20. The molecular weight of this mucopolysaccharide is in the region of 3×10^5.

The monosaccharide constituents of the echinoderm egg jelly mucopolysaccharides are most frequently found to be galactose and fucose, occurring separately or together in varying admixtures depending upon the species. The rule is not a general one, however.

The monosaccharides of the egg jellies of *E. cordatum*, *S. droebachiensis* and *P. lividus* were firmly identified by paper chromatography as fucose (Vasseur and Immers, 1949a) with small amounts of galactose in the *S. droebachiensis* material and glucose in the *P. lividus* jelly. Galactose was

TABLE III (see Table I)
Electrophoretic Mobilities of Echinoderm Egg Jellies[a, b]

Buffer	pH	Mobilities $\times 10^{-5}$ cm^2 sec^{-1} volt^{-1}			
		Strongylocentrotus purpuratus	*Lytechinus pictus*	*Arbacia punctulata*	*Echinarachnius parma*
Phosphate	2·00	18·8 ascending	—	—	—
HCl–KCl	2·00	—	—	20·2	—
Acetate	3·76	16·6 ascending			
		16·7 descending	—	—	—
Acetate	4·50	—	4·2	—	—
Acetate	4·84	—	—	15·7	—
Acetate	4·90	—	4·8	—	—
Acetate	5·65	17·8 ascending			
		17·4 descending	—	—	—
Phosphate	6·00	—	6·1	—	—
Phosphate	6·60	19·7 ascending			
		19·5 descending	—	—	—
Phosphate	7·02	—	—	19·9	18·2
Barbital	7·30	—	9·3	—	—
Phosphate	8·75	18·8 ascending	—	—	—
Barbital	8·58	16·6 ascending			
		16·7 descending	—	—	—

[a] Tyler (1949).
[b] Tyler *et al.* (1954).

found to be the sole monosaccharide constituent of *E. esculentus* egg mucopolysaccharide. Positive responses to tests for hexosamine in egg jelly hydrolysates (Tyler 1948a, b; Vasseur, 1948) were discounted by Vasseur and Immers (1949b) after they had shown that the hydrolysates of *S. droebachiensis* and *P. lividus* jellies gave a positive red colouration with the Ehrlich reagent following treatment with sodium carbonate solution and in the absence of acetyl acetone. The positive reaction presumably originated in the formation of amine-sugar condensation products formed during hydrolysis.

The galactan sulphate of *E. esculentus* egg coat has been characterized in greater detail by Vasseur (1950). Hydrolysis of the polysaccharide and isolation of the galactose showed that it had the unusual L-configuration. The sugar was separated on a cellulose column by elution with a butanol-water mixture (20 : 1); it had $[\alpha]_D^{19} -74\cdot5°$, in water and yielded a methyl-phenythydrazone with a melting point of 182°, compared with D-galactose 190°. The melting point was rather low, suggesting the presence of a small amount of the D-isomer.

Estimation of the galactan by the Somogyi method gave a reducing group content of one per 150 monosaccharide residues. Periodate oxidation studies giving a consumption of 1 oxygen per 60 residues coupled with the observation that mild acid treatment resulted in a parallel increase in reducing value and periodate consumption (the ratios between the two values remained constant) suggested that sulphate release did not give rise to a pair of adjacent free hydroxyls. In fact the periodate oxidation results suggested a maximum of three pairs of adjacent free hydroxyls per polysaccharide molecule, possibly less. The most probable structure would therefore be one similar to that of the seaweed polysaccharide carrageenan with 1 → 3 glycosidically linked galactose units sulphated at position 4 (Fig. 1). The chains could be cross-linked by divalent calcium bridges to

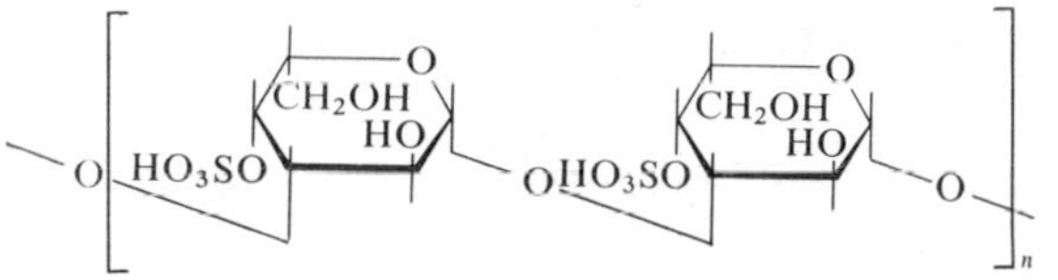

Fig. 1. Possible structure for the galactan sulphate from *Echinus esculentus* egg jelly.

give larger, more complex molecular networks. Vasseur (1949b) found the sperm agglutinating activity of the fucan sulphate, from *S. droebachiensis* egg jelly, to be strongly enhanced by the presence of isotonic calcium chloride solution; he attributed this effect, partly at least, to the formation of calcium bridges between the polysaccharide sulphate and acid groups on the sperm surface.

The structure of *E. cordatum* egg jelly fucan sulphate also has been investigated by Vasseur (1952). This polysaccharide was found to contain, in addition to 34·4% fucose (2·1 m moles/g), small amounts of galactose, mannose, and glucose totaling about 2%. The sulphate content of 20·6% (2·1 m moles/g) suggested a fucan sulphate in which each monosaccharide residue carried one esterified sulphate residue. The fucan sulphate content of the whole mucopolysaccharide was about 52% and a nitrogen content of 3·4% suggested a protein content of about 21%.

Treatment of the fucan sulphate with pH 3·5 periodate at 50° resulted in the consumption of two-thirds of the reagent after twelve hours. Disappearance of the periodate was at its most rapid during the first half hour of the reaction, concomitant with the loss of half the fucose residues; thereafter periodate was consumed only slowly and no further destruction of fucose took place. In fact it was shown that this latter periodate consumption could be accounted for by reaction with amino acids in the protein moiety of the mucopolysaccharide. If one considers the effect on the polysaccharide alone then it is apparent that the fucose sulphate residues are unlikely to be linked in an unbranched chain. In such a structure only one or both of the terminal residues could have unsubstituted α glycol groupings.

Several possible structures were suggested (Vasseur 1952). The most probable structure (Fig. 2) was considered to be that of a 1 → 4 linked

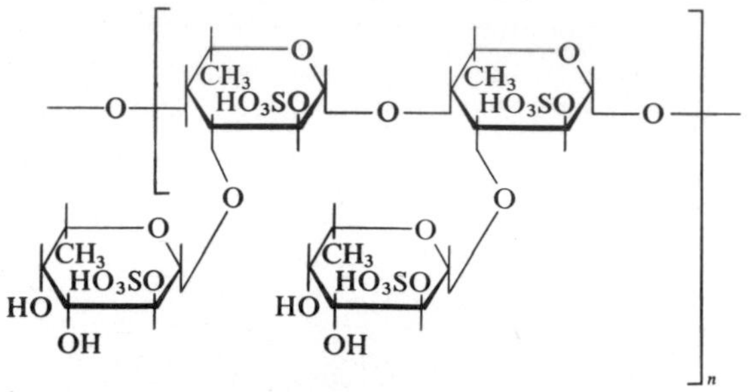

Fig. 2. Possible structure for the fucan sulphate of *Echinocardium cordatum* egg jelly.

chain with branch points at position 3 of each residue and with ester sulphate residues distributed at the 2 positions.

An alternative structure postulated was that of a main chain linked 1 → 3 by glycosidic bands and branching either at the 2 or 4 positions with ester sulphate residues at the 4 or 2 positions (Fig. 3).

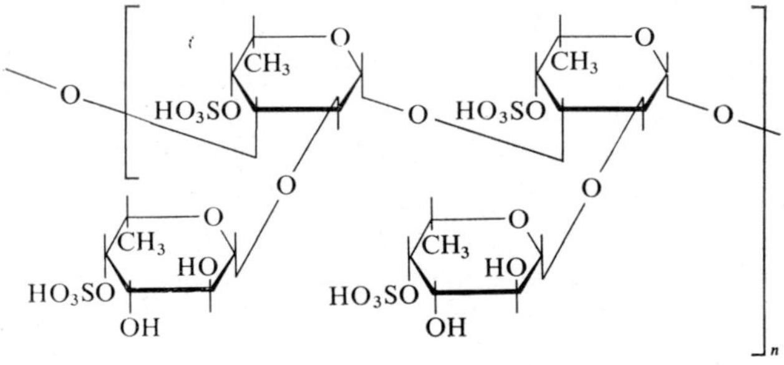

Fig. 3. Alternative structure for the fucan sulphate of *Echioncardium cordatum* egg jelly.

A rather different structure put forward was one in which the mucopolysaccharide comprised a protein backbone bearing sulphated fucose disaccharides.

During the course of periodate oxidation of this material the protein is precipitated and the unattached fucose sulphate residues are released with the supernatant.

A highly-branched structure was suggested for the egg jelly mucopolysaccharide of *P. lividus* by Minganti and Vasseur (1959). This polysaccharide had earlier been shown to consist mainly of fucose and a lesser amount of glucose (Vasseur and Immers, 1949) and was now demonstrated to contain also a small proportion of mannose, xylose, and hexosamine. The mucopolysaccharide comprised some 28·5% protein and 31·7% carbohydrate (fucose 24·3%, glucose 4·1%, mannose 2·8%, xylose 0·5% and hexosamine 2%) and sulphate 20·9%. Periodate oxidation resulted in the loss of one third of the fucose residues, most of the glucose residues, and half of the mannose residues.

The egg jelly mucopolysaccharides of *H. pulcherimus*, *P. depressus*, and *Heliocidaris crassispina* also have been shown to have fucose as their only sugar constituent (Nakano and Ohashi, 1953, 1954) while that of *Sphaerechinus granularis* has a hetero composition with glucose, mannose, and fucose residues (Minganti, 1958).

Reports of small quantities of monosaccharides other than the major constituent in these egg jelly mucopolysaccharides, in some instances at least, may represent heterogeneity of preparations rather than of polymerization. This point is illustrated by the egg jelly of *A. lixula*. Here the sea water extracts contain polymeric fucose, sulphate, and lesser amounts of galactose. Dialysed solutions however, when treated with the weakly anionic exchange resin Amberlite IR4B, resolve into two fractions, one bound to the resin and one not (Monroy *et al.*, 1954; Messina and Monroy, 1956). The bound component, which could be readily eluted from the resin, contained fucose, sulphate and a little galactose. This material had strong sperm agglutination activity while the neutral unbound component, which contained galactose alone, had only slight activity. Treatment of jelly coat solutions with protamine precipitated a fucose sulphate component which again contained a little galactose; a galactose containing component remained in the supernatant. Treatment of this supernatant with Amberlite IR120 cation exchange resin (to remove protamine), followed by dialysis against sea water, yielded a neutral, galactose-containing material which now had no sperm agglutination activity. Both of the egg jelly constituents were associated with proteins having similar amino acid compositions.

The *Arbacia* fucan sulphate had the property of binding strongly to sperm. Agglutination of sperm was shown to be prevented if the sperm amino groups were alkylated before exposure to the fertilizin substance (Metz and Donovan, 1951).

The fucan sulphate exhibited the viscous properties typical of polyelectrolytes: a 60% decrease in viscosity was noted on passing from distilled water to 0·6 M sodium chloride, 70% in the change to sea water. Divalent cations however did not affect the viscosity to such a great extent, presumably because of the formation of interchain salt bridges and the resulting stabilization of extended structure. As might have been expected for a molecular system whose physical properties would depend largely on ionic interactions, the viscosity was insusceptible to urea.

The jelly coat mucopolysaccharide of the "sand dollar", *E. parma*, has a rather unusual carbohydrate composition (Bishop, 1951; Bishop and Metz, 1952). This material contains neither fucose nor galactose but is instead a polymer of the keto sugar fructose. The polysaccharide is associated with a protein containing isoleucine, leucine, aspartic acid, cystine, methionine, tyrosine, threonine, glycine, arginine, lysine (or serine) and glutamic acid.

In a recent study of *Lytechinus variegatus* egg jelly, Stern (1966, 1967) has compared the physical properties and chemical compositions of univalent and multivalent fertilizins.

It should be explained here that in keeping with immunological doctrine since agglutination is species specific (there is believed to be an analogy with antibody agglutination of cells) Tyler (1948) proposed that agglutinating fertilizin must be multivalent in terms of combining groups or active sites and that fertilizin rendered non-agglutinating, but still having the capacity to combine with the specific complementing sites of the sperm, must be univalent.

Stern (1966, 1967) converted multivalent *L. variegatus* fertilizin to the univalent form by a 30 minute treatment with 2% hydrogen peroxide at room temperature and subsequent dialysis against water. Treatment in this manner resulted in a fall in the relative viscosity of a 0·6 mg/ml solution of multivalent fertilizin from 5·25 to 1·25, after 100 minutes. The viscosity change took place most rapidly in the first 30 minutes following the addition of the hydrogen peroxide. During this time the relative viscosity fell by 76%. The effect appeared to be dependent on peroxide concentration; a 0·4% solution of hydrogen peroxide only lowered the relative viscosity to 2·5 in 100 minutes. Similar results were obtained with *Tripneustes esculentus*, *Echinometra lucunter*, *Eucidaris tribuloides*, and *A. punctulata* fertilizins.

The difference in sedimentation properties, in the ultracentrifuge, between the multivalent and univalent fertilizins indicated a radical change from a highly asymmetric to a more compact and symmetrical molecular configuration on degradation to the univalent state. The univalent state seems to have the properties of a sub-unit of the multivalent form. Peroxide treatment causes the sedimentation constant to fall from 8·60s (at infinite dilution) to 3·05s, while the concentration dependence of sedimentation constant plots changes from ones showing extreme dependence to ones of almost complete independence.

On cellulose acetate electrophoresis, multivalent fertilizin gives one strong metachromatic band at the origin (pH 8·6 borate buffer), whereas the univalent material yields no band at the origin and four strong bands migrating towards the anode. At least two of these bands are capable of absorption by sperm.

Both multi and univalent fertilizins had similar amino acid compositions, at least 16 amino acids were identified and in addition peaks due to 2-amino-2-deoxy-D-glucose and 2-amino-2-deoxy-D-galactose were identified in both cases on the amino acid analyser chromatograms. There appeared to be of the order of twice as much 2-amino-2-deoxy-D-glucose as 2-amino-2-amino-2-deoxy-D-galactose; the total amino sugar content being about one per cent of the dry weight of the fertilizin. These two amino sugars were also identified in hydrolysates of *T. esculentens* and *E. lucunter* fertilizins. Comparison of the fucose contents of the multi and univalent fertilizins indicated that there was probably little or no change; the average percentage compositions of a number of samples were 26·7 and 25·5 respectively.

The echinoderm mucopolysaccharides discussed so far have all been derived from echinoids. In one case only the egg jelly of an asteroid has been investigated (Muramatsu, 1965).

Sea water extracts of *Asterias amurensis* eggs, acidified to pH 5·0 and filtered, were extracted with phenol (45% final concentration) and the aqueous layer dialysed and concentrated. The mucopolysaccharide thus obtained was completely precipitable by a one per cent protamine sulphate solution, gave a single hypersharp boundary in the ultracentrifuge and migrated as a single component on electrophoresis in 0·1 M acetate buffer at pH 5·5. The sedimentation properties were markedly concentration dependent; values for $S_{22,w}$ of 4·8 and 11 were obtained in pH 5·5 acetate buffer at concentrations of 0·3 and 0·15% respectively. The mucopolysaccharide seemed to be of a high degree of purity and free from contaminating neutral polysaccharides. The monosaccharide constituents were fucose, galactose, and 2-amino-2-deoxy-D-glucose, identified by paper chromatography after acid hydrolysis. Quantitatively the hexose content was 20·2% (phenol-sulphuric acid), fucose 15·7% (cysteine-sulphuric acid) and hexosamine 5·6% (Elson-Morgan). The ester sulphate content was 19·1% and the nitrogen content 3·4%. In common with most of the echinoid egg jelly mucopolysaccharides this material carries a protein moiety. The amino acid content, after hydrolysis, was 20·3%. No uronic or sialic acids could be detected. The general pattern in the asteroids thus appears to be similar to the echinoids.

Ishihara (1968) has recently presented some new data for the egg jelly of *A. punctulata*.

Ultracentrifugation showed the jelly to be practically homogeneous with a sedimentation coefficient, at infinite dilution, of 7s. The jelly contained 14.3% protein and the amino acid analysis of the jelly is given in Table IV. The N-terminal amino acids of the protein moiety were phenylalanine and arginine in the molar ratio 2 to 1. This would suggest that there were three distinct polypeptide chains present but some uncertainty was expressed as to the reproducibility of the results and it was suggested that some time-dependent-degradation might take place.

TABLE IV

Amino Acid Composition of *Arbacia punctulata* Egg Jelly[a, b]

Amino Acid	
Aspartic acid	1·65
Threonine	1·30
Serine	1·09
Glutamic acid	1·66
Proline	0·55
Glycine	0·68
Alanine	0·83
Cysteic acid	0·25
Cystine	—
Valine	1·23
Methionine	0·08
Isoleucine	0·93
Leucine	1·11
Tyrosine	0·55
Phenylalanine	1·16
Lysine	0·54
Histidine	0·33
Arginine	0·36
Total	14·3

[a] Ishihara (1968).
[b] Amino acids as percentages of total dry weight.

The major monosaccharide component of the jelly was fucose (36·5%), but a small proportion of mannose (not estimated), galactose (1·3%) and hexosamine (2·6%) were also present.

The sulphate content of the mucopolysaccharide was 19·0%. If the percentage compositions are converted to molar quantities this yields a ratio of approximately 1 : 1 of ester sulphate and fucose, suggesting possible mono substitution on each monosaccharide in the polysaccharide chain.

The infrared spectrum of the mucopolysaccharide was published (Fig. 4) and this shows the interesting feature of a quite strong C-O-S stretching band at 840 cm^{-1}. This would suggest an axially located ester sulphate

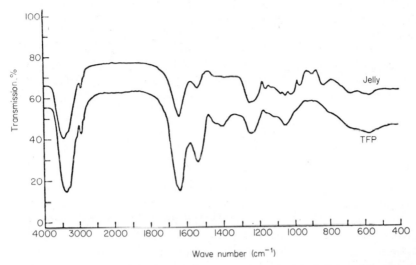

Fig. 4. Infrared spectra of the egg jelly mucopolysaccharide (Jelly) and fertilization product (TFP) from the eggs of *Arbacia punctulata* (from Ishihara, 1968).

TABLE V

Monosaccharide Components of Echinoderm Egg Jellies

Arbacia lixula	Fucose and some galactose
Arbacia punctulata	Fucose (major), mannose, galactose and hexosamine
Echinus esculentus	Galactose
Echinarachnius parma	Fructose
Echinocardium cordatum	Fucose
Hemicentrotus pulcherimus	Fucose
Heliocidaris crassispina	Fucose
Lytechinus anamesus	Galactose
Lytechinus variegatus	Fucose (major), 2-amino-2-deoxy-D-glucose and 2-amino-2-deoxy-D-galactose
Paracentrotus lividus	Fucose (major), glucose, mannose, xylose and hexosamine
Pseudocentrotus depressus	Fucose
Sphaerechinus granularis	Glucose, mannose, fucose
Strongylocentrotus purpuratus	Galactose
Strongylocentrotus pulcherimus	Fucose
Strongylocentrotus droebachiensis	Fucose (major) and galactose
Asterias amurensis	Fucose, galactose and 2-amino-2-deoxy-D-glucose

grouping and hence the localization of the ester sulphate residues at either carbon atoms 2 or 3 on the fucose molecules. These findings suggest that there may be structural similarities between this egg jelly mucopolysaccharide and that of *E. cordatum*.

The carbohydrate compositions of the various echinoderm egg jelly mucopolysaccharides which have been discussed here are summarized in Table V.

ANTI-COAGULANT PROPERTIES OF EGG JELLY MUCOPOLYSACCHARIDES

Apart from their sperm-agglutination properties the jelly coat substances have also been shown to have a rather strong anticoagulant activity (Immers, 1949). This particular biological property is to be expected since heparin and all other known strong anticoagulants are polysaccharide sulphates. Thus *E. esculentus* fertilizin had an antithrombic activity some 20 times less than that of heparin, sensitive to treatment with 0·5 N hydrochloric acid at 100° for 90 minutes or to periodate oxidation (Immers and Vasseur, 1949).

SURFACE ANTIGENS

The surface antigens of *P. lividus* eggs have been investigated by Perlmann (1957, 1958). The so-called J antigen is heat stable at pH 7–8 but is destroyed when heated to 100° at pH 1·0; it is also susceptible to periodate oxidation. This antigen appears to be an integral part of the main jelly coat polysaccharide.

Parthogenetic activation of the egg is brought about by the reaction between antibodies and the A antigen. The A antigen, which is readily soluble in water, is insusceptible to the action of trypsin and pepsin and is heat stable at pH 3·5–8·0. The antigen is slowly hydrolysed by heating to 100° at pH 1·0 and is partially susceptible to periodate oxidation. The electrophoretic mobility of A antigen in phosphate buffer (μ–0·1) at pH 7·7 is $-1\cdot6\times10^{-5}$ cm^2 volt^{-1} sec^{-1} (Perlmann, 1959). A antigen was assumed to have a polysaccharide character distinct from that of the J antigen; there was some evidence to suggest that it contained glucose, mannose, ester sulphate and amino acids.

ENZYMES DEGRADING EGG JELLY MUCOPOLYSACCHARIDES

Numanoi (1953 a, b, c) isolated, from the liver of the marine gastropod mollusc *C. lampas*, an enzyme with the ability to release ester sulphate from both the sperm and egg jelly of *H. pulcherimus*. Egami has found, however, that partially purified polysaccharide sulphatase preparations from the same source have only low activity towards sea urchin egg jellies (Egami and Takahashi, 1962; Takahashi and Egami, 1961).

These authors also obtained and purified a polysaccharase from the same source, capable of bringing about complete solubilization of the fucan sulphate from *H. pulcherimus* egg jelly (Takahashi *et al.*, 1960).

BIOSYNTHESIS

Little is known of the biosynthesis of the fertilizin substances. Administration of ^{35}S-labelled sodium sulphate to females of *A. punctulata* over a period of several days during ovogenesis yielded egg jelly mucopolysaccharides with ^{35}S-labelled sulphate residues (Tyler and Hathaway, 1958). Immers (1961a) demonstrated incorporation of ^{14}C-labelled amino acids and ^{35}S-labelled sulphate into protein-masked sulphated mucopolysaccharides present in the ovary, eggs and embryos of *P. lividus*.

OTHER RELATED MUCOPOLYSACCHARIDES OF ECHINODERM EGGS

Histochemical evidence has been obtained for the presence of acid mucopolysaccharides in the cortical granules of the eggs of *B. lyrifera*, *E. cordatum*, *P. lividus* and *P. miliaris* (Monné, 1949; Monné and Harde, 1951). These observations have been confirmed by Immers (1960a) using stratification of the cytoplasmic components of the unfertilized eggs of *A. lixula*, *E. cordatum*, *E. esculentus*, *P. lividus* and *P. miliaris* by centrifugation and histochemical identification of the fractionated material. Sulphated mucopolysaccharides were present in the cortex of all the eggs examined as well as in the internal granules.

Immers (1961b) has also demonstrated the presence of sulphated mucopolysaccharides in the perivitelline liquid of *E. esculentus* eggs. Transfer of fertilized eggs to 0·05% solutions of methylene blue, in sea water, yielded a precipitate in the perivitelline space. That the material actually carried ester sulphate groups was confirmed by labelling with ^{35}S sulphate. Hot water extracts of the dried, powdered eggs of *A. lixula* and *E. esculentus* contained two polysaccharides each consisting of galactose (Immers, 1960b).

Ishihara (1968) has detected two sulphated glycoproteins in the fertilization products of *A. punctulata* (see also Gregg and Metz, 1966). One of these diffuses out of the elevating fertilization membranes of jelly-free normal eggs. This material contained 44·7% protein, 6·3% hexose (made up of mannose, glucose and galactose), 3·2% fucose, 10·0% hexosamine and

6·4% ester sulphate. An amino acid analysis of this material is given in Table VI. N-terminal amino acid determinations suggested that the material was extremely heterogeneous probably being composed of a number of low molecular weight components.

The other sulphated glycoprotein-like material was released when the vitelline membrane was treated with trypsin following removal of the jelly. This preparation contains 78·6% protein, 6·3% hexose (made up of mannose, glucoce and galactose in the ratio 1 : 1 : 2), 3·4% fucose, 9·2% hexosamine and 5·0% sulphate. The amino acid analysis is given in Table VI. The infrared spectrum of this material clearly indicates that the sulphate is in the ester form (Fig. 4). N-terminal analyses again indicate considerable heterogeneity.

TABLE VI

Amino Acid Compositions of the Fertilization Product (FP) and Trypsin Released Fertilization Product (TFP) from *Arbacia punctulata* Egg Surface[a, b]

Amino Acid	TFP	FP
Aspartic acid	12·5	7·05
Threonine	6·05	4·04
Serine	5·13	2·85
Glutamic acid	12·0	5·50
Proline	5·48	3·10
Glycine	4·83	3·82
Alanine	3·41	1·95
Cysteic acid	3·24	2·74
Cystine	1·50	—
Valine	4·76	2·45
Methionine	0·25	—
Isoleucine	3·23	1·67
Leucine	4·36	2·28
Tyrosine	3·12	1·18
Phenylalanine	3·73	1·78
Lysine	1·27	1·15
Histidine	1·01	0·82
Arginine	2·74	2·35
Total	78·6	44·7

[a] Ishihara (1968). [b] Amino acids as percentages of dry weight.

REFERENCES

Ando, S. (1958). *Bull. biol. Stn. Asamushi*, **9**, 55.
Bishop, D. W. (1951). *Biol. Bull. mar. biol. lab.*, *Woods Hole*, **101**, 215.
Bishop, D. W. and Metz, C. B. (1952). *Nature, Lond.* **169**, 548.
Dorfman, W. A. (1963). "Physico-chemical Foundation of Fertilization". Academy of Sciences of the U.S.S.R., Moscow, U.S.S.R.

Egami, F. and Takahashi, N. (1962). *In* "Biochemistry and Medicine of Mucopolysaccharides" (F. Egami and Y. Oshima, eds), p. 53. University of Tokyo, Tokyo, Japan.
Gregg, K. W. and Metz, C. B. (1966). *Ass. Southeast Biol. Bull.* **13**, 34.
Immers, J. (1949). *Ark. Zool.* **42A**, No. 6.
Immers, J. (1960a). *Expl. Cell Res.* **19**, 499.
Immers, J. (1960b). *Ark. Kemi.* **16**, 63.
Immers, J. (1961a). *Expl. Cell Res.* **24**, 356.
Immers, J. (1961b). *Ark. Zool.* (2) **13**, 299.
Immers, J. and Vasseur, E. (1949). *Experientia,* **5**, 124.
Ishihara, K. (1968). *Biol. Bull. mar. biol. lab.,* Woods Hole, **134**, 425.
Lillie, F. R. (1913). *J. exp. Zool.* **14**, 515.
Lillie, F. R. (1914). *J. exp. Zool.* **16**, 523.
Messina, L. and Monroy, A. (1956). *Pubbl. Staz. zool. Napoli,* **28**, 266.
Metz, C. B. and Donovan, J. (1951). *Biol. Bull. mar. biol. lab.,* Woods Hole, **101**, 202.
Metz, C. B. and Monroy, A., eds (1967). "Fertilization". Academic Press, New York and London.
Minganti, A. (1958). *Boll. Zool.* **25**, 55.
Minganti, A. and Vasseur, E. (1959). *Acta Embryol. Morph. exp.* **2**, 195.
Monné, L. (1949). *Ark. Zool.* **1**, 101.
Monné, L. and Harde, S. (1951). *Ark. Zool.* (2) **1**, 487.
Monroy, A. (1965). "Chemistry and Physiology of Fertilization". Holt, Rhinehart & Winston, New York.
Monroy, A., Tosi, L., Giardini, G. and Maggio, R. (1954). *Biol. Bull. mar. biol. lab.,* Woods Hole, **106**, 169.
Muramatsu, T. (1965). *J. Biochem., Tokyo,* **57**, 223.
Nakano, E. and Ohashi, S. (1953). *Medicine Biol.* (in Japanese), **27**, 97.
Nakano, E. and Ohashi, S. (1954). *Embryologia,* **2**, 81.
Numanoi, H. (1953a). *Sci. Pap. Coll. Gen. Educ. Univ. Tokyo,* **3**, 55.
Numanoi, H. (1953b). *Sci. Pap. Coll. Gen. Educ. Univ. Tokyo,* **3**, 67.
Numanoi, H. (1953c). *Sci. Pap. Coll. Gen. Educ. Univ. Tokyo,* **3**, 71.
Perlmann, P. (1957). *Expl. Cell Res.* **12**, 454.
Perlmann, P. (1959). *Experientia,* **15**, 41.
Runnström, J. (1949). *Adv. Enzymol.* **9**, 241.
Runnström, J. (1951). *Harvey Lec.* **46**, 116.
Runnström, J. (1952). *Symp. Soc. exp. Biol.* **6**, 39.
Runnström, J. (1964). *Biol. Bull. mar. biol. lab.,* Woods Hole, **127**, 132.
Runnström, J., Tiselius, A. and Vasseur, E. (1942). *Arkiv. kemi Miner. Geol.* **15A**. No. 16.
Stern, S. (1966). *Physical and Chemical Comparison of Univalent and Multivalent Sea-Urchin Fertilizins,* Thesis, University of Miami, Coral Gables, Florida.
Stern, S. (1967). *Biol. Bull. mar. biol. lab.,* Woods Hole, **133**, 255.
Takahashi, N. and Egami, F. (1961). *Biochem. J.* **80**, 384.
Takahashi, N., Ishikawa, Y. and Egami, F. (1960). *Kagaku (Tokyo),* **30**, 591.
Tyler, A. (1940). *Biol. Bull. mar. biol. lab.,* Woods Hole, **78**, 159.
Tyler, A. (1948a). *Physiol. Rev.* **28**, 180.
Tyler, A. (1948b). *Anat. Rec.* **101**, 856.
Tyler, A. (1949). *Am. Nat.* **83**, 195.
Tyler, A. (1956). *Expl. Cell Res.* **10**, 377.

Tyler, A. and Fox, S. W. (1939). *Science N.Y.* **90**, 516.

Tyler, A. and Fox, S. W. (1940). *Biol. Bull. mar. biol. lab., Woods Hole*, **79**, 153.

Tyler, A., Burbank, A. and Tyler, J. S. (1954). *Biol. Bull. mar. biol. lab., Woods Hole*, **107**, 304.

Tyler, A. and Hathaway, R. (1958). *Biol. Bull. mar. biol. lab., Woods Hole*, **115, 369.**

Vasseur, E. (1947). *Arkiv. Kemi Miner. Geol.* **25B**, No. 6, 1–2.

Vasseur, E. (1948a). *Acta chem. scand.* **2**, 900.

Vasseur, E. (1948b). *Acta chem. scand.* **2**, 693.

Vasseur, E. (1949a). *Proc. 1st Int. Congr. Biochem.* (Cambridge), 281.

Vasseur, E. (1949b). *Ark. Kemi.* **1**, 105.

Vasseur, E. (1950). *Acta chem. scand.* **4**, 1144.

Vasseur, E. (1952). *Acta chem. scand.* **6**, 376.

Vasseur, E. (1954). "The Chemistry and Physiology of the Jelly Coat of the See Urchin Egg". Kihlstromstryk, Stockholm.

Vasseur, E. and Carlsen, I. (1948). *Ark. Zool.* **41A**, No. 16.

Vasseur, E. and Hogström, B. (1946). *Ark. Zool.* **37A**, No. 17.

Vasseur, E. and Immers, J. (1949a). *Ark. Kemi*, **1**, 39.

Vasseur, E. and Immers, J. (1949b). *Ark. Kemi*, **1**, 253.

Wallenfels, K. (1940). *Osterr Chem. Ztg.* **43**, 87.

CHAPTER 7

Polysaccharide Phosphates

INTRODUCTION

With very few exceptions, and if one also excepts the ribose and deoxy-ribose phosphate backbones of nucleic acids, polymers of sugar and phosphate have not been found to occur in higher animals, although various types of this class of macromolecule are found in microorganisms. In the latter case the teichoic acids, of the Gram-positive bacteria, are the most notable example. Here glycerol or ribitol residues, joined to each other by phosphate groups, carry sugars while D-alanine residues are ester linked to some of the hydroxyl groups (Baddiley, 1964). Phosphorylated mannans are found in some yeasts (Jeanes *et al.*, 1961).

ONUPHIC ACID

The major exception, mentioned above, of a sugar-phosphate polymer found in a higher animal occurs as the secretion of an annelid marine worm *Hyalinoecia tubicola* (previously *Onuphis tubicola*). Many of the phlychaete worms, of which *Hyalinoecia* is a typical example, secrete a protective tube in which they live. In the case of *Hyalinoecia* the secretion hardens into a tough but light quill-like tube of transluscent material.

In an analysis of the material of these tubes Schmiedeberg (1882) found an ash content of 37·5%, Mg 5·9%, Ca 2·4% and P 9·5%; he believed the tubes to be composed of protein ("albuminoid"), a carbohydrate material which he called "onuphin" and probably the acid phosphate salts of calcium and magnesium. Schmiedeberg noted however that preparation of onuphin freed from phosphate was difficult and accompanied by degradation of the carbohydrate substance. Clarke and Wheeler (1924) in an examination of the mineral deposits formed by marine invertebrates gave the results of the analyses for the tubes of *H. artifex*, a close relative of *H. tubicola*. These were MgO 8·57%, CaO 5·35% and P_2O_5 20·72%. Clarke and Wheeler found that there were discrepancies in the proportion of bases to acid in the tubes and suggested that the phosphate might be present as meta or polyphosphate.

In a detailed series of studies (Kelly, 1963; Pautard *et al.*, 1964; Graham *et al.*, 1965; Pautard and Zola, 1966, 1967a, b) it has been shown that the phosphate is in fact not present as inorganic salt but rather as a complex with carbohydrate. This sugar-phosphate polymer has been given the name onuphic acid.

The analyses of ash (37·2%) and phosphorus (9·3%) in washed and dried tubes closely agreed with those of Schmiedeberg. The X-ray diffraction patterns of the whole tubes contained two broad diffuse rings and no well-defined, oriented or continuous reflections as might have been expected if crystalline salts had been present. Moreover after demineralization in 1 N hydrochloric acid no ortho-, meta- or poly-phosphate could be detected in the supernatant. The demineralization process caused the tubes to swell to a gelatinous mass which, after washing, was found to contain only a small proportion of the total phosphate. Examination of the hydrochloric acid extract showed that the phosphate had in fact been extracted but in a bound form which only became detectable after calcining or after hydrolysis by severe reagents such as perchloric acid/sulphuric acid mixtures. The hydrochloric acid extracted phosphate was non-dialysable and the organic material containing it comprised 70% of the weight of the whole intact tube.

Onuphic acid is generally prepared by extraction with N hydrochloric acid at 2° for 24 hours or by treatment of the tubes with 10% EDTA solution at pH 7·4 for 24 hours at 2°. The polymer is precipitated from solution with cold ethanol in the presence of potassium chloride (1% concentration). After repeated reprecipitation and drying, the onuphic acid is dissolved in water and dialysed. Under these conditions a preparation is obtained which has the empirical formula $C_{12}H_{22}O_{24}P_3$ (Pautard and Zola, 1966). This material contains 14·4% P, while nitrogen appears to be absent in spite of the indication in the earlier papers of a low content (0·3–0·5%).

Onuphic acids prepared by either acid or EDTA extraction, under the conditions described above, had identical sharp sedimentation patterns, yielding S values, in 0·2M potassium chloride at 20° and infinite dilution, of 21·8. The critical ethanol solubility curves (critical ethanol concentration 40%) were also identical. Increasing the strength of acid, the duration of extraction, and the temperature of extraction caused a progressive increase in polydispersity reflected in diffuse sedimentation patterns and extended ethanol solubility curves. The polymer also appeared to be sensitive to mechanical degradation; crushed as opposed to whole tubes yielded more polydisperse preparations. A similar result was produced when homogeneous polymer was subjected to a period of shaking in a Mickle disintegrator.

Attempted purification of onuphic acid by precipitation with cetylpyridinium bromide, according to Scott (1960), failed to produce any significant alteration in the composition of the material. Deionization was effected by treatment with Amberlite IR120 (H^+ form) ion exchange resin. Drying onuphic acid *in vacuo* over phosphorus pentoxide at room temperature gives a product with 6% moisture content.

Although onuphic acid appears homogeneous by sedimentation, ethanol precipitation, detergent precipitation and also gel-filtration, moving boundary electrophoresis gives anomalous results. A single peak is produced on electrophoresis in veronal buffer pH 8·6 whereas in pH 6·4 phosphate buffer two peaks of similar mobility are produced. Pautard and Zola (1966) suggested the electrophoretic anomalies might be explained in terms of interactions rather than heterogeneity.

The primary cysteine-sulphuric acid reaction indicates that onuphic acid consists mainly of carbohydrate. Tests for hexuronic acid were negative. Although in an earlier report (Pautard et al., 1964), a weak maximum at 260 mμ was noted in the absorption spectrum of onuphic acid, a later study (Pautard and Zola, 1966) showed that the visible and near ultraviolet absorption spectrum was without maxima; the absorption increasing gradually with decreasing wave length in the 270–210 mμ region and thereafter rapidly towards 200 mμ.

The infrared spectrum had several characteristic phosphate bands as well as hydroxyl, methylene and methyl bands (Table I). The band at 11·9 μ suggests an α-D anomeric linkage although the absence of a band at 13 μ renders this conclusion doubtful. Onuphic acid has $[\alpha]_{5461}^{20} - 4 \pm 2°$ (water) (Pautard and Zola, 1966) or $[\alpha]_{5461}^{20} - 6·4 \pm 2°$ (in N HCl) (Pautard et al., (1964).

TABLE I

Infrared Spectral Characteristics of Onuphic Acid[a]

Wave length (μ)	Possible Correlation
2·9	O—H
3·4	P—O—H
4·3	P—O—H
6·10	P(O)—OH
6·85	—CH$_2$
7·30	C—CH$_3$
8·20	P=O
9·50 (very broad)	P—O—C
11·90	α-D anomeric linkage

[a] Pautard and Zola (1966).

Onuphic acid has been reported to have a molecular weight of 25,000 (Pautard et al., 1964) and of the order 10^5 when determined by approach to sedimentation equilibrium (Graham et al., 1965).

Determinations of Pi, after oxidative hydrolysis and after non-oxidative hydrolysis, gave identical results (Pautard and Zola, 1967a) indicating that onuphic acid contains no detectable P—C bonds. Mild acid hydrolysis does not release detectable Pi from onuphic acid, thus eliminating the possibility of polyphosphate or monoesterified pyrophosphate. It does not however eliminate the possibility of diesterified pyrophosphate groups since the mild acid treatment used caused depolymerization (measured by fall in viscosity) of the onuphic acid without Pi release. This observation might have been explained by the hydrolysis of diesterified pyrophosphate links, in a chain of sugar residues, to acid-stable monoesters:

$$R-O-P-(O_2H)-O-P(O_2H)-O-R' \rightarrow R-O-P(O_3H_2)$$
$$+R'-O-P(O_3H_2).$$

The results of potentiometric titrations indicated that approximately 10% of the phosphorus present was in the form of diester while the remaining 90% was in the monoester (presumably orthophosphate) form (Fig. 1, Table II). Since each phosphorus atom appeared to have at least one ionizable acid group, triesterified phosphate must have been absent.

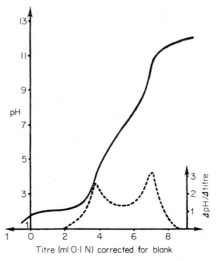

Fig. 1. Potentiometric titration curve (————) and differential titration curve (– – – – –) of onuphic acid (from Pautard and Zola, 1967).

The presence of diester phosphate was confirmed by a fall in the viscosity of onuphic acid solutions taking place on treatment with bovine spleen diesterase.

Treatment with alkaline and acid phosphatases, instead of releasing the

TABLE II

Potentiometric Acid-Base Titration Data for Onuphic Acid[a]

(1) Number of protons dissociated at the first inflexion per atom of P: $1 \cdot 1 \pm 0 \cdot 1$[b]

(2) Number of protons dissociated at the second inflexion per proton dissociated at the first inflexion: $1 \cdot 90 \pm 0 \cdot 02$

(3) Hence, % of P atoms having only one ionizable hydrogen atom: 10 ± 2

[a] Pautard and Zola (1967).

[b] The S.D. for (1) includes errors due to variable moisture content, weighing etc. These errors will not feature in (2) and (3) and hence the comparatively large error in (1) will not invalidate (2).

90% monoesterified orthophosphate as expected, in the case of the alkaline phosphatase released 4% Pi and in the case of acid phosphatase released $50 \cdot 6$% after repeated treatments. Zola (1967) has noted that wheat germ phosphomonoesterase released 42% of the available phosphorus of the whole tube as Pi in 48 hours.

Thus, although the evidence suggested that the phosphate was present as ester, it was not clear whether the diesterified phosphate was present as orthophosphate or as diesterified pyrophosphate similar to the pyrophosphate dinucleotides. Nor was it clear what was the form of the mono-esterified phosphate.

The major monosaccharide component released by acid hydrolysis was isolated by preparative thin layer chromatography. The monosaccharide was identified as D-glucose by chromatography, electrophoresis, colour reactions and by its infrared spectrum. A minor monosaccharide component detected on the chromatograms was identified as L-fucose.

Onuphic acid contains 59 ± 1% glucose and $2 \cdot 4 \pm 0 \cdot 4$% fucose.

Kinetic studies of the hydrolysis of onuphic acid by acid, under varying conditions, indicate that the polymer contains acid-stable groups as well as acid-labile linkages involving glucose.

Determination of the reducing power (Pautard and Zola, 1967b) showed that 1 g onuphic acid was equivalent to $1 \cdot 32 – 1 \cdot 51$ mg free glucose. Periodate consumption by onuphic acid was $44 \cdot 5 \times 10^{-5}$ moles per g onuphic acid, i.e., 1 mole periodate per 10 g atoms of onuphic acid phosphorus. Partial enzymic dephosphorylation caused no increase in periodate consumption.

Acid hydrolysis of borohydride-reduced, periodate-oxidized, onuphic acid, yielded glucose and inorganic phosphate. No fucose or non-reducing polyhydroxy compounds were detected but a minor component having the properties of a reducing disaccharide was found.

Treatment of onuphic acid with 2N hydrochloric acid at 25° or with $0 \cdot 2$ N potassium hydroxide at 25° reduces the viscosity markedly, the effect reaching a plateau at about 10 days. These hydrolysates, and hydrolysates produced by phosphodiesterase treatment, when examined in the ultra-centrifuge give a single, rapidly diffusing, broad peak in contrast to the sharp peak of the undegraded material. Fractionation of hydrolysates on columns of Sephadex G25 showed the presence of an excluded high molecular weight component and a retarded low molecular weight component. The high molecular weight fraction contained no fucose, the glucose-phosphate ratio being close to that of the parent molecule. The low molecular weight fraction varied in its composition, depending on the mode of degradation. In the case of enzymic hydrolysis it contained fucose while the material from acid hydrolysis contained fucose, glucose and inorganic phosphate. After alkaline hydrolysis a low molecular weight material was obtained which did not contain fucose but did contain glucose and inorganic phosphate.

A ratio of components glucose (22) : fucose (1) : phosphomonoester (25) : phosphodiester (3)—was proposed by Pautard and Zola (1967b) for onuphic acid. Since partial degradation resulted in macromolecular fragments similar in composition to the whole molecule, an even distribution of subunits and labile linkages throughout the parent molecule seemed to be indicated. It seemed possible also that fucose was bound to the rest of the molecule through the phosphodiester linkages although the action of phosphodiesterase in reducing the viscosity and in releasing fragments with molecular weights in excess of 5000 suggests that phosphodiester links may be distributed in the main chain also or in long branches.

The low reducing power suggests that the glycosidic groups of the glucose residues are blocked. The limited periodate attack seems to indicate substitution on the glucose residues at either position 3 or at both the 2 and 4 positions; the blocking groups do not appear to be phosphate since the action of periodate is not enhanced by removal of phosphate. It was suggested that the glycosidic link might be 1 → 3. Rapid degradation of onuphic acid by alkali (Pautard and Zola, 1967a) seems to indicate that the 2 position on the glucose molecule is free and adds further weight to the postulate of a 1 → 3 linked glucan structure for the main subunit (Pautard and Zola, 1967b). The periodate oxidation results suggest the glucose residues are in the pyranosyl form. The phosphate ester groups thus probably occupy position 4 or 6 on the glucose molecules while the excess of phosphate groups over sugar residues may indicate that in some instances both positions are substituted.

Zola (1967) has noted that the tube of *Chaetopterus variopedatus* also contains a phosphorus-rich polysaccharide similar in composition to onuphic acid. The polysaccharide, which was extracted with EDTA according to the procedure used for onuphic acid, contained 4·9% phosphorus and gave a single sharp boundary in the ultracentrifuge when dissolved in 0·2 M potassium chloride.

Hydrolysis of this material in 2N HCl at 80° for three hours yielded mannose, glucose and xylose, identified chromatographically, together with inorganic phosphate and carbohydrate phosphate. Specimens of *Spirorbis* sp. and *Spirorbis spallanzanii* did not contain significant quantities of phosphorus in their tubes.

Scott has examined the critical electrolyte concentration of complexes of various dyes with onuphic acid. Critical electrolyte concentration is defined as the value of the maximum concentration of electrolyte at which staining of a spot of the polyanion on paper is apparent. Onuphic acid yielded values of 0·3 in Alcian blue, 0·1 in Thioflavin T, 0·2 in Azure A and 0·2 in Acridine orange. The results quoted are in molarities of sodium chloride. These values may be compared with values of 0·5, <0·05, 0·3 and 0·2 respectively, for thymus DNA and 0·3, 0·1, 0·4 and 0·1 for RNA core (Scott and Willett, 1966).

OTHER POLYSACCHARIDE PHOSPHATES

Maki (1956) reported the presence of a water-soluble polyribose phosphate in the liver of the squid *Ommastrephes sloani* during the summer but not the winter months. This non-reducing, non-dialysable material was easily depolymerized by acid (0·5 N HCl), alkali (0·5 N NaOH) and by the enzyme ribonuclease. A polyribose phosphate is known to occur in the capsule of the microrganism *Hemophilus influenzae* type b. This substance consists of a chain of ribose phosphate units, such as are found in the backbone of the ribonucleic acids, but in which the place of the purines and pyrimidines is occupied by a second similar chain linked to the first in β-β glycosidic linkages (Zamenhof *et al.*, 1953, 1954; Rosenberg and Zamenhof, 1961; Rosenberg *et al.*, 1961). Rosenberg and Zamenhof (1962), believing that the polyribose phosphate found in squid by Maki might be similar in character to their *H. influenzae* material, attempted to repeat his work but totally failed to detect any ribose or significant phosphate in the polysaccharides isolated. The material used was frozen Japanese squid, obtained in Honolulu during July, fresh Atlantic coast squid obtained during November or August and Californian squid obtained during July. Since however many marine invertebrates are known to show sudden marked seasonal changes in metabolism, it is possible that isolation of the polyribose phosphate was attempted at the wrong time.

Monroy *et al.* (1954) have reported that *Arbacia lixula* (echinoderm) sperm in sea water, or in egg coat solution, release a fucose-containing phosphate ester.

A phosphate-containing polysaccharide appears to be present in *Helix pomatia* (gastropod mollusc) albumin. An aqueous extract of the gland, freed from galactan-containing mucins, was precipitated with alcohol at pH 7 to give a polysaccharide-protein complex (Geldmacher-Mallinkrodt, 1957). This was shown to contain three fractions which could be separated by electrophoresis. A mixture of polysaccharides was isolated from these fractions by boiling with alkali and subsequent precipitation with alcohol. This mixture was reported to be difficult to hydrolyse because of its high acid content.

REFERENCES

Baddiley, J. (1964). *Endeavour*, **23**, 33.

Clarke, F. W. and Wheeler, W. C. (1924). *Prof. Pap. U.S. geol. Surv.* No. 124.

Geldmacher-Mallinkrodt, M. (1957). *Hoppe-Seyler's Z. physiol. Chem.* **308**, 220.

Graham, G. N., Kelly, P. G., Pautard, F. G. E. and Wilson, R. (1965). *Nature, Lond.* **206**, 1256.

Jeanes, A., Pittsley, J. E., Watson, P. R. and Dimler, R. J. (1961), *Archs biochem. Biophys.* **92**, 343.

Kelly, P. G. (1963). *J. dent. Res.* **42**, 1085.

Maki, M. (1956). *Hirosaki med. J.* **7**, 211.

Monroy, A., Tosi, L., Giardini, G. and Maggio, R. (1954). *Biol. Bull. mar. biol. lab., Woods Hole*, **106**, 169.

Pautard, F. G. E. and Zola, H. (1966). *Carbohydrate Res.* **3**, 58.

Pautard, F. G. E. and Zola, H. (1967a). *Carbohydrate Res.* **3**, 271.

Pautard, F. G. E. and Zola, H. (1967b). *Carbohydrate Res.* **4**, 78.

Pautard, F. G. E., Kelly, P. G. and Zola, H. (1964). *Abs. 6th Int. Congr. Biochem. (N.Y.)* **6**, 84.

Rosenberg, E. and Zamenhof, S. (1961). *J. biol. Chem.* **236**, 2841.

Rosenberg, E. and Zamenhof, S. (1962). *J. Biochem. Tokyo*, **51**, 274.

Rosenberg, E., Leidy, G., Jaffee, I. and Zamenhof, S. (1961). *J. biol. Chem.* **236**, 2845.

Schmiedeberg, O. (1882). *Mitt zool. stn Neapel.* **3**, 373.

Scott, J. E. (1960). *Meth. biochem. Analysis* **8**, 145.

Scott, J. E. and Willet, I. H. (1966). *Nature, Lond.* **209**, 985.

Zamenhof, S. and Leidy, G. (1954). *Fedn. Proc. Fedn. Am. Socs exp. Biol.* **13**, 327.

Zamenhof, S., Leidy, G., Fitzgerald, P. L., Alexander, H. E. and Chargaff, F. (1953). *J. biol. Chem.* **203**, 695.

Zola, H. (1967). *Comp. biochem. Physiol.* **21**, 179.

MUCOPOLYSACCHARIDES
Neutral Mucopolysaccharides

CHAPTER 8

Chitin

INTRODUCTION

If quantity is any criterion then the most important structure polysaccharide of the animal kingdom is undoubtedly chitin. Chitin consists of linear, unbranched chains of $\beta1 \to 4$ linked 2-acetamido-2-deoxy-D-glucose residues. It thus resembles cellulose in its chemical structure, and at higher levels of organization, as well as in physical properties, the two polysaccharides have much in common. Both polysaccharides also serve rather similar functions, that is to say of support and defence. Chitin is most frequently found complexed with protein but the exact nature of this association is as yet only imperfectly understood. Chitin not only associates in nature with protein but frequently also is found in conjunction with insoluble salts of calcium, lipids and other organic substances.

So far as we know, chitin does not occur in the tissues of vertebrates, nor does it figure strongly, if at all, in the invertebrate evolutionary line which leads to the chordates. In the molluscan–arthropod evolutionary line however chitin shows an increasing significance as a skeletal substance which culminates in the role it plays in the organization of the arthropod cuticle or exoskeleton.

In the past the term "chitinous" has been used rather loosely to describe any invertebrate tissue, usually of skeletal or semi-skeletal origin, which had a hardened yet flexible character and as a result systems of tanned protein were frequently grouped with truly chitinous tissues. Chemical analyses of invertebrate tissues have done much to correct many of these earlier misconceptions.

In the literature the name "chitin" tends to be used liberally and rather uncritically for both the purified polysaccharide and for the chitin-protein complex in which the polysaccharide is most frequently found in the natural state. In fact we can probably no more regard the polymer of 2-acetamido-2-deoxy-D-glucose alone as a natural chemical entity than we can consider alkali treated, protein-free, chondroitin sulphate as an intact biopolymer. The polysaccharide chain of chitin, free from protein, may not exist in nature and it is therefore better if we can use a nomenclature which clearly distinguishes between the free polysaccharide as a degradation product of the chitin-protein complex and the intact complex. Hackman (1960) has

129

E

proposed that the term "chitin" should be reserved for the polysaccharide and that the complex with protein should be called "native chitin". In the author's view this latter term fails to describe the situation adequately, suggesting only changes in the macromolecular organization and conformation of the polysaccharide itself rather than association with other molecules. The more cumbersome, but more descriptive, "chitin-protein complex" will be used here.

OCCURRENCE

The distribution of chitin has been reviewed by several authors (Richards, 1951; Rudall, 1955; Foster and Webber, 1960; Jeuniaux, 1963; Brimacombe and Webber, 1964; Kent, 1964) and it is therefore unnecessary to repeat this in detail here.

Chitin is principally found in the arthropod phylum where it contributes to the structure of the exoskeleton, the lining of the gut, the tendons, wing coverings and the internal skeleton. In the annelid worms chitin does not form a part of the body wall itself but is found in the mouth parts, the lining of the gut and the chaetae. One of the purest forms of the polysaccharide occurs in the internal skeletons of squids (mollusc). Elsewhere in the molluscs we find chitin as a more minor component of the shell matrix and in the hardened teeth of the radula. Chitin also figures as an important constituent of the shell matrix in one of the major brachiopod groups, the Inarticulata. It also occurs as a component in the exoskeleta of certain other minor coelomate phyla, principally the Ectoprocta. Chitin seems not to be found in the adult sponge but is a constituent of the walls of the internal buds or gemmules. In the coelenterates the hydrozoan perisarc which is produced over the nutritive polyp contains chitin, while in the Anthozoa chitin occurs in the skeletal matrix. It is found in the egg shells of platyhelminths, nematodes and acanthocephalans but its occurrence in their body walls is uncertain. Chitin has been recorded in the copulatory bursae of rhabdocoels and in the hydatid cysts of cestodes. The exoskeletons of the pogonophores contains chitin (Fishenko, 1963; Brunet and Carlisle, 1958; Rudall, 1962) and it is therefore interesting to note that Florkin (1966), in spite of earlier reports to the contrary (see Carlisle, 1963; Manskaia and Drozdova, 1962; Foucart et al., 1965), categorically denies the presence of chitin in graptolites, a fossil group thought to be closely related to the pogonophores. The occurrence of chitin in the pogonophores is interesting also in view of its absence from other members of the chordate line and it may therefore bring into question the validity of their inclusion in the Deuterostomia. Its presence in echinoderms is doubtful and it is absent from the urochordates, the hemichordates, the cephalochordates and as we have already noted the vertebrates.

Chitin occurs also in lower plants where it is found in fungi, including yeasts, and green algae. It is not found in higher plants nor apparently in bacteria.

CHEMICAL STRUCTURE

The primary structure of chitin is by now reasonably well established and the steps leading to its elucidation need not concern us here; they have been particularly well reviewed recently by Brimacombe and Webber (1964) and Kent (1964).

As already mentioned, chitin is an unbranched polysaccharide composed of $\beta1 \rightarrow 4$ linked 2-acetamido-2-deoxy-D-glucopyranose residues (Fig. 1).

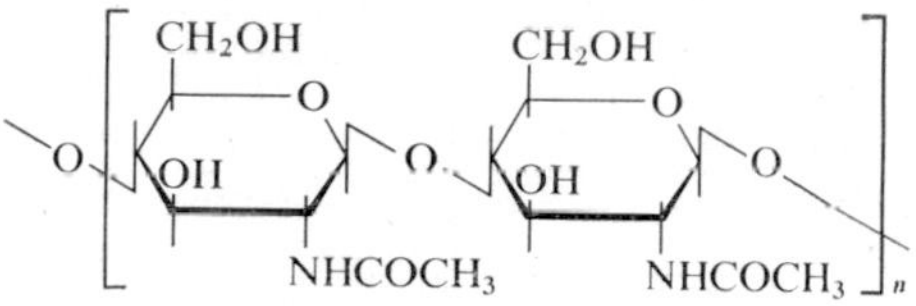

Fig. 1. Structure of Chitin.

Some doubt still exists as to whether all the hexosamine residues are in fact N-acetylated. Rudall (1963) has reviewed the evidence which suggests that a proportion of the residues do in fact occur as 2-amino-2-deoxy-D-glucose. Elemental analyses and histochemical properties suggest the presence of a proportion of free-amino groups, approximating to one every sixth or seventh residue in the chain. Hunt (unpublished) has failed to detect any free-amino groups in blowfly cuticle chitin by a dinitrophenylation technique.

PHYSICAL STRUCTURE

The probable secondary, tertiary and quaternary structures (if one may impertinently apply these protein chemistry terms to a polysaccharide) of chitin have been elucidated in considerable detail, largely as a result of numerous X-ray diffraction and infrared spectroscopy studies. This particular aspect of the molecular biology of chitin has been well reviewed by Rudall (1963), by Kent (1964) and, more briefly, by Brimacombe and Webber (1964).

Three molecular forms of chitin are now recognized. All are derived from the regular $\beta1 \rightarrow 4$ linked chain of 2-acetamido-2-deoxy-D-glucose units. The chitin of the arthropod cuticle occurs in the crystal form known as α, in the pen of the squid *Loligo*, the chaetae of the annalid *Aphrodite* and the tubes of pogonophores, is found the less common β form, while in the thick cuticle lining from the stomach of *Loligo* a third form, γ chitin, occurs (Rudall, 1962; Lotmar and Picken, 1950). *Loligo* does in fact possess all three forms, since α chitin occurs in a thin oesophageal cuticle lining the stomach.

Early X-ray diffraction studies of α chitin (Meyer and Pankow, 1935) indicated a fibre axis repeat almost identical with that of cellulose; a rhombic unit cell of eight 2-acetamido-2-deoxy-D-glucose residues, with the dimensions a = 9·40 Å, b = 10·46 Å (the fibre axis) and c = 19·25 Å, was proposed. As a result of the β linkage the chains were straight, with the acetamido groups lying alternately on either side of the chain. As with cellulose, equal numbers of chains ran in opposite directions, but in contrast to cellulose, paired chains in the same plane ran in the opposite sense.

Later, more refined studies by Carlström (1957) confirmed the general picture described above but simplified it with the adoption of an ortho-rhombic unit cell, having the dimensions a = 4·76 Å, b = 10·28 Å, c = 18·85 Å (fibre repeat). Two di-N-acetylchitobiose residues, in two polysaccharide chains, run anti-parallel through the unit cell. The chains, rather than being straight, have a bent conformation (Fig. 2), imposed partly by the bulky acetamido groups (Carlström, 1962) resulting from rotation in the glycosidic bonds. A similar type of structure has been proposed for cellulose (Liang and Marchessault, 1959). The chitin structure is stabilized by intermolecular full hydrogen bonding between acetamido side-chains of adjacent, parallel polysaccharide chains. Polarized infrared spectra provide good support for this structure (Carlström, 1957; Pearson *et al.*, 1960a, b).

Independently, Dweltz (1960) has derived a unit cell having dimensions—a = 4·69 Å, b = 19·13 Å, c = 10·43 Å (fibre repeat)—somewhat different to those put forward by Carlström. Moreover, Dweltz has suggested that the chains have a straight, rather than bent, conformation. While a feature of the structure proposed was intermolecular hydrogen bonding between the hydroxymethyl groups of the adjacent chains and bonding of C-3 hydroxyls to amide oxygens, no infrared evidence could be obtained in support of this structure (Pearson *et al.*, 1960a, b).

Rudall (1963) is of the opinion that the technique used by Carlström is likely to be more accurate than that used by Dweltz and that there may be, therefore, a greater experimental error in Dweltz's cell dimensions. Carlström (1962), in defence of his structure, has criticized that of Dweltz on the basis of unfavourable H– – – –H contacts at carbon atoms C-1 and C-4, adjacent to the glycosidic oxygen, and has also indicated errors in the atomic co-ordinates listed by Dweltz.

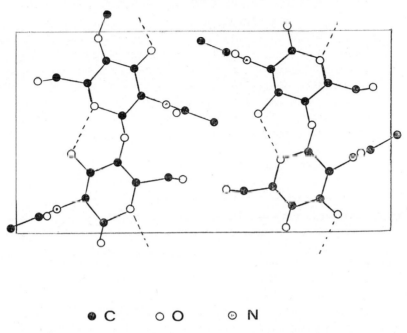

● C   ○ O   ☉ N

Fig. 2. Unit cell of α-chitin, viewed along the axis, according to Carlström, 1957.

Ramachandran and Ramakrishnan (1962) have examined the relative merits of the Carlström and Dweltz structures and have decided that on the basis of the X-ray diffraction data neither is preferable to the other. These authors note that the C-1, C-4 H‑ ‑ ‑H contact can be satisfied by either straight or bent chains. However, a re-examination of the fibre repeat distance of Dweltz's specimen yielded a value of 10·3 Å, comparing more favourably with that of Carlström (10·28 $\pm$ 0·03 Å) than with that obtained originally by Dweltz (10·43 Å).

In an X-ray diffraction study of chitin, from the pen of the squid *Loligo* and from the chetae of the worm *Aphrodite*, Lotmar and Picken (1950) were able to define a unit cell having the dimensions a = 9·32 Å, b = 10·17 Å and c = 22·15 Å. This therefore provided concrete evidence for the existence of a second crystallographic form of the chitin molecule. Density measurements indicated unequivocally that this crystalline form could not have arisen from the alpha form of chitin by hydration and resultant swelling along the c-axis. The alpha and beta forms differ only at the macromolecular level; both yield 2-amino-2-deoxy-D-glucose only, on hydrolysis, and solutions of β-chitin prepared in formic acid give rise to the alpha form of the molecule when the chitin is reisolated.

A smaller unit cell for *Loligo* β-chitin has been proposed by Dweltz

(1961). This cell, which is monoclinic with the dimensions a = 4·7 Å, b = 10·5 Å, c = 10·3 Å (fibre repeat), contains two 2-acetamido-2-deoxy-D-glucose units in a single straight polysaccharide chain. Thus all the chains in a crystallite have the same direction. Dweltz in his model places one water molecule per amino sugar residue, and earlier studies of β-chitin have also indicated that water has a significant role to play in the structure of the system (Rudall, 1955). The water molecule lies bound between the C-3 hydroxyl of the amino sugar on one chain and the hydroxymethyl group of the hexosamine on the adjacent chain along the b-axis (Dweltz, 1961). Thus β-chitin is a monohydrate, in the dry state. Changes in humidity lead to alterations in the b-repeat distance confirming these observations; the implication being that introduction of further molecules of water would lead to separation of the chains.

Carlström (1962) has suggested that before a definite structure for the beta form can be deduced from the X-ray data, a consideration also of stereochemical requirements in the system should be undertaken. Rudall (1963) also believes that Dweltz may well be wrong in maintaining the straight chain structure although he agrees that the deduction that all the chains are parallel is probably correct. It seems to Rudall that the configurations of the chains themselves is likely to be similar to that in α-chitin.

Just as consideration of the X-ray diffraction diagrams of β and α-chitins leads to their structures being considered as parallel and anti-parallel arrangements of chains respectively, by the same arguments consideration of the diffraction patterns for γ-chitin, from the thick cuticle which lines the stomach of *Loligo*, leads to the necessity for proposing a completely novel arrangement of chains in the molecular system. Along the c-axis of this structure the chains repeat, not in groups of one as in β-chitin, nor in groups of two as in α-chitin, but rather in groups of three (Fig. 3) (Rudall, 1962, 1963). The simplest way in which this structure can be explained is by the folding of single chains upon themselves (Fig. 4). Folding, incidentally, can account for the β to α transition, which is accompanied by a

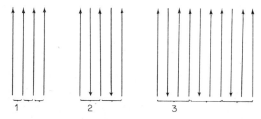

Fig. 3. Representations of chitin chains viewed in the plane of the sugar rings. Chains in groups of 1, 2 and 3 (according to Rudall, 1963).

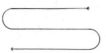

Fig. 4. Folding of a chain upon itself to give three parallel segments as envisaged by Rudall (1963).

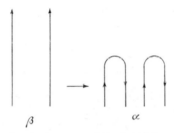

Fig. 5. Possible mechanism for the β to α transition in chitin according to Rudall (1963).

50% contraction, which occurs when β-chitin is placed in 6 N hydrochloric acid (Fig. 5).

Since γ-chitin possesses in one molecular structure characteristics of both α and β-chitins, i.e. anti-parallel and parallel arrangements of chains, it might be expected that this molecular type should also form hydrates and X-ray diffraction data indicate that hydration does indeed occur (Rudall, 1963).

CHITIN-PROTEIN COMPLEXES

The arthropod cuticle, while frequently referred to as chitinous, actually contains considerable quantities of other substances among which proteins figure strongly (see Chapter 15 also). Indeed chitin may be present in certain instances in relatively insignificant quantities. It is unlikely that chitin, in animals at least, occurs naturally in an absolutely pure form un-complexed with protein. References in the literature to the identification of protein accompanying chitin are numerous; too numerous, in fact, to be listed here. Even in such cases as the squid internal skeleton, where chitin occurs in perhaps its purest natural form, there is still a certain amount of accompanying protein, and we can say with a fair measure of certainty that this also applies in varying degrees to all chitins *in vivo*. The question is to what degree is the accompanying protein a part of a true macromolecular complex with the chitin, in the same sense that we define the polysaccharide protein complexes found in other connective tissues (e.g. chondroitin sulphate-protein complex of cartilage tissue).

It is important to remember that the X-ray crystallographic data considered in the previous section was obtained from materials often subjected to harsh deproteinization procedures (e.g. strong alkali for long periods at elevated temperatures) and that the structures deduced may not reflect with complete accuracy the configurational state of the polysaccharide chains when complexed with protein in the natural state.

In Chapter 15 we note that a progressive extraction of the soft, insect cuticle, with solvents of increasing harshness, results in removal of several distinct groups of proteins (Hackman and Goldberg, 1958). Thus, some 14% of the total protein in the cuticle could be removed merely by treatment with cold water and it is obvious that the association of this material with other cuticle components, including chitin, can be only by the weakest bonds. A further 2% of the cuticular protein is soluble in salt solution, suggesting some degree of ionic association with the remaining constituents of the cuticle. Some 25% of the protein can be removed by treatment with 7 M aqueous urea; this protein may thus be associated with chitin, and other materials, by hydrogen bonds. Treatment of the residue with cold dilute aqueous sodium hydroxide solubilizes a further 3% of the total protein, a perhaps significant finding in view of the now well-known and well-characterized alkali labile protein-polysaccharide linkages found in many glycoproteins and connective tissue mucopolysaccharides. After such a series of extractions 56% of the total protein of the soft cuticle still remains and must be considered as being more closely, or firmly, associated with the chitin component of the cuticle than are the other proteins. This apparently firmly bound protein (35% of the cuticle) is approximately equal in quantity to the chitin present (37% of the cuticle). It should be noted however that soft cuticles from other arthropod larval sources have been observed to have lower protein contents and the ratio of protein to chitin in the firm complex may not always be the convenient 1 : 1 found in *Agrionome spinacollis* cuticle. Thus, for example, in the larval cuticle of the blowfly *Sarcophaga falculata* the proportion of the total protein in the cuticle, remaining after all extractions, is far less than the chitin content of the cuticle.

While this remaining protein gives the impression of being bound by some strong, probably covalent, type of bond to the chitin, it is worth pointing out that Picken (1960) has suggested that the vigorous conditions which are required to deproteinize sclerotized cuticle need not necessarily indicate tenacious chitin-protein interactions but may be due to the great stability of a tanned protein constituent.

Hackman (1955a, b) has investigated the interaction of 2-acetamido-2-deoxy-D-glucose as well as chitin, with amino acids, peptides and proteins. The amino sugar combined with arthropodin (the water-soluble cuticular protein), peptides and amino acids, including tyrosine. The complexes dissociated at low pH but were stable at higher values. Chitin formed temperature stable complexes with proteins binding up to 8% of its own weight of protein. These complexes were sensitive to ionic strength and pH, dissociation being complete at pH 9. Thus weak complexes of chitin and arthropodin may be possible. Theoretically the extended β configuration of arthropodin would favour interaction with chitin since the molecular spacing of the extended protein chains agree identically with the lattice spacings of the α-chitin chains (Fraenkel and Rudall, 1947; Rudall, 1950).

Evidence does exist for chemical bonding of protein to chitin although this has recently been disputed. Chitin from the soft cuticle of *S. falculata* and the decalcified cuticle of the edible crab *Cancer pagurus* can be dissolved in saturated solutions of lithium thiocyanate and reprecipitated by addition of acetone (Foster and Hackman, 1957). This chitin still retains protein which is bound to it, or at least associated with it, in some firm manner. Chitin isolated from the crab cuticle, by mild decalcification, contained some 5% protein and on hydrolysis yielded several acid, basic and neutral amino acids; notable among these were glutamic and aspartic acid. Jeuniaux (1959) has described a membrane in the cuticle of crustaceans which is resistant to the action of chitinase and which contains 73% chitin and 12 to 15% protein. Removal of protein by alkaline digestion permits the chitin to be degraded by the enzyme.

Hackman (1960) has subjected the whole problem to a more extensive study isolating α and β-chitins from insect and crustacean cuticles, from cuttlefish shells and squid pens. The methods of preparation were mild, avoiding the use of alkali. A series of chitin-protein complexes were obtained. Significantly the ratio of chitin to protein showed considerable variation from about 1 : 1 (cuttlefish shell) to 20 : 1 (*A. spinacollis* larval cuticle). Dissolving the complexes in lithium thiocyanate solutions and reprecipitation with acetone yielded fractions whose amino acid analyses were essentially similar to those of the original materials. Treatment of these complexes with hot alkali (N NaOH at 100° for 60 hours) removed all protein except for those amino acids supposed to be immediate to the carbohydrate-protein linkage region. In all cases small amounts of aspartic acid and histidine only could still be detected on acid hydrolysis of the alkali treated materials (Hackman, 1960, 1962). This is particularly interesting also since one is dealing here with chitins in more than one molecular form. Of the order of two histidines and one aspartic acid residue were present for each 400 hexosamine residues (giving a polysaccharide chain about 4000 Å long) in the *A. spinacollis* chitin-peptide complex.

The amino acid analyses of the whole complexes were odd in certain respects. For example the chitin-protein complexes of the arthropod seemed to be deficient in glycine, cystine/cysteine, lysine, serine and methionine. Hydroxyproline was also absent. An almost complete absence of glycine in a protein is a sufficiently unusual circumstance to be worthy of comment. The molluscan chitin-protein complexes also lacked methionine and hydroxyproline.

It is not clear from Hackman's studies whether we are to regard aspartic acid or histidine as the amino acid actually involved in linkage to the polysaccharide or whether both are bound to the polysaccharide chain as expressions of two distinct protein chains. Precedent, at least, places aspartic acid in a favourable position for consideration as the amino acid linked directly to the polysaccharide. Histidine, on the other hand, has not hitherto been detected in this role. The types of linkage usually occurring in stable covalent complexes of protein and carbohydrate have been reviewed elsewhere in this book. Apart from the absence of serine and threonine in the chitin-peptide complex preparations obtained by Hackman, the considerable stability of the linkage region to alkali argues against the type of O-glycosidic bond found in submaxillary gland glycoproteins and the connective tissue acid mucopolysaccharides. Where aspartic acid has been demonstrated to be involved in linkage to hexosamine the bond seems to be of the glycosylamine type, that is to say between the reducing group of the sugar and the β-amido group of asparagine. Such bonds have been noted to be stable to alkali (Andrews et al., 1967), a property of this chitin-protein bond. Alternatively it would be feasible to invoke a bond of the amide type involving an amino group on a hexosamine residue and a β-carboxyl group on aspartic acid; such bonds have not been demonstrated in other situations however and it seems at present that protein-carbohydrate links invariably involve the anomeric carbon atom of the sugar molecule. This is not to say however that such a situation is necessarily impossible or that because natural complexes of histidine and carbohydrate have not been detected that these too are unlikely. If we do accept linkage to the reducing groups of the carbohydrate, only then must we accept also that the bound protein is found only at the extremities of the chitin chains since these are unbranched. This may possibly prove to be an unattractive hypothesis when the physical data concerning the in vivo macromolecular organization of the complex comes to be considered.

Considerable doubt has been cast on Hackman's results in a paper by Attwood and Zola (1967). These workers have attempted to repeat Hackman's studies using, in their case, Calliphora larval cuticle and Loligo pen. The tissues were sequentially extracted in water (24 hours at 2°), aqueous sodium sulphate (0·17 M for 72 hours at 2°), urea (7 M for 48 hours at 2°), 0·01 N sodium hydroxide (20° for 5 hours), 1·0 N sodium hydroxide (50° for five hours) and 1·0 N sodium hydroxide at 100° for 72 hours. In the case

of the maggot cuticle two different chitin-protein complexes, obtained at
the end of the initial alkaline treatment from two different samples of
cuticle, had chitin-protein ratios of 1·7 : 1 and 1·2 : 1 while two preparations from *Loligo* pens had ratios of 0·6 : 1. These results are at vari-
ance with those of Hackman (1960). Hydrolysates of preparations which
had been subjected to the extended alkaline extraction procedure were
examined for their amino acid compositions. The chitin-protein complex
from maggot cuticle contained aspartic acid, threonine, serine, glutamic
acid, glycine, alanine, valine, methionine, isoleucine, leucine, tyrosine,
phenylalanine, ornithine, lysine and histidine while that from *Loligo* pen
contained valine, methionine, tyrosine, phenylalanine, ornithine, lysine and
histidine. All these amino acids were present in trace amounts only and there
was no evidence to suggest that either aspartic acid or histidine might
predominate.

Differences in the behaviour of these chitin-peptide preparations, compared with Hackman's materials, were noted towards lithium thiocyanate
as a solvent. *Loligo* pen chitin-peptide complex dissolved in saturated
LiSCN when heated at 130° for 18 hours but the corresponding *Calliphora*
material would not dissolve at all even when heating was prolonged for
five days or when the temperature was raised to 170° for 12 hours. The
Loligo complex could be reprecipitated on dilution and the precipitate was
then insusceptible to re-solution in lithium thiocyanate. The LiSCN solution was examined in the ultracentrifuge and its behaviour proved to be
incompatible with that of true macromolecular solution, being far more
like a dispersion of particulate aggregates or coacervates. Thus reprecipitation of chitin-protein complexes intact from solution need not necessarily
imply the presence of covalent links between the polysaccharide and the
peptide or protein. Obviously the entire problem is in need of a further and
more detailed re-examination.

Stegeman (1963) in a study of the internal supporting structures of the
cuttlefish *Sepia officinalis* and *Loligo vulgaris*, the octopus *Octopus vulgaris*
and the dwarf calamary *Alloteuthis subulata*, found protein components
(conchagens) which could be solubilized or gelatinized by autoclaving at
143° or by treatment with 0·01 N sodium hydroxide. The insoluble residue
of chitin which remained contained some 2% protein and smaller amounts
of sugars not identical with 2-amino-2-deoxy-D-glucose one of them being
either 2-amino-2-deoxy-D-galactose or possibly 2-amino-2 deoxy-D-man-
nose.

Brunet and Carlisle (1958) report that the highly crystalline chitin of
the Pogonophoran tube is accompanied by a very stable protein which
lacks sulphur-containing amino acids.

Glutamic acid has been identified in hydrolysates of a chitin-protein complex preparation from the epicuticle of the arachnid *Palamnaeus swammerdami* (Krishnan, *et al.*, 1955).

Voss-Foucart (1968) removed soluble nacrine from the shell-matrix of *Nautilus pompilus* and solubilized the remaining material in saturated LiSCN solution (130° for 4 hours). Dialysis of this solution against buffered (pH 5·6 acetate) 8M urea gave two fractions, one soluble, one insoluble. The insoluble fraction had a much higher 2-amino-2-deoxy-D-glucose content than the soluble fraction and it was suggested that it was a chitin-protein mucopolysaccharide. The amino acid composition is given in Table I.

TABLE I

Amino Acid Composition of a Chitin-Protein complex
from *Nautilus pompilus* Shell-Matrix[a]

Amino acid	Molar percentages	
	Sample 1	Sample 2
Aspartic acid	6·3	6·2
Threonine	2·0	2·1
Serine	10·3	11·0
Glutamic acid	6·3	6·3
Proline	Tr	Tr
Glycine	25·4	24·4
Alanine	29·3	29·7
Valine	1·6	1·8
Methionine	Tr	0·7
Isoleucine	1·6	1·5
Leucine	2·0	1·8
Tyrosine	0·8	0·6
Phenylalanine	5·2	5·0
Lysine	1·6	1·5
Histidine	Tr	Tr
Arginine	7·3	7·2

[a] Voss-Foucart (1968).

PHYSICAL EVIDENCE FOR THE EXISTENCE OF CHITIN-PROTEIN COMPLEXES

As early as 1940 it had been suggested that the differences noted between the X-ray diffraction diagrams of chitinous cuticles and the purified chitins might be interpreted as being due to modifications in the chitin lattice resulting from complexing with protein (Fraenkel and Rudall, 1940, 1947). Picken and Lotmar (1950) observed extra reflections in the X-ray diffraction photographs of *Aphrodite* bristles as well as in other chitinous structures impregnated with tanned protein. They suggested that these reflections might arise from oriented protein.

Rudall (1963) reports a widely occurring type of X-ray diagram given by chitin-protein systems. This is found in the soft plastic cuticles of dipteran, lepidopteran and coleopteran larvae and in the pliable intersegmental membranes of adult insects and crustacea (Fig. 6).

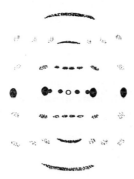

Fig. 6. X-ray diffraction diagram of stretched intersegmental cuticle of lobster (redrawn from Rudall, 1963).

A second and equally widely distributed structure is also described by Rudall (1963, 1965a). This occurs in the cuticles of adult true Orthopterans, in all adult Hymenopterans and in the chaetae of the annelid *Aphrodite* (Fig. 7). Here the X-ray diffraction data indicate a structure repeating along

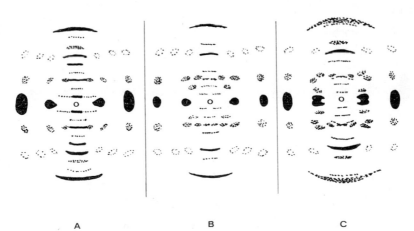

A   B   C

Fig. 7. X-ray diffraction diagrams of chitin-protein complexes in A, Hymenopteran cuticle (*Sirex* ovipositor), B, Orthopteran cuticle (locust leg-segment cuticle) and C, *Aphrodite* chaetae (redrawn from Rudall, 1963).

the chitin chains at a period of 31 Å or at intervals of six hexosamine residues. It is of interest to note that this phenomenon is observed in the case of both α (the insect cuticles) and β (annelid chaetae)- chitin arrangements. Definite proof is lacking as to whether this structural repeating unit is protein or some other compound. Lipids would seem to be eliminated since the macroperiod survives boiling in benzene for 40 hours (Rudall, 1965a). Rudall (1965a) has compared the intensities in the axial subperiods of 31 Å system with diffraction data for other proteins of known conformation. The closest correspondence was with feather keratin. Lack of special intensities at 1·5, 1·15 or 1·11 Å suggested the absence of oriented systems of α or β protein chains.

Electron micrographs of hymenopteran ovipositor (Ichneumon fly, *Megarhyssa*) show two main fine structures (Rudall, 1965a), a superficial (0·5 μ thick) regular system of layers, each about 250 Å thick, and, for the bulk of the ovipositor wall (when seen in cross section) a regular system of non-staining chitin cores surrounded by thin uniform layers of highly staining material interpreted as protein. This system of rods is, in places, in a regular hexagoral array.

A third type of chitin-protein X-ray diffraction diagram is given by the oesophageal cuticle of the squid *Loligo* (Fig. 8) (Rudall, 1963, 1965b). Here an axial period of 41 Å is found which corresponds to repeats of eight amino sugar residues. Consideration of the X-ray diffraction diagram of the intact

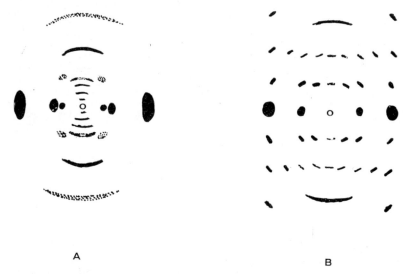

A

B

Fig. 8. X-ray diffraction diagrams of A, natural *Loligo* oesophageal cuticle (stretched) and B, purified α-chitin from the same source (redrawn from Rudall, 1963).

cuticle indicated that the chitin chains were not packed in the usual three-dimensional array but were intimately associated with protein molecules in some specific way. Removal of the protein caused the chitin chains to aggregate spontaneously to give the usual α-chitin structure.

Rudall (1963, 1965a, b, 1968) has suggested that the type of chitin (α, β, or γ) conformation and the particular arrangement of chitin and protein in the chitin-protein complex are probably related to the mechanical function which the structure has to fulfil. He has also noted that α-chitin appears to be incompatible with collagen and that where the collagenous cuticle of Annelids is replaced by a chitinous cuticle, as in Arthropods, the type of chitin involved is alpha. On the other hand chitin in the beta or gamma form appears to be able to coexist with collagen. Surely this incompatibility of the alpha form of chitin and collagen may provide some clue to the organization of the chitin system itself.

BIOSYNTHESIS AND BIODEGRADATION OF CHITIN

In common with most other polysaccharides the mode of biosynthesis of chitin is as yet only imperfectly understood.

Using an enzyme prepared from *Neurospora crassa*, Glaser and Brown (1957a, b) were able to bring about *in vitro* incorporation of ^{14}C-labelled 2-acetamido-2-deoxy-D-glucose from uridine 5-(2-acetamido-2-deoxy-D-glucosyl dihydrogen pyrophosphate) (UDPNAG) into chitin by transfer to an insoluble primer. No synthesis could be detected in the absence of primer. The synthesis was stimulated by the presence of 2-acetamido-2-deoxy-D-glucose, although neither this sugar nor its 1 or 6 phosphate derivatives could replace the nucleotide derivative. This is in a fungal system but there is evidence to suggest that synthesis may follow a similar course in invertebrates. Kent and Lunt (1958, 1961) have isolated UDPNAG from the hypodermis of the spider crab *Maia squinado* and the hepatopancrei of the shore crab *Carcinus maenas* and the lobster *Homarus vulgaris*. The nucleotide sugar also occurs in moth haemolymph (Carey and Wyatt, 1960) and in the wings of the desert locust *Schistocerca gregaria* (Candy and Kilby, 1962). The enzyme bringing about incorporation of 2-acetamido-2-deoxy-D-glucose from UDPNAG into chitin, chitin synthetase, has recently been identified in several invertebrate tissues: in the epidermis of moulting blue crabs (*Calinectus sapidus*), the larvae of *Artemia salina* (Carey, 1965) and in cell-free extracts of the caterpillar of *Prodenia eridania* (Jaworski *et al.*, 1963).

Chitin synthesis of course requires 2-acetamido-2-deoxy-D-glucose and the pathway of synthesis for this from glucose has been well documented in bacteria and vertebrates although not in invertebrates (Kent, 1960;

Kent and Draper, 1968; Allen and Kent, 1968; Kent and Allen, 1968) (see Chapter 2, Fig. 6).

There is evidence to suggest that glucose for 2-acetamido-2-deoxy-D-glucose, and thence chitin, synthesis is derived in the arthropods from glycogen. Unlike many other structural macromolecules chitin, in the arthropods, is capable of being both synthesized and degraded at very high rates; cuticle is largely digested, reabsorbed and rebuilt at each moult. Thus, following a rapid degradation during the moult there is a period of intense chitin synthesis, probably at the expense of glycogen (Candy and Kilby, 1962; Zaluska, 1960; Meenakshi and Scheer, 1961). To what extent the degradation products of chitin breakdown, i.e. 2-acetamido-2-deoxy-D-glucose and 2-amino-2-deoxy-D-glucose, are returned directly for biosynthesis is not clear. They may in fact be degraded further to glucose and either incorporated into glycogen or passed directly to the hexosamine synthesis path. There is evidence to suggest that in the pre-moult period glycogen synthesis occurs at the expense of the cuticular chitin (Bade and Wright, 1962). In *Cancer pagurus* there is a six-fold increase of glycogen in the hepatopancreas and hypodermis before the moult (Renaud, 1949) and a similar change has been observed in the case of the lobster *Panulurus argus* (McWhinnie and Scheer, 1958). The hexosamine content of the body fluid of *C. pagurus* rises sharply during pre-moult while an increase in blood glucose is noted in *Hemigraspus nudus* during inter-moult. There appears to be extensive utilization of fat-body glycogen during moulting in insects also, while the epidermal cells of *Bombyx mori* have been observed to lose glycogen during the period when new cuticle is being built up (Paillot, 1938; Billard, 1944).

The occurrence, biological significance and phylogenetic significance of chitinolytic enzymes have been reviewed in detail by Jeuniaux (1963, see also Florkin, 1966). Chitinases are found in the moulting fluids of insects especially during the larval nymph stages (Jeuniaux, 1955). Surprisingly the activity of chitinolytic enzymes is not apparent throughout the moult and in *Hyalophora*, for example, chitinase is found only during the latter third of the development (Passonneau and Williams, 1953). Chitinase has been detected in the hypodermis of *C. maenas* together with N-acetyl-β-glucosaminidase (Lunt and Kent, 1960).

Chitinolysis is the result of action by two distinct enzymes, chitinase and chitobiase. There is thus a strong resemblance to the process of the enzymological breakdown of cellulose which also requires two enzymes, cellulase and cellobiase. The similarity is interesting in view of the close structural relationship between the two polysaccharides. An attractive hypothesis would be the suggestion that the two chitin-degrading enzymes are homologues of their cellulose-degrading counterparts. Chitinase or more correctly poly-β-1,4-(2-acetamido-2-deoxy-)-D-glucoside glycanohydrolase, degrades chitin to chitobiose and chitotriose as almost the only products of hydrolysis. A small percentage of these products (about 1%) may be monosaccharide. Chitobiase then further degrades the di and trisaccharides to 2-acetamido-2-deoxy-D-glucose.

Jeuniaux (1963) has examined the variation of chitinase and chitobiase levels in the epidermis of crabs (*Maia squinado* and *C. pagurus*) during the moult cycle. The epidermis synthesizes chitinase permanently but chitobiase passes through a period of intensified synthesis during the pre-exuvial stages, falling off immediately following exuviation. In *B. mori*, at the time of moulting, the epidermis retracts and secretes an exuvial fluid rich in chitinolytic and proteolytic enzymes (Jeuniaux, 1963). Chitinase activity is absent during the inter-moult period, appearing at the onset of the moult and disappearing immediately prior to ecdysis.

Chitinases are common among unicellular organisms (bacteria and protozoa) and occur in the fungi. In the animal kingdom they appear widely distributed among chitin eaters and are retained in the Deuterostomia as well as in the Protostomia. The ability to synthesize chitinolytic enzymes does however tend to be lost in those species of vertebrates which do not include arthropods, zooplankton or fungi in their diet. Jeuniaux (1961) has noted the curious fact that the ability to break down chitin, using their own enzymes, has been retained by many vertebrates while the ability to degrade the structurally similar cellulose, often also included in the vertebrate diet, by their own enzymes is absent.

REFERENCES

Allen, A. and Kent, P. W. (1968). *Biochem. J.* **107**, 589.
Andrews, A. T. de B., Herring, G. M. and Kent, P. W. (1967). *Biochem. J.* **104**, 705.
Attwood, M. M. and Zola, H. (1967). *Comp. Biochem. Physiol.* **20**, 993.
Bade, M. L. and Wright, G. R. (1962). *Biochem. J.* **83**, 470.
Billard, L. (1944). *Bull. Inst. océanogr.* **866**, 4.
Brimacombe, J. S. and Webber, J. M. (1964). "Mucopolysaccharides", p. 18. Elsevier, Amsterdam.
Brunet, P. C. J. and Carlisle, D. B. (1958). *Nature, Lond.* **182**, 1689.
Candy, D. J. and Kilby, B. A. (1962). *J. exp. Biol.* **39**, 129.
Carey, F. G. (1965). *Comp. Biochem. Physiol.* **16**, 155.
Carey, F. G. and Wyatt, G. R. (1960). *Biochem. biophys. Acta*, **41**, 178.
Carlisle, D. B. (transl. and ed.) (1963). *In* "Pogonophora" by A. V. Ivanov. Academic Press, London and New York.
Carlström, D. (1957). *J. biophys. biochem. Cytol.* **3**, 669.
Carlström, D. (1962). *Biochim. biophys. Acta*, **59**, 361.
Dweltz, N. E. (1960). *Biochim. biophys. Acta*, **44**, 416.
Dweltz, N. E. (1961). *Biochim. biophys. Acta* **51**, 283.

Fishenko, N. V. (1963). *Cited in* "Pogonophora", by A. V. Ivanov (D. B. Carlisle, transl. and ed.) Academic Press, London and New York.

Florkin, M. (1966). "A Molecular Approach to Phylogeny". Elsevier, Amsterdam.

Foster, A. B. and Hackman, R. H. (1957). *Nature, Lond.* **180**, 40.

Foster, A. B. and Webber, J. M. (1960). *Adv. Carbohyd. Chem.* **15**, 371.

Foucart, M. F., Bricteux-Grégoire, S., Jeuniaux, C. and Florkin, M. (1965). *Life Sciences*, **4**, 467.

Fraenkel, G. and Rudall, K. M. (1940). *Proc. R. Soc.* B **129**, 1.

Fraenkel, G. and Rudall, K. M. (1947). *Proc. R. Soc.* B **134**, 111.

Glaser, L. and Brown, D. H. (1957a). *Biochem. biophys. Acta*, **23**, 449.

Glaser, L. and Brown, D. H. (1957b). *J. biol. Chem.* **228**, 729.

Hackman, R. H. (1955a). *Aust. J. biol. Sci.* **8**, 83.

Hackman, R. H. (1955b). *Aust. J. biol. Sci.* **8**, 530.

Hackman, R. H. (1960). *Aust. J. biol. Sci.* **13**, 568.

Hackman, R. H. (1962). *Aust. J. biol. Sci.* **15**, 526.

Hackman, R. H. and Goldberg, M. (1958). *J. Insect Physiol.* **2**, 221.

Jaworski, E., Wang, L. and Marco, G. (1963). *Nature, Lond.* **198**, 790.

Jeuniaux, C. (1955). *Mém. Soc. entomol. Belg.* **27**, 312.

Jeuniaux, C. (1959). *Archs. int. Physiol. Biochem.* **67**, 516.

Jeuniaux, C. (1961). *Nature, Lond.* **192**, 135.

Jeuniaux, C. (1963). "Chitine et Chitinolyse". Masson, Paris.

Kent, P. W. (1960). *Biochem. Soc. Symp.* **20**, 90. Cambridge University Press, Cambridge.

Kent, P. W. (1964). *In* "Comparative Biochemistry" (M. Florkin and H. S. Mason, eds). Vol. 7, p. 93. Academic Press, New York and London.

Kent, P. W. and Allen, A. (1968). *Biochem. J.* **106**, 645.

Kent, P. W. and Draper, P. (1968). *Biochem. J.* **106**, 293.

Kent, P. W. and Lunt, M. R. (1958). *Biochim. biophys. Acta*, **28**, 657.

Krishnan, G., Ramachandran, G. N. and Santanam, G. S. (1955). *Nature, Lond.* **176**, 557.

Liang, C. Y. and Marchessault, R. H. (1959). *J. Polymer. Sci.* **37**, 385.

Lotmar, W. and Picken, L. E. R. (1950). *Experientia*, **6**, 58.

Lunt, M. R. and Kent, P. W. (1960). *Biochim. biophys. Acta*, **44**, 371.

Lunt, M. R. and Kent, P. W. (1961). *Biochem. J.* **78**, 128.

Manskaia, S. M. and Drozdova, T. V. (1962). *Geokhimiya*, 952.

McWhinnie, M. A. and Scheer, B. T. (1958). *Science, N.Y.* **128**, 90.

Meenakshi, V. R. and Scheer, B. T. (1961). *Comp. Biochem. Physiol.* **3**, 30.

Meyer, K. H. and Pankow, G. W. (1935). *Helv. chim. Acta*, **18**, 589.

Paillot, A. (1938). *C.r. Séanc. Soc. Biol.* **127**, 1502.

Passonneau, J. V. and Williams, C. M. (1953). *J. exp. Biol.* **30**, 545.

Pearson, F. G., Marchessault, R. H. and Liang, C. Y. (1960a). *J. Polymer. Sci.* **43**, 101.

Pearson, F. G., Marchessault, R. H. and Liang, C. Y. (1960b). *Biochim. biophys. Acta*, **45**, 499.

Picken, L. E. R. (1960). "The Organization of Cells". Oxford University Press, London and New York.

Picken, L. E. R. and Lotmar, W. (1950). *Nature, Lond.* **165**, 599.

Renaud, L. (1949). *Annls Inst. océanogr.* **24**, 259.

Ramachandran, G. N. and Ramakrishnan, C. (1962). *Biochim. biophys. Acta*, **63**, 307.

Richards, A. G. (1951). "The Integument of Arthropods". University of Minnesota Press, Minneapolis, Minnesota.

Rudall, K. M. (1950). *Prog. Biophys. biophys. Chem.* **1**, 39.

Rudall, K. M. (1955). *Symp. Soc. exp. Biol.* **9**, 49.

Rudall, K. M. (1962). *In* "Comparative Biochemistry" (M. Florkin and H. S. Mason, eds). Vol. 4, p. 397. Academic Press, New York and London.

Rudall, K. M. (1963). *Adv. Insect Physiol.* **1**, 257.

Rudall, K. M. (1965a). *In* "Aspects of Insect Biochemistry" (T. W. Goodwin, ed.), p. 83. Academic Press, London and New York.

Rudall, K. M. (1965b). *In* "Structure and Function of Connective Tissue", p. 191. Butterworth, London.

Rudall, K. M. (1960). *In* "Treatise on Collagen" (B. S. Gould, ed.). Vol. 2A, p. 83. Academic Press, London and New York.

Stegemann, H. (1963). *Hoppe-Seyler's Z. physiol. Chem.* **331**, 269.

Voss-Foucart, M. F. (1968). *Comp. Biochem. Physiol.* **26**, 877.

Zaluska, H. (1960). *Acta Biol. exp., Lodz,* **19**, 339.

CHAPTER 9

Cellulose

INTRODUCTION

Prior to 1845 the polysaccharide cellulose had been thought to be exclusive to the plant kingdom. The discovery by Schmidt (1845) of cellulose in ascidians was therefore an important event, doubly so not only because of the plant associations but also since it marked the first definite indentification of an invertebrate connective tissue polysaccharide.

CHEMICAL CHARACTERIZATION OF ANIMAL CELLULOSE

Following Schmidt's discovery, the material and its tissue source were studied in detail and the identity of the polysaccharide with cellulose generally confirmed by a number of workers (Loewig and Koelliker, 1846; M. Berthelot, 1858, 1859; A. Berthelot, 1859, 1872; Schulze, 1863; Schäffer, 1871; Hertwig, 1873; Semper, 1875; Winterstein, 1894). Notable amongst these workers, M. Berthelot (1858) coined the name tunicin for this first animal cellulose, deriving the name from that particular ascidian group, the tunicates, in which it had first been identified.

The improved techniques which came with the twentieth century permitted more critical comparisons of plant, bacterial and animal cellulose. Abderhalden and Zemplén (1911) showed, by partial acid hydrolysis techniques, that the tunicin of *Phallusia mammilaris* and phytocellulose were very closely related and probably identical. The tunicin was converted into octa-acetylcellobiose by the action of acetic anhydride and concentrated sulphuric acid in the cold. The derivative formed had m.p. 225° and $[\alpha]_D^{20}$, $+42\cdot19$; values in good agreement with the properties of authentic acetylated cellobiose. An ozazone derivative and free cellobiose were also prepared. Confirmation of this result, using tunicin from the same source, was obtained by Zechmeister and Toth (1933). Hydrolysis by hydrochloric acid (saturated at 0°) and fractionation with ethanol yielded cellohexose, cellotetraose and cellotriose all having the same optical properties, copper numbers, iodine numbers, molecular weights and acetyl derivatives as similar preparations obtained from cotton cellulose. Octa-acetyl-cellobiose was obtained by acetylation of the tunicin.

More recently Tsuchiya and Suzuki (1963) have isolated tunicin from the test of the ascidian *Cynthia roretzi* v. Drasche by alkali treatment. The tunicin gave only glucose on acid hydrolysis while partial hydrolysis yielded cellobiose, cellotriose, cellotetraose and cellopentaose. Although this strongly indicates a cellulose-like structure, variation from phyto-cellulose seemed to be suggested by slight solubility differences. Thus, while the tunicin dissolved in Schweizer's reagent, acid zinc chloride solution, and super-concentrated hydrochloric acid, there was a greater resistance to solution than exhibited by cotton or wood celluloses.

Viel (1939) in a study involving thermal fractionation of the gaseous products formed during the pyrolysis at 300–400° of a variety of natural celluloses, concluded that tunicin could not be a typical cellulose (i.e. a phyto or bacterial cellulose) since cotton and *Acetobacter xylinum* cellu-loses yielded 20–30% volatile material in contrast to tunicin which yielded 60–70% volatile products with a correspondingly lower yield of residual carbon.

Grassmann *et al.* (1932, 1933) found bacterial β-cellulose to be distinct from tunicin in its response to a cellulase derived from *Aspergillus niger*. Bacterial cellulose was not degraded by the enzyme whereas tunicin and plant cellulose were.

COMPARATIVE AND STRUCTURAL INFORMATION DERIVED BY PHYSICAL MEANS

The amount of structural information which one can obtain regarding a complex insoluble substance with a high molecular weight such as cellulose, using chemical techniques, tends to be limited. Often the application of physical methods, with the reduced tendency to destructive degradation of the molecule, will yield a more detailed view than might otherwise be obtained.

The period following 1920 marked an increased awareness of the useful-ness of physical techniques for the identification and characterization of biological macromolecules. Apart from structural characterization, which we shall consider shortly, physical methods, applied to earlier studies of tunicin, helped to confirm the chemical data which had indicated a close correspondence with plant and bacterial celluloses. In this respect the technique of X-ray diffraction proved of immense importance.

Several studies showed that the X-ray diffraction patterns produced by tunicate celluloses (*P. mammillata* and *P. nigra*) were identical with those produced by phytocelluloses (Herzog and Gonell, 1925; Meyer and Mark, 1930; Spence and Richards, 1940). In 1929 Mark and Susich were able to demonstrate that *P. mammillata* tunicin and β-cellulose (bacterial) had respectively a ring fibre structure (101), parallel to the foil plane, and a pure fibre structure (010) parallel to the direction of tension, closely related to, and producing X-ray diagrams in harmony with, the monoclinic quadratic form of cellulose.

Sutra (1932) found no difference between the X-ray diffraction patterns of *Polycarpa varians*, *P. mammillata* and *Ascidia mentula* and cotton linters. Moreover nitration of tunicin yielded X-ray diagrams identical with nitro-cotton; denitration restored the cellulose diagram. Acetyl tunicin exhibited similar properties to those of acetyl cellulose; forming films having similar X-ray diffraction diagrams, and dissolving in chloroform to give solutions with $[\alpha]_D - 21°.5'$.

Gross and Clark (1938) found that the X-ray diffraction patterns of a number of oriented celluloses examined, including bacterial, *Valonia ventricosa* (algal) and tunicin, were all consistent with the Meyer and Mark cellulose unit cell.

Once the general identity of animal cellulose with its phyto and bacterial counterparts had been reasonably established, attention naturally turned to a more detailed characterization of this molecule from the point of view of size and shape, primary structure [i.e. departures, if any, from the simple linear $\beta 1 \rightarrow 4$ linked structure typified by plant cellulose (Fig. 1)], tertiary

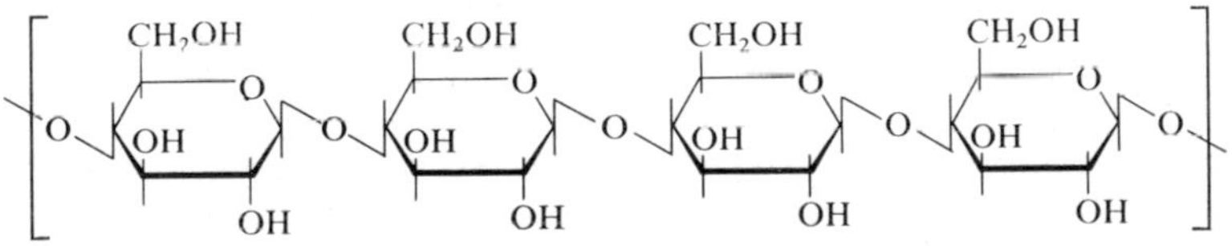

Fig. 1. Structure of cellulose.

structure (confirmation of the chains and intermolecular stabilization via hydrogen bonding), and the mode of formation and organization of higher structural orders resulting ultimately in such macromolecular systems as fibres and sheets visible at the microscopic level.

Since, like plant cellulose, tunicin is a water-insoluble polymer, the determination of its molecular weight presents considerable problems. Indeed it is in all probability unlikely that the molecular weight of the molecule, *in vivo*, can be set within any very definite limits. Purified preparations, however, may have suffered some degradation during isolation, such that the chain lengths do fall within a certain definable range.

An early estimate of the degree of polymerization of tunicin (Schmidt *et al.*, 1931) placed it in the same range as plant (cotton) and bacterial celluloses with a minimum of 96 $C_6H_{11}O_5$ (glucose anhydride) groups. This value was based on the occurrence of carbonyl groups and would seem in the light of later results to be a considerable underestimation.

Krässig (1954) determined the intrinsic viscosities, in Schweizer's reagent, of a series of polymer homologous tunicins, derived from *P. mammillata*, having degrees of polymerization of the order 700–3500. Samples were converted by polymer analogue nitration to nitrates and their intrinsic viscosities and osmotic molecular weights measured. The relation between molecular weight and the viscosity of the tunicin solutions was the same as that between the same parameters of vegetable cellulose nitrates. Mass distribution curves of nitrates from hydrolytically degraded tunicin showed pronounced maxima at degrees of polymerization of about 500 and 1000 glucose units.

Krässig deduced that the tunicin consisted of unbranched linear chains of glucose anhydrides and that with regard to the distribution of the more easily decomposed bonds it was equivalent to phytocellulose. Such a conclusion was in complete contrast to the results obtained by Wyk and Schmorak (1953), also based on viscometric data. Here a highly ramified structure for tunicin, comparable at the molecular level to that of glycogen, was indicated.

Rånby (1952a), estimating the degree of polymerization of tunicins from two species of ascidian, *Corella parallelogramma* and *Ascidia* sp., found values of approximately 4000. Again these results were obtained from viscometric measurements on nitrate derivatives in solution. The tunicin used had been isolated by treating the tunic with one per cent sodium hydroxide at 100° for 20 hours. This extraction procedure had the effect of lowering the nitrogen content of the tunicin from between 2·5 to 3·0% to approximately 0·07%.

Carbohydrates exhibit bonds between 4000 and 3000 cm^{-1}, in their infrared spectra, which arise from OH stretching vibrations. In the case of polysaccharides these bonds are complex, particularly so if a crystalline form of the molecule is examined. Mann and Marriman (1958) obtained the infrared spectra, in both polarized and unpolarized radiations, of a series of crystalline modifications of cellulose, including *P. mammillata* tunicin.

The infrared spectrum of tunicin in the OH stretching region shows some notable differences from other celluloses. It differs considerably from hydrate cellulose (cellulose II; a crystalline form regenerated from oriented films of cellulose acetate) and from cellulose III (a form prepared by treating stretched water-swollen films of bacterial cellulose with liquid ammonia or ethylamine), as well as from the spectra of algal (*V. ventricosa*) and bacterial celluloses, although these are all grouped under a common crystallographic category, cellulose I. Tunicin gives five bands between 3600 and 3000 cm^{-1}, whereas the algal and bacterial celluloses both show six.

Spectra determined in liquid nitrogen show the same bands as are observed at room temperature, but shifted 16 to 29 cm^{-1} towards lower frequencies (Fig. 2).

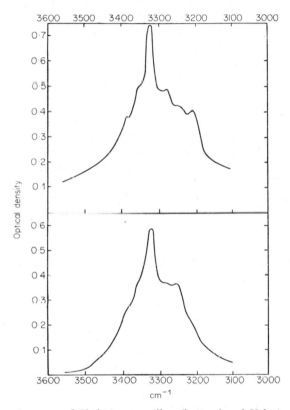

Fig. 2. Infrared spectra of *Phalusia mamaillata* (bottom) and *Valonia ventricosa* (top) celluloses at the temperature of liquid nitrogen (from Mann and Marriman, 1958).

The spectra of molecules in the crystalline state are complicated by the possibility of bands arising from the following factors: (a) coupling of an intramolecular vibration with a vibration of the crystal lattice, (b) appearance of extra bands due to sum or difference frequencies of the vibrations described in (a), (c) coupling between similar vibrations in different molecules giving rise to more than one frequency, (d) appearance of bands due to sum and difference frequencies or overtone frequencies of intramolecular vibrations, and (e) appearance of multiple peaks in hydrogen-bonded crystals, the origins of which are unclear.

Spectra of crystalline regions of polysaccharide containing 100% OD groups will show whether the complications due to the factors cited in (a), (b) and (d) are present. If the 100% OD spectrum shows the same bonds as the 100% OH spectrum, but shifted to lower frequencies by a factor of 1·35, these complications are absent and (e) also is probably not a factor to be considered.

Thus the spectra of tunicin and bacterial cellulose, 100% substituted by OD groups, are effectively the same as for those having 100% OH groups. The bond shift factor is of the order of 1·35 to lower frequencies. The two bonds having the highest frequencies are not however resolved in the deuterated spectra.

The presence or absence of coupling of intramolecular vibrations (factor c) can be studied by comparing 100% OD and 10% OD substituted spectra. If coupling occurs the 10% OD spectrum will differ from the 100% OD spectrum, since in the specimen containing 10% OD groups a given OD group will have only a very small chance of having another OD group as its nearest neighbour.

The spectrum of tunicin having 10% OD groups in its crystalline regions is identical with 100% substituted tunicin, suggesting that the coupling of vibrations does not occur in this cellulose. Coupling does however occur in bacterial cellulose.

Plane-polarized radiation studies assist in the determination of the direction of orientation of the OH groups within the crystalline regions of the molecule. The spectra obtained with polarized radiation from the three crystalline modifications of cellulose (I, II, and III), all indicate the presence of OH groups which are orientated so that they make an angle much smaller than 55° with the chain direction, while yet other OH groups present make angles greater than 55° with this direction. The former OH groups can be attributed to intramolecular hydrogen bonds which join successive glucose residues in the cellulose chain. It would appear that ether oxygen atoms must participate in the hydrogen bonding systems in all these modifications. Examination of the spectra of tunicin and bacterial cellulose (Fig. 3) indicate the presence of at least three infrared absorption bands with independent dichroic behaviour; two are of the parallel type and one perpendicular. Mann and Marrinan consider the most reasonable interpretation of the infrared results on tunicin, bacterial and algal cellulose, to be that of a structure based on a cellobiose residue having a two-fold screw axis with two intramolecular hydrogen bonds and one or two intermolecular hydrogen bonds. The frequencies of the bands due to intramolecular H bonds suggest that the chains in cellulose I have a molecular form different from those in II and III. Differences may not be very large however, since the H-bonded OH frequencies are particularly sensitive to the distance between the two oxygen atoms involved.

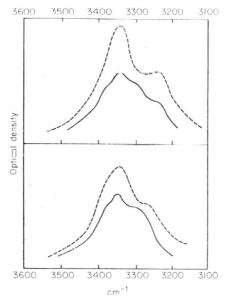

Fig. 3. Infrared spectra in polarized radiation of tunicate cellulose (bottom) and bacterial cellulose (top). Both are forms of cellulose I (from Mann and Marriman, 1958). (– – – –) Electric vector parallel to chain direction. (———) Electric vector perpendicular to chain direction.

QUATERNARY ORGANIZATION

Development of the electron microscope made possible the direct examination of celluloses at the macromolecular level.

Frey-Wyssling and Frey (1950), examining tunicate cellulose in the electron microscope, found the molecules to be organized into microfibrils having a diameter of about 250 Å; similar in fact to organization noted in plant celluloses. Also observed was a tendency of the microfibrils to associate by lateral aggregation into systems of ribbons. In this latter respect the tunicate cellulose represented the type of bacterial cellulose characteristically produced by *Bacterium xylinium*. Production of ribbons such as these allows the formation of "higher" three-dimensional orientations; a property incompatible with the "individualized" microfibrils of plant fibres. Schmidt (1924) had previously noted that in the polarization microscope the tunicate test appeared to be composed of layers of interwoven anisotrophic fibrils.

Results comparable with those of Frey-Wyssling were obtained by Meyer and his co-workers (Meyer *et al.*, 1951) who showed that the tunicin of *P. mammillata* was composed of fibres having a diameter 700 Å and bearing cross-striations at a 200 Å periodicity. These fibres were continuous, lacking any apparent branching.

Rånby (1952a, b) has investigated the fine structure and crystallinity of tunicate (*C. parallelogramma* and *Ascidia* sp.) celluloses in some detail. Mercerized tunicate, cotton and wood celluloses all gave qualitatively similar X-ray diffraction patterns; the tunicate material differed however in the sharpness of definition in its diagrams, suggesting a greater degree of order. Here again electron microscopy showed the tunicin to have a fine structure comparable to plant materials. The micelle strings of tunicin, which had diameters of 120 ± 20 Å, were however somewhat thicker than those observed in the phytocelluloses where the widths were of the order 75–80 Å for wood cellulose and 90 Å for cotton. Bacterial cellulose fibres examined had diameters of 100 Å. Fibrils of 100 Å diameter appeared to contain approximately 240–250 cellulose chains. Treatment of tunicin with boiling 2·5 N sulphuric acid for four hours had the effect of subdividing the strings into micelles larger than those which resulted from a similar treatment of plant celluloses. Resistance to both alkali and acid hydrolysis was in fact greatest in animal cellulose, decreasing through bacterial and cotton celluloses to wood celluloses where resistance was least. This probably suggests a higher degree of order and crystallinity in the animal material, since attack is usually thought to commence at amorphous rather than crystalline regions.

THE RELATION OF INVERTEBRATE TO VERTEBRATE CELLULOSE

For a long period animal cellulose was considered to be an evolutionary quirk confined to the urochordates. The identification by Hall (Hall *et al.*, 1958; Hall and Saxl, 1960) of cellulose in mammalian tissues caused not a little excitement, excusable when one considers that this discovery provided a molecular link between the tunicates and the vertebrates. Zoologists had already considered this link to exist on embryological and morphological grounds; it places the urochordates at the junction where the vertebrates spring from the chordate evolutionary line.

Mammalian cellulose takes the form of anisotrophic fibres, associated with protein, in dermis and aortic tissues of humans. Because of their evolutionary significance a detailed comparison of tunicate and mamalian celluloses was made (Hall *et al.*, 1960; Hall and Saxl, 1960, 1961).

Chemical examination of *Ascidiella aspersa* tests confirmed that the fibrous material present in the tunic was identical with plant cellulose. The nature of the protein with which the polysaccharide is associated was determined after collagen had been removed by treatment with collagenase and elastase.

Polysaccharide isolated from the mid-section of the tunic and treated with these enzymes yielded a protein residue still associated with the cellulose and having a characteristic amino acid composition similar in all respects to that associated with the cellulose of the mammalian anisotrophic fibres. This protein resembled elastin in having a low hydroxyproline content (2%) and a high concentration of proline and glycine (50–60%) coupled with a relatively high concentration of valine (ca. 8%). It differed however from elastin in its basic and acidic amino acid contents, ca. 11% and 9% respectively. The amino acid composition was in many respects reminiscent of both collagen and elastin, although it did not consist of a mixture of these proteins, unattacked by the enzymes, since treatment of separate samples of the whole tissue with the individual enzymes indicated that both substrates were in fact present and susceptible to their respective enzymes. The collagenase treatment lowered the hydroxyproline content of the tissue by 80%, while elastase reduced the valine content by 40%. It was therefore concluded that, in common with aged human aorta, the tunic contained a cellulose-protein complex in which the protein moeity had a composition intermediate between elastin and collagen. In the light of more recent discoveries it would now be of interest to know whether this tunicate tissue contains both or either of the demosines (Chapter 15) and whether these, if present, are removed from the cellulose-protein complex by elastase treatment or whether a proportion remains with the complex itself.

Electron diffraction studies of plant cellulose, the human connective tissue anisotrophic fibres and tunicin produced patterns all essentially identical with regard to the position of the reflexions, although differences in intensities were observed, probably reflecting differences at higher levels of organization.

Evidence was put forward, and a hypothesis advanced, relating tunicin (referring here not to the polysaccharide alone but to the fibrous complex with protein) and the mammalian anisotrophic cellulose-containing fibres to the reticular elements of connective tissues and to the phylogenetic development of reticulin.

OTHER POLYSACCHARIDES ASSOCIATED WITH TUNICIN

Arabinose, xylose and galactose were found in hydrolysates of tunicin by Adams and Bishop (1955), in addition to the glucose derived from the cellulose. The pentose content did however fall on purification of the cellulose and cellulose fractions obtained by fractional precipitation from cuprammonium hydroxide solutions were pentose free. These purified preparations still contained considerable quantities of galactose. Whether this galactose is glycosidically linked into the cellulose chain itself, or whether it forms a part of the protein moeity observed by Hall and Saxl will have to be established; certainly a firm link to the cellulose either directly or through some intermediary molecule seems to be indicated by the survival of the association through a moderately severe purification process. The pentose sugars may have been linked to collagen associated with tunicin in the test.

HISTOLOGY AND HISTOCHEMISTRY

Early histological and histochemical studies of the ascidian test were reviewed by Saint-Hilaire (1931). Saint-Hilaire found the "cartilaginous" test of *A. mentula* to contain 90% water while the membranous test of *Pyura papillosa* contained only 65%. The fibres of the test were found to be richer in nitrogenous material than the ground mass, while a similarity in histological appearance between the tunicate test and vertebrate connective tissue was noted in the respect that cells were present in a ground mass enmeshed by fibres. *A. mentula* had a test in which there was more ground mass than fibres, while the test of *P. papillosa* had a predominantly fibrous structure. In some species the fibres were found to stain with resorcin, fuchsin and orcein, in a similar manner to vertebrate connective tissue fibres, while in others the fibres lacked these properties. Perhaps rather surprisingly both the ground mass and the fibres appeared to contain tunicin.

The tunicin in the test of *Clavelina lepadiformis* appears to be a part of a complex containing protein and acid mucopolysaccharide (Pérès, 1948a, b); the tunic exhibits γ metachromasia with toluidine blue. A similar observation has been made for the test of *P. stolonifera* (Endean, 1955a) where the acid mucopolysaccharide appears to be closely related to chondroitin sulphate. Endean found that the gelatinous test resolved under the microscope into a network of fibres with variable thickness; the larger ones, and probably the intermediate sizes also, being aggregates of the smaller $0 \cdot 3 – 0 \cdot 5 \mu$ diameter fibrils. In some cases the fibres were oriented in definite directions. The amorphous interfibrilar material seemed to be mainly aqueous as the test contained over 97% plasma.

Histochemically the test resembled vertebrate connective tissue. The fibres were composed of a highly insoluble polysaccharide resembling plant cellulose in many respects and presumably constituted of tunicin. Some other factor must be operative, however, if this is tunicin, since it did not dissolve in Schweitzer's reagent or stain with Lugol's solution followed by

1 : 3 $H_2SO_4-ZnCl_2/KI$ solution. The fibres seemed to be cemented together by mucopolysaccharide. This material, which stained positively with toluidine blue, alcian blue and Hale's stain, was sensitive to the action of bovine testicular hyaluronidase. Treatment with the enzyme left the fibres in a disordered state. There seemed to be very little protein associated with the fibres.

Defatted test was hydrolysed with 6 N hydrochloric acid for eight hours and yielded both Molisch and Elson-Morgan positive materials. Removal of plasma, acetone-soluble materials, and mucopolysaccharide from the test left a residue of highly insoluble polysaccharide (C, 40·7%; H, 6·2%) with no detectable protein (Millon reaction). This polysaccharide was assumed to be cellulose.

Endean (1955b) also examined the "biogenesis" of the test. Certain blood cells, termed ferrocytes, were found to produce fibres when cytolysed and similar fibres were produced when the cells were maintained in "hanging drop" preparations. In both instances histological and histochemical examination indicated that the fibres were identical with those present in the test. Under normal conditions the ferrocytes seemed to migrate across the blood vessel walls and enter the test matrix. Here, after a period of movement in the fibrilar matrix already present, the cells produced fibres by one of three different methods observed. A cytoplasmic centre from which fibres subsequently arose might be isolated from the main corpuscular mass; alternatively fibres might arise directly from the surface of the ferrocyte; or fibres might arise from material secreted by the ferrocyte into its immediate vicinity. The process of fibre formation resembles the manner in which connective tissue is laid down by vertebrate fibroblasts. During regeneration, ferrocytes were observed to migrate directly to the test-sea water interface where, as a result of cytolysis, fibres were laid down from released fibre-producing intraglobular material.

Acid mucopolysaccharides have been demonstrated by histochemical analysis and fluorescent microscopy in the tunic of *Ciona intestinalis* (Bierbauer and Vagas, 1962).

CELLULOSE FROM OTHER INVERTEBRATE SOURCES

Cellulose-like materials have been reported to be present between the fibroin and the seracin in natural silk from *Bombyx mori* (Shimizu, 1955; Buonocore, 1958). Accounts of cellulose in the tubes of Pogonophora (Hyman, 1958) and in the cuttlebones of Sepia (see Muller, 1934) have not been substantiated and probably arise from mistaken identification of chitin as cellulose.

REFERENCES

Abderhalden, E. and Zemplén, G. (1911). *Hoppe-Seyler's Z. physiol. Chem.* **72**, 58.

Adams, G. A. and Bishop, C. T. (1955). *TAPPI*, **38**, 672.

Berthelot, A. (1859). *J. Physiol, Paris*, **2**, 577.

Berthelot, A. (1872). *Bull. Soc. chim. Paris*, 18.

Berthelot, M. P. E. (1858). *C.r. hebd. Séanc Acad. Sci.*, *Paris*, **47**, 227.

Berthelot, M. P. E. (1859). *Annls chim. Phys.* **56**, 149.

Bierbauer, J. and Vagas, E. (1962). *Acta biol. hung.* **13**, 139.

Buonocore, C. (1958). *Annali Sper. agr.* **12**, 681.

Endean, R. (1955a). *Aust. J. mar. Freshwat. Res.* **6**, 139.

Endean, R. (1955b). *Aust. J. mar. Freshwat. Res.* **6**, 157.

Frey-Wyssling, A. and Frey, R. (1950). *Protoplasma*, **39**, 657.

Grassmann, W., Zechmeister, L., Toth, G. and Stadler, R. (1932). *Naturwissenschaften*, **20**, 639.

Grassmann, W., Zechmeister, L., Toth, G. and Stadler, R. (1933). *Justos Liebigs Annln. Chem.* **503**, 167.

Gross, S. T. and Clark, G. L. (1938). *Z. Kristallogr. Kristallgrom* **99**, 357.

Hall, D. A. and Saxl, H. (1960). *Nature Lond.* **187**, 547.

Hall, D. A. and Saxl, H. (1961). *Proc. R. Soc.* B**155**, 202.

Hall, D. A., Lloyd, P. F. and Saxl, H. (1958). *Nature, Lond.* **181**, 470.

Hall, D. A., Happey, F., Lloyd, P. F. and Saxl, H. (1960). *Proc. R. Soc.* B**151**, 497.

Hertwig, O. (1873). *Jena Z. Naturw.* **7**, 46.

Herzog, R. O. and Gonell, H. W. (1924). *Z. physiol. Chem.* **141**, 63.

Hyman, L. H. (1958). *Biol. Bull. mar. biol. lab.*, *Woods Hole* **114**, 106.

Krässig, H. (1954). *Makromolek. Chem.* **13**, 21.

Loewig, C. and Koelliker, A. (1846). *C.r. hebd. Séanc Acad. Sci.*, *Paris*, **22**, 38.

Mann, J. and Marriman, H. J. (1958). *J. Polym. Sci.* **32**, 357.

Mark, H. and Susich, G. (1929). *Z. phys. Chem.* **4**, 431.

Meyer, K. H. and Mark, H. (1930). *In* "Der Aufbau der hochpolymeren organischen Naturstoffe." Vol. 100, p. 117. Leipzig.

Meyer, K. H., Huber, L. and Kellenberger, E. (1951). *Experientia*, **7**, 216.

Muller, S. (1934). *Magy. Kém. Foly.* **40**, 112.

Pérès, J. M. (1948a). *Archs Anat. Microse, Morph. exp.* **37**, 230.

Pérès, J. M. (1948b). *C.r. hebd. Séanc Acad. Sci. Paris*, **226**, 1220.

Rånby, B. G. (1952a). *Ark. Kemi.* **4**, 241.

Rånby, B. G. (1952b). *Bull. Ass. tech. Ind. pap.* **6**, 67.

Saint-Hilaire, K. (1931). *Zool. Jber. Neapel*, **54**, 435.

Schäffer, A. (1871). *Justos Liebigs Annln. Chem* **160**, 312.

Schmidt, C. (1845). *Justos Liebigs Annln. Chem.* **54**, 284.

Schmidt, E., Simson, W. and Schnegg, R. (1931). *Naturwissenschaften*, **19**, 1006.

Schmidt, W. J. (1924). "Die Bausteine der Tierkörpers im polarisiertem Licht," p. 226. F. Cohen, Bonn.

Schulze, F. E. (1863). *Z. wiss. Zool.* **12**, 175.

Semper, C. (1875). *Archs. Zool., Zootec. Inst. Würzburg*, 2.

Shimizu, M. (1955). *C.r. Séanc Soc. Biol.* **149**, 853.

Spence, J. and Richards, O. W. (1940). *Pap. Tortugas Lab.* **32**, 165.

Sutra, R. (1932). *C.r. hebd. Séanc Acad. Sci. Paris*, **195**, 181.

Tsuchiya, Y. and Suzuki, Y. (1963). *Tohoku. J. agric. Res.* **14**, 39.
Viel, G. (1939). *C.r. hebd. Séanc Acad. Sci. Paris*, 208, 1689.
Winterstein, E. (1894). *Hoppe-Seyler's Z. physiol. Chem.* **18**, 43.
Wyk, A. J. A. and Schmorak, J. (1953). *Helv. chim. Acta*, **36**, 385.
Zechmeister, L. and Toth, G. (1933). *Hoppe-Seyler's Z. physiol. Chem.* **215**, 267.

F

CHAPTER 10

Galactan

INTRODUCTION

While galactans are not in general of common occurrence in animal tissues they are found, accompanying heparin, in mammalian lung tissue and are also quite widely distributed in the reproductive secretions of pulmonate gastropod molluscs. Their function in molluscs, at least, would appear to be as storage polysaccharides provided for the embryo. Why galactan should be more suitable is not clear; glycogen would appear to be the principal storage polysaccharide of the adult animal (Goddard and Martin, 1966).

Mammalian galactan, or pneumogalactan as it is called, is not a homopolysaccharide but contains one glucuronic acid residue for every 35–40 galactose units. Moreover it has been suggested that pneumogalactan is in fact a mixture of molecular species containing different galactose: glucuronic acid ratios. In contrast to the majority of molluscan galactans all the hexose units are D-galactose there being no L-galactose present. The structure shown in Fig. 1 has been proposed, on the basis of good experimental

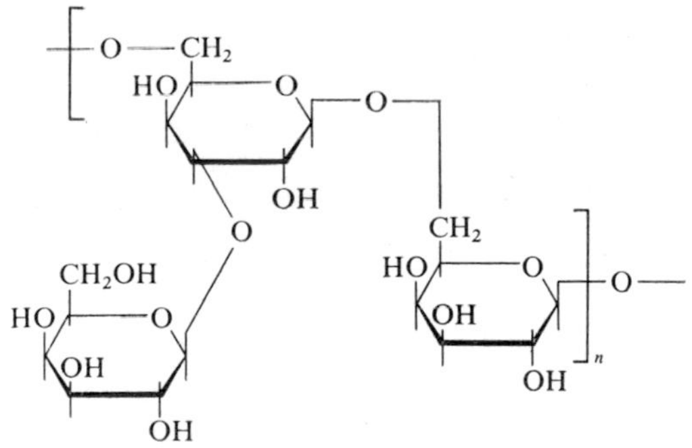

Fig. 1. Structure of pneumogalactan from beef lung.

evidence, for pneumogalactan. Further details of this polysaccharide will be found in Stacey and Barker (1962).

163

INVERTEBRATE GALACTANS

HELIX POMATIA GALACTAN

The first accounts of a mucilaginous polysaccharide, contained in the eggs and albumin glands of the pulmonate *Helix pomatia* can be attributed to Eichwald (1865) and later to Landwehr (1882). This polysaccharide was later given the name sinistrin by Hammarsten (1885) and following his studies of the material, in which he found polysaccharide to be associated with nitrogen, a considerable period elapsed before it was further investigated. Levene (1925) failed to recognize the presence of a galactan in the secretion of the albumin gland, considering it to be a glycoprotein. However May (1931, 1932a, b, 1934a, b, c, d; May and Weinbrenner, 1938) was able to show that sinistrin was in fact a true galactan. A purified preparation of the polysaccharide, isolated by digestion of the mucin with 30% potassium hydroxide and precipitated as the copper complex, yielded galactose as the sole product of acid hydrolysis. Galactan prepared by Baldwin and Bell (1938) showed $[\alpha]_D - 16 \cdot 1°$ (water). Hydrolysates gave crystalline D-galactose but there was rotational evidence as well for the presence of some of the L form of the sugar. A tri-O-acetate derivative of the polysaccharide was methylated to give a methyl ether with 43·1% OMe and having $[\alpha]_D - 20°$. This derivative gave 2 : 3 : 4 : 6-tetra-O-methyl galactopyranose and 2 : 4-di-O-methyl galactopyranose on hydrolysis (Schlubach *et al.*, 1937; Baldwin and Bell, 1938). Tetra-O-methyl galactose prepared in this way and isolated as the anilide was a mixture of the D and L forms but the di-O-methyl derivative appeared to consist of the D isomer alone (Bell and Baldwin, 1940, 1941). Optical rotation data suggested a D to L ratio of 6 : 1. The absence of a trimethyl derivative coupled with the distribution of the D and L forms in the methyl derivatives suggested a highly branched structure (Fig. 2) having a $\beta 1 \rightarrow 3$ linked backbone of D-galactose units

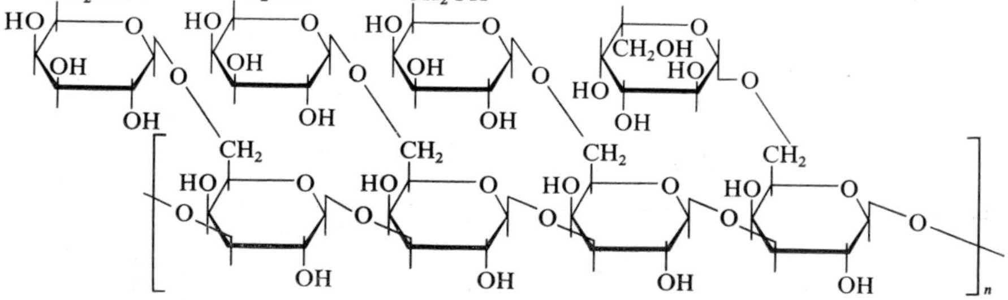

Fig. 2. Structure for *Helix pomatia* galactan based on methylation evidence.

with single residue side chains linked $\beta 1 \rightarrow 6$ to the main chain. The L-galactose residues would constitute 25% of the side chains. Treatment however of the galactan with periodate and phenylhydrazone (Barry degradation) indicated a dichotomously branched structure of $1 \rightarrow 6$ and $1 \rightarrow 3$ linkages rather than the simpler structure postulated earlier (O'Colla, 1953). Identification of a highly branched structure does of course agree well with the concept of the galactan as a storage polysaccharide.

Separation of a partial hydrolysate of the galactan yielded 3-O-β-D-galactosyl D-galactose, 6-O-β-D-galactosyl D-galactose and traces of 6-O-α-L-galactosyl D-galactose (Weinland, 1956). Weight was added to the idea that the L-galactose residues might constitute single side chains, or at least terminate side chains, on noting their more rapid rate of release on acid hydrolysis (May and Weinland, 1953, 1954, 1956).

Apart from the presence in *Helix* albumin gland of a galactan identical with Hammarsten's sinistrin the presence also has been noted of a second strongly laevorotatory polysaccharide having a high L-galactose content. (Geldmacher-Mallinkrodt and May, 1957). In fact, Geldmacher-Mallinkrodt (1958; Geldmacher-Mallinkrodt and Traexler, 1960) has suggested that the galactan sinistrin is itself not homogeneous. Galactan obtained by alkaline digestion at 100° for several hours, and isolated as the copper complex, was found to contain at least three components when examined in the analytical ultracentrifuge. Geldmacher-Mallinkrodt and Traexler (1961) have demonstrated the presence of low-molecular weight galactans, by ultrafiltration, in the eggs of *H. pomatia*.

Horstmann (1964) has isolated *H. pomatia* egg galactan by aqueous extraction and ultracentrifugation. The polysaccharide was polymolecular with a molecular weight of 4×10^6 at 20° in 0·1 M potassium chloride solution. This value was calculated from sedimentation and diffusion data. Hot alkali split the galactan into smaller units precipitable by dilute ethanol.

BIOMPHALARIA GLABRATA GALACTAN

A galactan having a rather similar composition and structure to that of *H. pomatia* has been isolated from the albumin gland of *Biomphalaria* (*Australorbis*) *glabrata* (Duarte *et al.*, 1964; Corrêa *et al.*, 1967).

The glands of parasite-free snails (*B. glabrata* is the intermediary host for *Schistosoma mansoni*) were inactivated by storage at 80° for ten minutes, homogenized and treated with *Bacillus subtilis* protease for 15 hours. After centrifugation at 18,000 g the supernatant was precipitated with ethanol and the precipitate redissolved in water. Following three Sevag treatments and reprecipitation with ethanol, polyanions (mostly nucleic acid) were removed by precipitation with cetyltrimethylammonium bromide at pH 7. On adjusting the supernatant to pH 8·5, with borate buffer, a precipitate of polysaccharide was obtained which contained no detectable protein. Hydrolysis and paper chromotography of the polysaccharide indicated that galactose was the only sugar component. Neither uronic acid nor sulphate could be detected. Estimation with D-galactose oxidase indicated a 2 : 1 ratio of D to L galactopyranose.

Fractionation of the products of partial hydrolysis containing 70% reducing carbohydrate, on charcoal-celite, yielded free galactose, disaccharides and a small amount of oligosaccharide. The reducing disaccharides, following treatment with sodium borohydride, had electrophoretic mobilities on paper, in molybdate buffer, identical with those of standard disaccharide alcohols containing $1 \rightarrow 3$ and $1 \rightarrow 6$ linkages and it was concluded that the disaccharides also contained these glycosidic links, in equal proportions.

Treatment of the intact polysaccharide with D-galactose oxidase resulted in reaction of 27% of the galactopyranose residues. It appeared therefore that the D- and L-galactopyranose residues were present as end groups to the extent of 27% and 23% respectively.

The galactan, which had $[\alpha]_D^{22} + 19 \cdot 5$ in water before hydrolysis and $[\alpha]_D^{22} + 56°$ after 40% hydrolysis, showed a moderately intense infrared absorption band at 890 cm^{-1}. The glycosidic linkages thus appeared to be, in part at least, of the β type. Periodate oxidation confirmed the presence of $1 \rightarrow 6$ linkages.

Subjection of the galactan to Smith degradation (periodate oxidation, borohydride reduction and hydrolysis) yielded galactose and glycerol (identified by paper chromatography and by gas liquid chromatography of the triacetate). The ratio of glycerol to galactose was $1 : 0 \cdot 9$, thus confirming a $1 : 1$ ratio of $1 \rightarrow 6$ and $1 \rightarrow 3$ linkages.

Successive Barry degradations applied to periodate oxidized polysaccharide suggested a structure in which there were the same number of non-reducing end groups as inner residues of the branched chain structure. It was considered that for each pair of terminal residues, one was linked to C-3 and the other to C-6 of the same inner residue (Fig. 3). The structure would thus be similar to that suggested for *H. pomatia* galactan by O'Colla.

GALACTANS IN OTHER MOLLUSCAN SPECIES

As Bayne (1966) has pointed out, the observation by May (1934) that galactan is limited to the secretion of the albumin gland in *H. pomatia* contrasts with Meenakshi's report (1954) of the occurrence of galactan in the uterus of *Pila virens*. In *Pila* the content of the polysaccharide was found to decline during the sexually active period from 28% to between 14% and 15% after oviposition, increasing again after egg laying to 25% and then declining once more during aestivation. Identification of galactan was

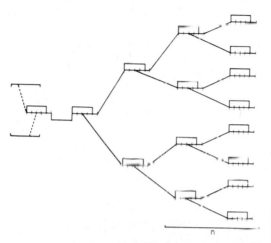

Fig. 3. Structure proposed for *Biomphalaria glabrata* galactan (from Corrêa *et al.*, 1967). This structure probably more closely resembles that of the *Helix pomatia* galactan than that shown in Fig. 2. The exact distribution of D- and L-galactose residues is not known but the 64% D and 36% L-galactopyranose units appear to be linked $\beta 1 \rightarrow 3$ and $\beta 1 \rightarrow 6$ in approximately the same proportion.

based on the detection of galactose in hydrolysates; this might perhaps be considered to be somewhat insufficient evidence. The albuminous fluid of *Pila* eggs was also reported to contain a galactan as were the albumin glands of *Viviparus* and *Ariophanta*.

McMahon *et al.* (1957) detected galactans in several species of gastropod, including *Otala lactea*, *Biomphalaria glabrata* and *Helix aspersa*. It was noted that while the galactans of terrestrial species are composed of galactose alone, the polysaccharide derived from the uterii of two fresh water operculate species contained both galactose and fucose; in this respect it is of interest to note that a similar polysaccharide has been found in *Fissurella volcano* by Emmerson (1966). Male specimens of *Lanistes bottenianus* and *Pomacea zetebi* contained glycogen only and no galactan. Young egg clutches of pulmonate and operculate snails contained non-glycogen polysaccharides only but glycogen appeared towards the end of development. During development some polysaccharide was consumed. Freshly hatched *Lymnea* still contained large amounts of galactan.

Galactan has been identified in *Achatina fulica* (Ghose, 1963), *Oxichilus cellarius* (Rigby, 1963), *Agriolimax columbianis* (Meenakshi and Scheer, 1968), in the egg albumen of *A. reticulatus* (Bayne, 1966) and in *Lymnea stagnalis* (Holm, 1946; Horstmann, 1956). Extraction of *L. stagnalis* eggs yielded a practically phosphate-free galactan ($[\alpha_D^{20} + 23°]$) which appeared to consist entirely of D-galactose. The galactan was neutral and homogeneous electrophoretically but showed inhomogeneity in the ultracentrifuge (Horstmann and Geldmacher-Mallinkrodt, 1961).

OTHER INVERTEBRATE SOURCES

So far as we can say at the present, galactans appear to be confined to the molluscs and mammals with perhaps the exception of a small amount of polysaccharide which was detected in the helminth *Macrocanthorynchus hirudinaceus* (Von Brand and Saurwein, 1942).

BIOSYNTHESIS

The enzymatic synthesis of galactan has been studied in *H. pomatia* by Goudsmit and Ashwell (1965). Albumin glands were homogenized in 0·25 M sucrose—0·001 M EDTA and cellular debris removed at 18,000 g. Precipitation at pH 7·0 with cold saturated ammonium sulphate yielded a protein fraction at between 20% and 50% saturation. Suspension of the precipitate in 0·05 M *tris* buffer, pH 7·0, and dialysis against the same buffer yielded an active enzyme preparation. Incubation of this preparation with UDP-(^{14}C)D-galactose, galactan and *tris* buffer yielded labelled galactan. The reaction was terminated by adding KOH and heating to 100° for 15 minutes. The galactan was precipitated with ethanol and further purified by repeated reprecipitations from aqueous solution.

The enzyme preparation exhibited an absolute requirement for galactan primer and also showed a species primer specificitly; incorporation was far lower when *L. stagnalis* galactan (which contains D-galactose only) was substituted for *H. pomatia* galactan. Glycogen had no primer properties.

Apart from primer specificity there was a substrate specificity also. The specificity was for UDP-D-galactose, neither the free monosaccharide nor its 1-*P* derivative being incorporated. As an alternative nucleotide-sugar precursor, TDP-D-galactose was less than half as useful as the UDP derivative. The ability of the enzyme preparation to utilize, to an appreciable extent, UDP-D-glucose was of some interest; this would suggest the presence of a specific epimerase of the type which has been detected in bacterial preparation where NAD$^+$ exists as an obligatory cofactor. Wherever the incorporation of radioactive material was found to have taken place its identity with galactose was always confirmed by characterization of the products of hydrolysis. Although the authors suggested that the donor of L-galactose might prove to be the TDP derivative, it was later shown that this aspect of galactan synthesis was mediated via the GDP-sugar (Goudsmit and Neufeld, 1966). GDP-L-galactose was identified in extracts of the albumen gland.

The GDP-L-galactose is formed from GDP-L-mannose by a unique double epimerization not involving formation of an intermediate deoxy sugar (Goudsmit and Neufeld, 1967). Albumen gland extracts from *H. pomatia* brought about the conversion of labelled derivative in a 30% to 50% yield. The GDP-L-galactose produced was isolated by paper chromatography and characterized by comparing its R_f values with authentic samples, by examining its ultraviolet absorption spectrum and by identifying L-galactose in the products of acid hydrolysis. The conversion seemed not to require DPN nor could conversion of GDP-D-glucose to GDP-L-galactose be demonstrated.

Sawicka (1967) has observed the presence of glucose-1-phosphate uridyltransferase (E.C.2.7.7.9.) in the albumen gland of *H. pomatia*, while Sawicka and Chajnacki (1968), using labelled precursors, have been able to demonstrate that extracts of the albumen gland will catalyse the formation of galactan from glucose-6-phosphate via glucose-1-phosphate, UDP-glucose and probably UDP-galactose. They were also able to show that the gland extracts would catalyse the incorporation of labelled phosphate from galactose-1-phosphate into nucleotides, in the presence of UTP or UDP-glucose.

COMPLEXES WITH PROTEIN

The possibility of snail galactans being complexed with protein remains uncertain and unresolved. Earlier reports of such complexes probably arose from the presence of other glycoproteins. However the methods of purification frequently employed might be expected to break certain types of covalent protein-carbohydrate bonds and would certainly destroy weaker interactions; we can therefore deduce little at the present time from reports of purity which depend upon low or zero nitrogen contents.

Protein, associated with the galactogen globules, was found in the albumen glands of *H. pomatia* by Grainger and Shillitoe (1952). With the onset of the reproductive season the protein decreased and disappeared.

REFERENCES

Baldwin, E. and Bell, D. J. (1938). *J. chem. Soc.* 1461.
Bayne, C. J. (1966). *Comp. Biochem. Physiol.* **19**, 317.
Bell, D. J. and Baldwin, E. (1940). *Nature, Lond.* **146**, 559.
Bell, D. J. and Baldwin, E. (1941). *J. chem. Soc.* 125.
Correâ, J. B. C., Dmytraczenko, A. and Duarte, J. H. (1967). *Carbohydrate Res.* 3, 445.
Duarte, J. H., Fokama, G., Correâ, J. B. C. and Bacila, M. (1964). *Ciênc. Cult., S. Paulo*, **16**, 189.
Eichwald, E. (1865). *Justus Liebigs Annln Chem.* **134**, 177.

F*

Emmerson, D. N. (1966). *Rep. Am. malac. Soc.* 1966.
Geldmacher-Mallinkrodt, M. (1958). *Hoppe-Seyler's Z. physiol. Chem.* **309**, 190.
Geldmacher-Mallinkrodt, M. and May, F. (1957). *Hoppe-Seyler's Z. physiol. Chem.* **307**, 191.
Geldmacher-Mallinkrodt, M. and Traexler, G. (1960). *Hoppe-Seyler's Z. physiol. Chem.* **322**, 112.
Geldmacher-Mallinkrodt, M. and Traexler, G. (1961). *Hoppe-Seyler's Z. physiol. Chem.* **325**, 116.
Ghose, K. C. (1963). *Proc. zool. Soc. Lond.* **140**, 681.
Goddard, C. K. and Martin, A. W. (1966). *In* "Physiology of Mollusca" (K. M. Wilbor and C. M. Yonge, eds). Vol. 2, p. 275. Academic Press, New York and London.
Goudsmit, E. M. and Ashwell, G. (1965). *Biochem. biophys. Res. Commun.* **4**, 417.
Goudsmit, E. M. and Neufeld, E. F. (1966). *Biochem. biophys. Acta.* **121**, 192.
Goudsmit, E. M. and Neufeld, E. F. (1967). *Biochem. biophys. Res. Commun.* **26**, 730.
Grainger, J. N. R. and Shillitoe, A. J. (1952). *Stain Technol.* **27**, 81.
Hammarsten, O. (1885). *Pfluger's Arch. ges. Physiol.* **36**, 373.
Holm, L. W. (1946). *Trans. Am. microsc. Soc.* **65**, 45.
Horstmann, H. J. (1956). *Biochem. Z.* **328**, 342.
Horstmann, H. J. (1964). *Biochem. Z.* **340**, 548.
Horstmann, H. J. and Geldmacher-Mallinkrodt, M. (1961). *Hoppe-Seyler's Z. physiol. Chem.* **325**, 251.
Landwehr, H. A. (1882). *Hoppe-Seyler's Z. physiol. Chem.* **6**, 74.
Levene, P. A. (1925). *J. biol. Chem.* **65**, 683.
May, F. (1931). *Z. Biol.* **91**, 215.
May, F. (1932a). *Z. Biol.* **92**, 319.
May, F. (1932b). *Z. Biol.* **92**, 325.
May, F. (1934a). *Z. Biol.* **95**, 277.
May, F. (1934b). *Z. Biol.* **95**, 401.
May, F. (1934c). *Z. Biol.* **95**, 606.
May, F. (1934d). *Z. Biol.* **95**, 614.
May, F. and Weinbrenner, H. (1938). *Z. Biol.* **99**, 199.
May, F. and Weinland, H. (1953). *Z. Biol.* **105**, 339.
May, F. and Weinland, H. (1954). *Hoppe-Seyler's Z. physiol. Chem.* **296**, 154.
May, F. and Weinland, H. (1956). *Hoppe-Seyler's Z. physiol. Chem.* **305**, 75.
McMahon, P., Von Brand, T. and Nolan, M. O. (1957). *J. cell. comp. Physiol.* **50**, 219.
Meenakshi, V. R. (1954). *Curr. Sci.* **23**, 301.
Meenakshi, V. R. and Scheer, B. T. (1968). *Comp. Biochem. Physiol.* **26**, 1091.
O'Colla, P. (1953). *Proc. R. Tr. Acad.* **B55**, 165.
Rigby, J. (1963). *Proc. zool. Soc. Lond.* **141**, 311.
Sawicka, T. (1967). *Bull. Acad. pol. Sci., Cl. II Sér. Sci. Biol.* **15**, 521.
Sawicka, T. and Chajnacki, T. (1968). *Comp. Biochem. Physiol.* **26**, 707.
Schlubach, H. H., Loop, W. and Schmidt, H. (1937). *Justus Liebigs Annln Chem.* **532**, 228.
Stacey, M. and Barker, S. A. (1962). "Carbohydrates of Living Tissues." Van Nostrand, London and New York.
Von Brand, T. and Saurwein, J. (1942). *J. Parasitol.* **28**, 315.
Weinland, H. (1956). *Hoppe-Seyler's Z. physiol. Chem.* **305**, 87.

CHAPTER 11

Glycogen

INTRODUCTION

The glucan, glycogen, occurs in all animals as the major energy reserve and carbohydrate store. Glycogen is a branched polymer of glucopyranose units joined by $\alpha 1 \rightarrow 4$ glucosidic linkages with the actual branch points at $\alpha 1 \rightarrow 6$ links (Fig. 1). Whilst this overall structural picture appears to apply

Fig. 1. Structure of a section of the glycogen molecule. ○ are D-glucose residues while ● is the D-glucose residue having a free reducing group—are $\alpha 1 \rightarrow 4$ bonds and $\rightarrow \alpha 1 \rightarrow 6$ bonds.

largely to glycogens throughout the animal kingdom the fine structural detail is subject to a greater variation than was originally thought, both from species to species and from tissue to tissue.

The literature relating to the structure and the biochemistry of glycogen is vast and to no little extent concerned with invertebrate glycogens, particularly those of the mollusca and the parasitic worms. It would be both difficult and unneccessary, for the theme of this book, to review this literature, particularly so since there have been several recent reviews dealing with all the various aspects of glycogen chemistry and biochemistry. The reader is therefore referred to the accounts by Stacey and Barker (1962), Chefurka (1965), Von Brand (1966), Goddard and Martin (1966) and Manners (1962).

The differences between the various glycogens appear largely to reside in the length of the $\alpha 1 \rightarrow 4$ linked glucose chains and the degree of branching. Differences in molecular weight may be more apparent than real in view of the variety of extraction procedures used by different authors. *In vivo*, glycogen probably has a very high molecular weight and may suffer chemical and mechanical degradation during the extraction and purification procedures.

SIZE, SHAPE AND STRUCTURE OF INVERTEBRATE GLYCOGENS

Ascaris lumbricoides (nematode) glycogen has an average chain length of 12 (Kjölberg *et al.*, 1963). This is a fairly typical value comparable with rabbit liver glycogen (13) or the glycogen of the lamellibranch mollusc *Mytilus edulis* (13). However these average values may be misleading; Bell and Manners (1952) and Manners (1953) isolated glycogens from *Mytilus edulis* with chain lengths of 5, 9, 12 and 17 units. The chain lengths of *Helix pomatia* glycogen appear to vary between 7 and 12 although the application of different techniques to the estimation of chain length, i.e. periodate oxidation and methylation, seemed to yield different results (Bell and Manners, 1952; Manners, 1953; Baldwin and Bell, 1940). The glycogen isolated from the liver and pancreas of *Mytilus galloprovinciallis* had an average chain length of five glucose units while that obtained from the mantle tissue had an average chain length of 10 units. The glycogens isolated from six species of fresh-water clam (*Lampsillis siliquoidea, Obovaria obliquaria, Amblema costata, Fusconaia undulata, Elliptio dilatus* and *Anodenta grandis*) had average chain lengths of 11 to 13 glucose units (Bahl and Smith, 1966).

Bathgate and Manners (1966) have shown by the use of the debranching enzyme, pullalanase, that in common with vertebrate glycogens the invertebrate glycogens of *A. lumbricoides, M. edulis* and oyster contain a high proportion of A chains, i.e. side chains linked to the molecule only by the reducing group, in contrast to B chains (main chains) which are not only linked by the reducing group but also have other chains attached to them. The proportion of A chains demonstrated was equal to or approached that calculated for multiple branched molecules containing equal members of A and B chains.

Molecular weights of ca. 1×10^6 to as high as 71×10^6 have been reported for vertebrate glycogens (see Stacey and Barker, 1962). In contrast, values in excess of $1–2 \times 10^8$ were reported for cold-water-extracted glycogen of the trematode *Fasciola hepatica*, while for *A. lumbricoides* the cold-water-extracted glycogen was divided between two varieties, one with an average molecular weight of $4 \cdot 5 \times 10^8$ and the other lighter component with an average molecular weight of 5×10^7 (Bueding, 1962). In *Hymenolepis diminuta* 30% of the glycogen is of a very high molecular weight

(19×10^0) very sharply differentiated from a low molecular weight form (6×10^7) (Orrell et al., 1964).

COMPLEXES WITH PROTEIN

The occurrence in vivo of glycogen in two distinct states has been frequently discussed and disputed. Thus lyoglycogen, soluble in trichloracetic acid, is in the free form, whilst desmoglycogen is conceived as being bound to proteins and extractable only by alkali treatment.

The existence of glycogeno-protein complexes in the cestode *Moniezia expansa* was demonstrated by Kent and Macheboeuf (1947). Evidence that the glycogen might be more than loosely associated with protein came from the observation that precipitation of a dialysed aqueous extract of the cestode with 44·6% ammonium sulphate brought glycogen out of solution as well as protein; glycogen might have been expected to remain in solution under these conditions. Electrophoresis showed two distinct and homogeneous fractions, both containing protein, and one with 60% and the other 11% glycogen. The glycogen-rich protein complex was named baerine and the glycogen-poor one moniezine. Glycogen-protein complexes of the baerine type were also demonstrated in *Taenia saginata* (Kent, 1947; Kent and Macheboeuf, 1947, 1949) in *Hymenolepsis diminuta* and in *Raillietina cesticillus* (Kent, 1957a, b) and in *Echinococcus granulosus* (Kent, 1963).

Fractionation of the blowfly (*Calliphora erythrocephala*) pupa glycogen into four types can be achieved by a combination of phenol-water extraction, trichloracetic acid and La^{3+} precipitation, and papain digestion (Lind, 1967). Fraction 1 does not precipitate in La^{3+} solution, fractions 2 and 3 are precipitated with La^{3+} and of these, Fraction 2 is soluble in ice-cold 5% trichloracetic acid. Fraction 4 precipitates in phenol-water and is soluble in water only after proteolytic digestion. These glycogens appear to differ only in their peptide content and possibly in the degree of branching. Fraction 1 is an almost pure glucan with a low nitrogen content between 0·01 and 0·02 μgN/mg glucan. The peptide content of the other glucans varies with pupal age but is highest in Fraction 4 achieving ca. 69 μg N/mg glucan at its greatest. The nitrogen contents of Fractions 2 and 3 are lower, the maximum values noted being 4·4 and 4·7 μg N/mg glucan respectively. However, since Fractions 3 and 4 had been subjected to a period of exposure to N KOH, during the purification process, the nitrogen contents may not be representative of the in vivo state. The amino acid compositions of the four glycogen-peptide complexes were quantitatively different (Table I) and showed an increasing acidity from Fractions 1 to 4. No amino sugars were detected.

Lindberg (1945) has reported that the greater part of the glycogen of the sea urchin egg is in a bound form, combined with protein.

TABLE I

Percentage Amino Acid Compositions of Glucan Fractions from *Calliphora erythrocephala* Pupal Glycogen[a]

Amino Acid	Fraction			
	1	2	3	4
Alanine, threonine	12·8	9·2	12·0	10·0
Arginine	15·4	11·1	7·9	11·2
Aspartic acid	3·4	5·9	5·5	10·7
Glutamic acid	9·2	14·1	13·4	18·1
Glycine, serine	24·8	27·7	24·7	17·0
Histidine	7·9	2·7	6·8	5·4
Phenylalanine, leucine	12·6	12·6	12·4	9·1
Proline	5·3	6·6	12·4	2·1
Tyrosine	2·1	3·3	2·4	4·8
Valine, methionine	2·7	4·8	2·3	7·1
Lysine	—	—	—	3·0
Unidentified	3·7	1·7	1·0	1·2

[a] Lind (1967).

REFERENCES

Bahl, O. P. and Smith, F. (1966). *J. org. Chem.* **31**, 1479.
Baldwin, E. and Bell, D. J. (1940). *Biochem. J.* **34**, 139.
Bathgate, G. N. and Manners, D. J. (1966). *Biochem. J.* **101**, 3C.
Bell, D. J. and Manners, D. J. (1953). *J. chem. Soc.* 3641.
Bueding, E. (1962). *Fedn. Proc. Fedn. Am. Socs exp. Biol.* **21**, 1039.
Chefurka, W. (1963). *A. Rev. Ent.* **10**, 345.
Goddard, C. K. and Martin, A. W. (1966). *In* "Physiology of Mollusca" (K. M. Wilbur and C. M. Yonge, eds). Vol. 2, p. 275. Academic Press, New York and London.
Kent, N. H. (1947). *Bull. Soc. Neuchâtel. Sci. nat.* **70**, 85.
Kent, N. H. (1957a). *Expl. parasit.* **6**, 351.
Kent, N. H. (1957b). *Expl. parasit.* **6**, 486.
Kent, N. H. (1963). *Am. J. Hyg., Monogr. Ser.* **22**, 30.
Kent, N. H. and Macheboeuf, M. (1947). *C.r. hebd. Séanc. Acad. Sci., Paris* **225**, 602.
Kent, N. H. and Macheboeuf, M. (1949). *Schweiz Z. Path. Bakt.* **12**, 81.
Kjölberg, O., Manners, D. J. and Wright, A. (1963). *Comp. Biochem. Physiol.* **8**, 353.
Lind, N. O. (1967). *Comp. Biochem. Physiol.* **20**, 209.
Lindberg, O. (1945). *Arkiv. Kemi, Miner., Geol.* B20, No. 1.
Manners, D. J. (1953). *Biochem. J.* **55**, xx.
Manners, D. J. (1962). *Adv. Carbohyd. Chem.* **17**, 371.

Orrell, S. A., Bueding, E. and Reissig, M. (1964). *In* "Control of Glycogen Metabolism", (W. J. Whelan and M. P. Cameron, eds), p. 29. J. and A. Churchill Ltd., London.

Stacey, M. and Barker, S. A. (1962). "Carbohydrates of living Tissues". Van Nostrand, London and New York.

Von Brand, T. (1966). "Biochemistry of Parasites", p. 48. Academic Press, New York and London.

GLYCOPROTEINS

CHAPTER 12

Glycoproteins

INTRODUCTION

Our knowledge of glycoproteins stems in the main from intensive studies of materials derived from vertebrate (principally mammalian) sources. The following general introductory remarks are made therefore in this context and do not necessarily apply to invertebrates.

A glycoprotein is now usually defined as a conjugated protein where the prosthetic group or groups are constituted by covalently bound hetero-saccharide units. A glycoprotein may have one or many of these groups, whose molecular weights may range between 500 and 4000, distributed along the length of the protein chain; often all the prosthetic groups are identical, but in many cases there may be two or more different types of heterosaccharide present. As a general rule the heterosaccharides of these vertebrate glycoproteins contain hexosamine; either 2-amino-2-deoxy-D-glucose or 2-amino-2-deoxy-D-galactose or both, usually N-acetylated. The linkage of the heterosaccharide to the protein chain is usually mediated via one of the hexosamine residues and these covalent links seem to fall roughly into two major classes, although other types perhaps may also exist. Thus O-glycosidic bonds may form between the amino sugar and one or other of the hydroxy aliphatic amino acids serine and threonine, or a glycosylamine bond may be formed involving the amide groups of asparagine or glutamine (Fig. 1).

Apart from hexosamine, some or all of the neutral monosaccharides mannose, galactose and fucose (6-deoxy-L-galactose) are frequently present, while more occasionally glucose may also be found. An important constituent of many glycoprotein prosthetic groups is sialic acid present as one

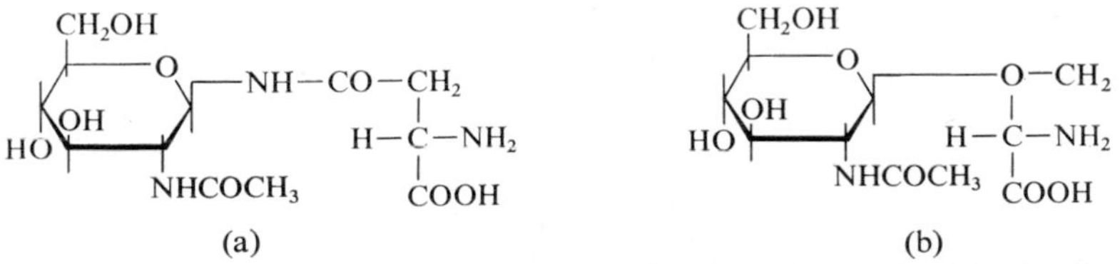

Fig. 1. Types of carbohydrate-protein covalent linkage found in vertebrate glycoprotein; (a) N-(β-aspartyl)glycosylamine, (b) O-seryl glycoside.

or other of the neuraminic acid derivatives (see Chapter 13), occupying a terminal position or positions on the heterosaccharide and linked ketosidically to it. Sialic acids tend, if they are present in any quantity, to markedly affect the physical properties of glycoproteins, notably their viscosity. Occasionally vertebrate glycoproteins are encountered where one or other of the monosaccharide residues carries an ester sulphate group; these are however rather uncommon.

The differences therefore between vertebrate glycoproteins and vertebrate mucopolysaccharides (Chapter 2) are quite well defined. On the one hand we have a macromolecule in which protein is quantitatively the most important constituent but linked to a smaller, often highly-branched heteropolysaccharide containing perhaps up to seven different mono-saccharides. Neutral monosaccharides, amino sugars and sialic acids are present but uronic acids are not. The heterosaccharides usually lack clearly-defined sequences of short oligosaccharide repeats. On the other hand we have the mucopolysaccharides, where in the isolated molecule at least, protein is quantitatively less important than the polysaccharide which is of high molecular weight, usually linear and built up from a regularly repeating sequence of oligosaccharides (often disaccharides) containing a limited variety of sugars which are often hexosamines and uronic acids and less often neutral sugars in conjunction with hexosamines and/or uronic acids. Sialic acid is rarely encountered but sulphate esterification is common.

DISTRIBUTION AND FUNCTION IN THE VERTEBRATES

Glycoproteins are ubiquitously distributed in vertebrate tissues and secretions; it is probably true to say that more vertebrate proteins are glycoproteins than are not. The mucous secretions of the respiratory, gastrointestinal and urinary tracts serve both a lubricatory and protective function. Many glycoproteins are biologically active molecules; these include many enzymes, some of the hormones, and the blood group substances.

INVERTEBRATE GLYCOPROTEINS

Although a moderately extensive literature refers to invertebrate glyco-proteins most of it is rather vague, and our knowledge of the occurrence and chemistry of these substances in invertebrates is in fact somewhat limited in comparison with vertebrate glycoproteins.

Gottschalk and Graham (1966) reviewing the field, comment: "Glyco-proteins are widely distributed throughout the animal world. They are regular components of mucous secretions and mucus is found almost universally in soft-bodied animals and vertebrates". While this is definitely true of vertebrates it is less true of invertebrates. Gottschalk and Graham seem to equate mucus with glycoprotein, an assumption which one would feel to be unacceptable. They go on to say, "Since glycoproteins are essential components of all mucous secretions and endow these secretions with their characteristic physiochemical properties, the functions of mucous secretions may be taken as a reflection of the functions of glycoproteins. In soft-bodied animals, mucus lubricates the surface of the body, thus facilitating locomotion, and sometimes it also protects the animal against the environment". In fact, so far as we can tell at the present time, acid mucopolysaccharides of various types seem to be far more important functional constituents of invertebrate mucus secretions than do glyco-proteins, although the latter may be present also. In spite of this there is a fairly considerable body of evidence to suggest that glycoproteins do occur widely in invertebrate phyla even though this evidence in the main lacks the detail we might wish. In seeking to compare vertebrate and invertebrate glycoproteins on the basis of the more obvious physical functions, i.e. as viscous lubricatory or protective agents, we would be falling into rather obvious error. We should remember that the functions of the heterosaccharide prosthetic groups in the biologically active vertebrate glycoproteins, i.e. enzymes, hormones and blood group substances, are understood imperfectly, if at all. This function or functions, understood or not, may be of greater fundamental importance than the promotion of viscosity which may well have arisen later in response to secondary demands upon organisms. We should therefore be prepared to search for invertebrate glycoproteins not only in mucin secretions but in the metabolically active proteins of the tissues themselves.

We shall now deal with what is known at the present time of invertebrate glycoproteins, taking the phyla not in any particular evolutionary order but in order of decreasing knowledge. Molluscs, perhaps because of their slimyness, perhaps because of their accessibility, would seem to have received the greatest attention and yielded the most detailed information.

MOLLUSCA

1. The Crystalline Style: In many lamellibranches and a few gastropods there occurs in a diverticulum of the stomach an organ known because of its appearance as the crystalline style. The elongated rod-shaped style juts out from a ciliated sac into the anterior end of the stomach where it comes into contact with the gastric shield. Rotation of the style by the cilia (Fig. 2) causes it to fray at the end and a mucinous material becomes dispersed, entangling food particles in the stomach. The style is continually renewed by epithelial cells lining the style sac. The mucin immobilizes food particles but the function of the style is primarily a digestive one, since, although it contains no proteases, amylase, lipase, cellulase, and ortho-phenolase activities have been detected (Coupin, 1900; Mitra, 1901; Yonge, 1926; George, 1952; Hozumi, 1959; Newell, 1953; Fish, 1955; Nair, 1955; Johansson, 1945).

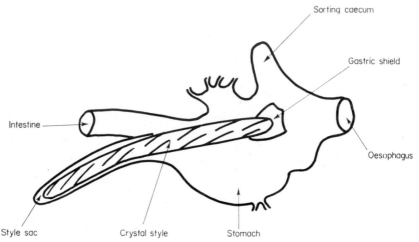

Fig. 2. Semi-diagrammatic view of the generalized lamellibranch stomach showing the location of the crystalline style.

Several of the earlier authors (Coupin, 1900; Barrois, 1899; Mackintosh, 1925) considered the crystalline style to be composed of mucous substances impregnated with the digestive fluid. The first definite indications to this effect however came much later from Berkeley (1935) who was able to demonstrate the presence, in the styles of several species of lamellibranch including *Mya arenaria*, of glucuronic acid and hexosamine. The style of *Pinna nobilis* was investigated in some detail by the late Kenneth Bailey (1958; Bailey and Worboys, 1960) and shown to contain glycoprotein of a very similar type to vertebrate materials.

Pinna styles were homogenized by Bailey at pH 8 (bicarbonate solution) and the small amount of insoluble residue which remained was removed by centrifugation. The solution was precipitated successively with dilute acetic acid at pH 5 and then with two volumes of acetone. Both precipitates were recombined, dissolved at pH 8, reprecipitated as before and freeze-dried to give two fractions (Fraction 2, the pH 5 precipitate and Fraction 3, the acetone precipitate). In spite of the separation of the material into two fractions Bailey considered it unlikely that the two fractions differed substantially since the pH 5 precipitate could, after redispersing, be again divided into soluble and insoluble fractions. An extract of the whole style was also investigated by dispersing in the manner described above, dialysing against water, and after removal of the slight sediment, precipitated with acetone after bringing to pH 5 with acetic acid ("whole style fraction"). When a style extract was made two per cent with respect to trichloracetic acid, little precipitation occurred unless the solution was heated. This slight precipitate, thought to represent the enzymes of the style, was dried in acetone and ether ("enzyme fraction"). The supernatant remaining after

TABLE I

Analysis of *Pinna nobilis* Crystalline Style and Style Extracts[a]

	Whole style fraction	Fraction 2	Fraction 3	Enzyme	"Mucoprotein"
N (g/100 g dry material	11·2	12·0	12·4	—	—
Hexosamine	11·2	12·4	10·8	6·2	13·5
Hexose	12·2	11·5	8·3	—	—
Fucose	5·7	6·5	—	—	—
Total carbohydrate as anhydro residues	26·2	—	—	—	—
2-amino-2-deoxy-D-glucose (% of total hexosamine)	—	49	—	—	—
2-amino-2-deoxy-D-galactose (% of total hexosamine)	—	51	—	—	—
Acetyl as percentage of free hexosamine					
Theoretical	24·0	—	—	—	—
Found	20·7	—	—	—	—
Total carbohydrate as anhydro residues + acetyl (% of dry wt.)	28·6	—	—	—	—
Hence protein (% of dry wt.)	71·4	—	—	—	—
Protein from amino acid analysis (Table II) (% of dry wt.)	67	—	—	—	—

[a] Bailey and Worboys (1960).

such treatment, expected to contain the style glycoproteins, was precipitated at pH 5 with acetone to yield the "mucoprotein" fraction.

Analytical data for the various style fractions are given in Tables I and II. Tests for uronic acid and sialic acid were negative. Paper chromatography

TABLE II

Amino Acid Analysis of *Pinna nobilis* Crystalline Style[a]

Amino Acid	Whole Style		Enzyme Fraction	
	b	c	b	c
Aspartic acid	10·87	95	14·0	122
Threonine	8·91	81	8·85	88
Serine	5·29	61	5·1	59
Glutamic acid	10·51	82	11·0	86
Proline	8·47	87	12·0	124
Glycine	4·38	77	6·35	111
Alanine	3·20	45	3·5	49
Cystine	4·28	42	0·9	9
Valine	5·33	54	5·5	56
Methionine	1·90	14	1·8	14
Isoleucine	4·19	37	3·9	35
Leucine	5·09	45	4·6	41
Tyrosine	10·12	62	6·1	37
Phenylaeanine	4·82	33	4·1	28
Lysine	5·24	41	6·2	49
Histidine⌉	1·84	13	1·7	12
Arginine⌋	4·07	26	4·1	26
Tryptophane	2·22	12	not measured	—
NH₃	(1·51)	108	—	—
	100·7		99·8	

[a] Bailey and Worboys (1960).
[b] Wt. of anhydro amino acid/100 g of protein excluding the carbohydrate portion.
[c] Residues/10^5 g of protein.

of hydrolysates of Fractions 2 and 3 indicated that galactose and fucose were present while ion-exchange chromatography showed the hexosamine in Fraction 2 to be made up of equal quantities of 2-amino-2-deoxy-D-glucose and 2-amino-2-deoxy-D-galactose.

Bailey was surprised to note that the so-called "enzyme fraction" (trichloracetic acid precipitate) still contained considerable hexosamine and suggested that the true enzyme content must be very small indeed. This observation was based, of course, on the now known to be erroneous notion that enzymes and glycoproteins were incompatible. It is likely that the carbohydrate content of this material was either too low or too unevenly distributed to protect against the denaturing effect of the trichloracetic acid. Fraction 2 was tested for blood group activity (A, B, H and Le^a); the tests were negative.

The results of amino acid analyses are unremarkable. Initially Bailey inferred that the style glycoproteins might possess alkaline iso-electric points since the base plus amide group contents exceeded the total dicarboxylic acid contents by 11 groups. Since however the pH of minimum solubility was in the region of five, the true iso-electric point appears to be acid and it was concluded that the amide values must be too high in spite of corrections having been made for ammonia arising from the decomposition of hexosamine. Really serine and threonine decompositions should also have been taken into account. The amino acid analyses of the "whole style" material indicated a minimum protein content for the style of 67% as opposed to a value of 71·4% deduced from the total carbohydrate and acetyl contents.

Some of the physical parameters of the style proteins were also determined (Table III). In the ultracentrifuge the whole style extract gave rise to

TABLE III
Sedimentation Coefficients and Relative Proportions of Components in the Ultracentrifuge Schlieren Diagrams (Fig. 5) for *Pinna nobilis* Crystalline Style^a

	Whole style extract		Fraction 2	
	$S_{20, w}$	Relative Proportion (%)	$S_{20, w}$	Relative Proportion (%)
Component 1	12·4	11	9·3	42
Component 2	9·3	47	5·6	58
Component 3	5·6	42	—	—

^{a} Bailey and Worboys (1960).

three clearly-defined peaks while in Fraction 2 the heaviest component had disappeared (Fig. 3). Electrophoresis (moving boundary) presented a more complex picture of the whole style extract: four distinctly separated components were present (Fig. 4).

Thus Bailey's results would appear to indicate clearly the existence in an invertebrate of a vertebrate type glycoprotein or glycoproteins. Little can be deduced from the amino acid composition as to the possible character of the protein-carbohydrate linkage region. Aspartic acid, glutamic acid, serine and threonine contents are all high; any of these might be involved in the bond to carbohydrate.

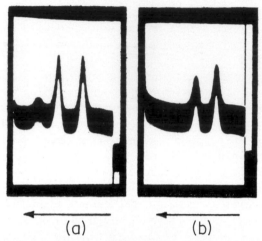

(a)   (b)

Fig. 3. Ultracentrifuge patterns for: (a) whole style extract of the crystalline style from *Pinna nobilis*, concentration 1·5%; (b) fraction 2 of the style (see text), concentration 0·9%. The solvent in both cases was sodium phosphate (I 0·028) − KCl (I 0·1), pH 7·5. Patterns are at 269,000 g, 20°, 64 minutes after reaching full speed (from Bailey and Worboys, 1960).

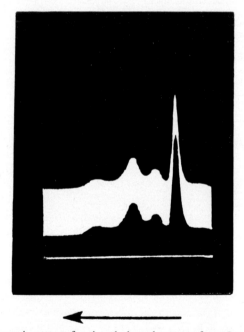

Fig. 4. Electrophoresis pattern for the whole-style extract from *Pinna nobilis* crystalline style. Concentration 1·5%, buffer sodium phosphate (I 0·028) − KCl (I 0·1), pH 7·5, ascending limb 100 V, 13 mA 270 minutes (from Bailey and Worboys, 1960).

The crystalline style of *Mactra sulcataria* has also been shown to contain glycoproteins, similar in some properties to vertebrate materials but exhibiting at the same time atypical characteristics (Hashimoto *et al.*, 1954; Hashimoto and Sato, 1955, 1956).

Defatted and dried style powders gave positive reactions to the biuret, xanthoproteic, ninhydrin and Millon tests. As is the case with many vertebrate glycoproteins the proteins in solutions of the style powders were not precipitable by heat treatment or by trichloracetic acid. Phosphotungstic and tannic acids did however precipitate the proteins. They could also be precipitated in the usual way with saturated ammonium sulphate solutions or with solutions of sodium sulphate. The authors suggested that it was unlikely that globulin was present. Hydrolysis of the style with hydrochloric acid yielded phenylalanine, methionine, tyrosine, alanine, threonine, serine, glutamic acid, aspartic acid, glycine and cystine (identified by paper chromatography). The style solutions were Molisch positive but only gave positive Fehling's reactions after acid hydrolysis; they thus contained some bound carbohydrate. Paper chromatography of acid hydrolysates indicated the presence of a number of monosaccharides; fucose, deoxyribose, xylose, mannose, galactose, 2-amino-2-deoxy-D-glucose, 2-amino-2-deoxy-D-galactose and glucuronic acid were identified. The presence of deoxyribose suggests contamination by nucleic acids, an opinion strengthened by its low concentration relative to the other sugars. The monosaccharides, characterized as their phenylhydrazones, methylphenylhydrazones, or in the case of the amino sugars as their hydroxynapthylidene derivatives, were all, with the exception of fucose, in the D form. The quantitative compositions of the styles, in terms of carbohydrate, are given in Table IV.

TABLE IV

Carbohydrate Composition of
Mactra sulcataria Crystalline Style[a]

Hexosamine (%)	14·3–15·2
Glucuronic acid (%)	10·7
Xylose (%)	3·1
Mannose (%)	3·9
Galactose (%)	13·2
Fucose (+ unknown; deoxy ribose?) (%)	7·3

[a] Hashimoto and Sato (1955, 1956); Hashimoto *et al.* (1954).

The occurrence of xylose in a glycoprotein is unusual but not without precedent in vertebrate glycoproteins (Moret *et al.*, 1964). The presence of glucuronic acid is perhaps the most atypical feature; however since no attempt was made to fractionate the style constituents it is impossible to say whether the hexuronic acid and the other more typical glycoprotein sugar constituents were in fact a part of the same molecule.

Doyle (1965, 1966) has identified a range of bound monosaccharides similar to those of *Mactra* in the styles of *Mya arenaria* and *Cardium edule*; galactose, glucose, mannose, xylose, fucose, deoxyribose, 2-amino-2-deoxy-D-glucose, 2-amino-2-deoxy-D-galactose and hexuronic acid were present. Approximately half the dry weight of the style in both species was protein; quantitative analyses are given in Table V. Although no amino acid analyses were carried out the style proteins had ultraviolet absorption spectra which indicated the presence of aromatic residues. Gel-filtration of papain digested *Mya* style on Sephadex-G75 yielded two carbohydrate-containing components with considerably reduced protein contents. These replaced the single protein-carbohydrate peak obtained with untreated style. The style materials did not exhibit metachromasia below pH 2 with thionin dyes and Doyle considered therefore that the sulphate detected was unlikely to be derived from ester sulphate residues on the carbohydrate. This assumption may be correct but is not wholly justified since isolated ester sulphate residues on non-linear heterosaccharides might not be expected to give rise to a metachromatic effect.

TABLE V

Analysis of *Mya arenaria* and *Cardium edule* Crystalline Styles[a]

	Mya	*Cardium*
	Dry style powders	
N (%)	8·99	9·06
Sulphate μg/mg	3·20	3·46
Phosphorus μg/mg	0·48	0·35
Hexosamine μg/mg	111·0	84–84·8
2-amino-2-deoxy-D-glucose (% total hexosamine)	45·3	43·0
2-amino-2-deoxy-D-galactose (% total hexosamine)	54·7	57·0
Uronic acid μg/mg	4·46	7·83
Glucose μg/mg	16·7	7·6
Deoxyribose μg/mg	1·85	1·20
Mannose μg/mg	7·98	7·98
Fucose μg/mg	22·60	38·87
Xylose μg/mg	2·73	0·77
N (%) Corrected for hexosamine nitrogen (7·8%)	8·18	8·28
Hence protein (%)	51·2	51·8

[a] Doyle (1966).

Using paper electrophoresis Inoue (1960) has found the crystalline styles of several molluscan species to have a complex composition. Styles of *Meretrix lusoria* contained four electrophoretically distinct protein components, those of *Venerupis philippinarium* six, *Mactra veneriformis* four and those of *Anadar inflata* four. One of the protein components of *A. inflata* style was associated with a polysaccharide as was one of the *M. lusoria* proteins, two of the *M. veneriformis* proteins and no less than four of the *V. philippinarium* proteins.

2. *The Egg and Reproductive System:* The secretions of the reproductive tracts of gastropod molluscs have been the subject of wide study (see also Chapter 15). Indeed the first account of a protein-polysaccharide complex was that made by Eichwald (1865), a Russian doctor working in Scherer's laboratory at Winzberg, during a study of *Helix pomatia* egg albumin. Eichwald regarded mucin as a conjugated single compound consisting of a moiety with all the properties of a genuine protein and a moiety released under certain conditions as a sugar, a definition of a glycoprotein still valid today. He was however probably mistaken in his interpretation of the albumen as a glycoprotein, the reducing substance he detected after acid hydrolysis almost certainly being galactose from the galactan in the albumen (Chapter 10). Hammarsten (1885) also mistakenly felt that the albumen was principally a glycoprotein as did Levene in 1925.

Geldmacher-Mallinkrodt and May (1957) believed the protein-polysaccharide complexes in the *H. pomatia* albumen gland not to be genuine "mucosubstances"; however, apart from galactogen these authors detected at least three other polysaccharide-protein complexes containing, besides galactose, glucose and other sugars. Other polysaccharides, possibly present as glycoprotein, have been detected on several occasions incidental to studies of the galactogen. Horstmann (1964) and Geldmacher-Mallinkrodt and Traexler (1961) have referred to a glycoprotein in *H. pomatia* albumen. Geldmacher-Mallinkrodt and Horstmann (1963) isolated a low molecular weight heteropolysaccharide from *H. pomatia* albumen gland. The glands were homogenized, boiled down, dissolved in water and precipitated with alcohol; twice in alkali, six times in acid, and twice at neutral pH. Ultrafiltrates of an aqueous solution were concentrated and precipitated with 20 volumes of alcohol, washed with alcohol-ether and dried. The heteropolysaccharide obtained was regarded as having been derived from one of three original protein-polysaccharide complexes present. These complexes were soluble in one per cent acetic acid. The polysaccharide had $[\alpha]_D^{20}$ $-68 \cdot 5$, contained L-galactose and 2-amino-2-deoxy-D-glucose (32% based on N), and had an ash content of $8 \cdot 6 \%$ with $1 \cdot 92 \%$ Pi. Although it separated into two zones on electrophoresis the polysaccharide was homogeneous in the ultracentrifuge and had a molecular weight of about 7500. In view of the monosaccharide composition, the relatively low molecular weight and the drastic extraction conditions it seems possible that this material constitutes the prosthetic group or groups of a conventional glycoprotein.

Bayne (1966) has investigated in some detail the composition of the layers of the eggs of a pulmonate gastropod *Agriolimax reticulatus* (see also Chapter 15). The egg jelly, which lies between the shell and the perivitelline sac (Fig. 5) gave on hydrolysis 2-amino-2-deoxy-D-glucose and

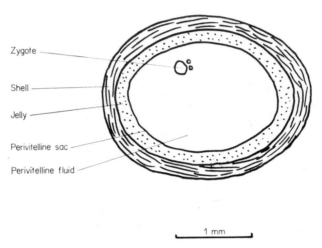

Zygote

Shell

Jelly

Perivitelline sac

Perivitelline fluid

1 mm

Fig. 5. Layers of the egg of a typical land pulmonate gastropod mollusc (redrawn from Bayne, 1966).

galactose, identified chromatographically. Identical results were obtained with the perivitelline fluid and the perivitelline membrane on hydrolysis and chromatographic analysis. The hydrolysed fluid contained 2-amino-D-deoxy-D-glucose and galactose together with alanine, glycine, aspartic acid, glutamic acid, lysine, histidine, arginine, serine, threonine, tyrosine, phenylalanine, leucine/isoleucine, cystine, methionine, proline, valine and nor-valine. In addition to galactose two other unidentified components were present, one of which stained positively with both silver nitrate and aniline hydrogen phthalate and one which stained only with silver nitrate. These unknowns were not thought to be sugars, however. Agar gel electrophoresis of the perivitelline fluid separated it into seven components; one of these was almost pure polysaccharide and probably contained galactan. Of the remaining protein bands one gave a weak periodic acid-Schiff reaction. Gel-filtration of the perivitelline fluid on Sephadex G25 and G100 failed to separate protein from carbohydrate.

Morrill *et al.* (1964) separated the egg albumen of *Limnaea palustris* into several components by electrophoresis and noted that four of these were periodic acid-Schiff positive.

McMahon *et al.* (1957) reported the presence, in the females of two species of aquatic pulmonate snail, of glycoproteins or mucopolysaccharides containing galactose and fucose. They suggested that the galactose-fucose containing polysaccharides found in the eggs of the snails were derived from the uterus.

George and Jura (1958), in a histochemical study of the egg capsule fluid from the land snail *Succinea putris*, found that a fractionation into three components could be achieved by centrifugation. The outer transparent layer immediately beneath the membrane did not sediment and was presumed to be a glycoprotein on the basis of its positive reactions to protein stains and periodic acid-Schiff reagent. A "centripetal" fraction also contained glycoprotein while a sedimenting fraction appeared to contain polysaccharide only.

Smith (1965) has investigated the secretions of the reproductive tract in the garden slug *Arion ater* using a rather extensive range of histochemical tests; in spite of this no definite conclusions were drawn with regard to the presence of glycoproteins.

3. Hypobranchial Mucins: The hypobranchial glands of marine gastropod molluscs produce a viscous "mucinous" secretion which apparently is involved in defence and in the promotion of mantle hygiene. These secretions are characterized by the presence of highly-sulphated polysaccharides (Chapter 5); however, there is always a protein component present although the polysaccharide has usually been the primary object of interest.

Hunt and Jevons (1963, 1965) separated the protein from its associated glucan sulphate in the hypobranchial mucin of the whelk *Buccinum undatum* and showed it to be a glycoprotein. The glycoprotein was obtained, in a denatured state, by treatment of the mucin with 40% (w/v) phenol solution at 60°. The glycoprotein contained eight per cent hexose and 4·5% hexosamine while paper chromatography of acid hydrolysates indicated the presence of galactose, mannose, glucose and fucose. Ion-exchange chromatography identified the hexosamine as containing equal proportions of 2-amino-2-deoxy-D-glucose and 2-amino-2-deoxy-D-galactose. No sialic acid could be detected. An amino acid analysis of the glycoprotein is given in Table VI. The acid amino acid content is high; aspartic and glutamic acids contribute almost 23% of the total amino acid content. The overall ratio of aspartic plus glutamic acid to basic amino acids is 3 : 2, but if the amide content is taken into account this becomes 1 : 1·3, suggesting a slightly basic isoelectric point sufficient to account for salt linking to the polysaccharide sulphate moiety of the mucin. Dinitrophenylation and

TABLE VI

Amino Acid Composition of the Glycoprotein
Moiety of *Buccinum undatum* Hypobranchial Mucin[a, b]

Amino acid	
Cysteic acid	1·73
Aspartic acid	11·10
Threonine	6·45
Serine	6·27
Glutamic acid	11·34
Proline	5·19
Glycine	8·27
Alanine	7·67
Valine	6·91
Methionine	0·31
Isoleucine	4·27
Leucine	8·16
Tyrosine	2·85
Phenylalanine	3·66
Ornithine	0·45
Lysine	7·29
Histidine	2·45
Arginine	5·54
Amide Ammonia[c]	51·00

[a] Hunt and Jevons (1965).
[b] μ moles of amino acid/100 μ moles of total amino acids.
[c] μ moles/100 μ moles acidic amino acid.

chromatography of the DNP derivatives, extracted from acid hydrolysates of the DNP-glycoprotein, yielded DNP alanine, serine and aspartic acid in the molar ratios 1 : 2 : 1 respectively. Thus there were probably at least four chains presumably linked by disulphide bridges; peptides sensitive to performic acid oxidation were detected by paper electrophoresis of proteolytic digests of the glycoprotein.

Solutions of the denatured glycoprotein could be prepared by adjusting aqueous suspensions briefly to pH 11 and then readjusting to pH 8. Analytical ultracentrifugation of a 1% (w/v) solution, prepared in this manner and then dialysed against pH 8 phosphate buffer, gave a single sedimenting component, 16·6 s at 20°. The infrared spectrum of the glycoprotein had well-defined peptide bond bands at 1650 and 1550 cm^{-1} with a sharp N—H stretch at 3250 cm^{-1} having only traces of the adjacent O—H stretch. A residual broad absorption, characteristic of glycoproteins, extended over the range 950–1150 cm^{-1}. No sulphate absorptions could be detected.

Moving boundary electrophoresis or analytical ultracentrifugation of whole mucin in phosphate buffer, pH 8 (Chapter 5: Figs. 5 and 6) showed a separation of the glycoprotein from glucan sulphate. Under these conditions the glycoprotein had a mobility of $-9 \cdot 3 \times 10^{-5}$ cm^2 v^{-1} sec^{-1} and a sedimentation coefficient 9·0 s ($S_{20, w}$). The lower sedimentation coefficient is probably a result of sedimentation in the presence of the highly-viscous polysaccharide.

In contrast to these latter studies with *B. undatum* mucin, Shashoua and Kwart (1959) found there to be no carbohydrate bound to the protein which they separated from polysaccharide sulphate present in the hypobranchial mucin of the gastropod *Busycon canaliculatum*. The amino acid compositions of the *Buccinum* and *Busycon* proteins however are closely similar (Chapter 5: Table V), the chief difference being a higher proportion of basic amino acids in the *Busycon* material.

Unfortunately no data is available for the protein associated with glucan sulphate in *Charonia lampas* mucin (Chapter 5) (Iida, 1963) while although Doyle (1964) detected some glucose and fucose residues in the mucin of the whelk *Neptunea antiqua* it seems likely that these were associated, not with the protein but with the polyhexosamine sulphate (Chapter 5).

4. Serum Glycoproteins: Vertebrate serum and plasma are rich in glycoproteins and an attempt has been made to detect similar materials in molluscan blood (Yamaguchi *et al.*, 1961). The hexose contents of hydrolysates of total blood, Winzler's seromucoid fraction and a 93 % ethanol precipitate from *Gryphea gigas* blood, were found to be 125 % (av.), 52 % (av.) and 106 % (av.) respectively. The seromucoid fraction, after acid hydrolysis and paper chromatography, was found to contain maltose, galactose and glucose together with four unidentified components.

5. Salivary Glycoproteins: Studies of vertebrate salivary gland glycoproteins have done much to advance our knowledge of the chemistry of these substances. Arvy (1963) has reviewed the comparative histo-enzymology of salivary glands in the invertebrata. The toxicity of octopus saliva has been noted by Ghirett (1960) who suggested that it might be a glycoprotein.

Gennaro *et al.* (1965) isolated glycoproteins from the anterior salivary gland of the large Pacific Coast octopus. Glands were treated with two per cent sodium hydroxide in the cold for 48 hours in order to bring about solubilization. After dialysis and removal of water-insoluble material by filtration through Celite, a viscous solution remained. Glycoprotein was precipitated from this solution by the addition of four volumes of 95 % ethanol. The precipitable and ethanol soluble fractions constituted 0·665 % and 0·58 % of the weight of the wet gland tissue. Sulphate, total sugar, nitrogen, uronic acid and hexosamine contents were estimated; these are given as molar ratios in Table VII. The two fractions were

G

TABLE VII

Molar Ratio Composition with Respect to
Hexosamine of Octopus Salivary Gland Fractions[a]

	Ethanol soluble	Ethanol precipitable
% wet tissue	0·58	0·665
Sulphate	0·09	0·01
Total sugar (phenol)	2·13	1·06
Uronic acid (Carbazole)	0·15	0·14
Hexosamine	1·00	1·00
Nitrogen	16·07	8·30

[a] Gennaro et al. (1965).

hydrolysed and fractionated by ion-exchange chromatography and thin layer chromatography. In neither case could uronic acid be detected, although the hydrolysis conditions had been mild. In the alcohol precipit-able fraction the hexosamine was identified as being primarily 2-amino-2-deoxy-D-galactose with traces of 2-amino-2-deoxy-D-glucose, while the ethanol soluble material contained equal quantities of these amino sugars. Both glycoproteins contained an approximately equal amount of fucose and galactose; in addition the ethanol soluble fraction contained an unidentified sugar component. There appears to be some unexplained discrepancy in the results obtained by these authors on different occasions. In their article they do not make clear whether they are working with the same species on all occasions and in the work described above say only that they used the large Pacific Coast octopus; in an earlier part of the article however they describe work using the anterior salivary glands of *Octopus vulgaris*. In this instance the method of preparation was similar with the addition of a precipitation by cetylpyridinium chloride from the dialysed and filtered extract. This precipitate was solubilized in 4 M sodium chloride solution and reprecipitated with 95% ethanol. The resultant material was dialysed against distilled water to bring it into solution. Although this material was not analysed it separated into two components on disc-gel electrophoresis one of which had a mobility similar to hyaluronic acid and stained orthochromatically with toluidine blue, and one which had a similar mobility to chondroitin-4-sulphate and stained metachroma-tically. The whole fraction gave a weak carbazole reaction for uronic acid. These findings, in contrast to the later ones, suggest the presence of acid mucopolysaccharides rather than glycoproteins. It is unfortunate that no evidence was given for the presence or absence, in these secretions, of a sialic acid.

6. *Epithelial Mucins:* In spite of the slimy nature of the epithelia of molluscs little attention has been paid to the nature of the mucous substances involved. The hypobranchial secretions we have already described are in fact derived from a modified epithelium and may perhaps give us some sort of indication of the properties of the general epithelium in this respect.

Papain-digested preparations of *Pecten maximus* mantle epithelium have been fractionated by ion-exchange chromatography on DEAE-Sephadex A50 using a stepwise elution with sodium chloride solutions (Doyle, 1967). This yielded two glycoprotein fractions. One of these was eluted from the column by 0·1 M sodium chloride and contained hexosamine, galactose, glucose, fucose, small amounts of mannose and a pentose (probably xylose). The other glycoprotein fraction was eluted by 0·5 M sodium chloride and had a similar composition, apart from being almost glucose-free. The glycoproteins were associated with a glucan sulphate component (Chapter 5) and we can therefore see affinities with hypobranchial gland secretions.

Arcadi (1963, 1965) has conducted histochemical studies of the mucus-producing cells in the integument of the garden slug *Lehmania poirieri*. There appear to be two types of structure involved in the secretion of epithelial mucin, the so-called "basket" cell complex and the "granular" cell complex. The latter apparently secretes an acid mucopolysaccharide (Chapter 5) onto the dorsal and lateral integument and the foot pad. The contents of the basket cell complex stain with the periodic acid-Schiff reagent (PAS) and the staining reaction is partially blocked by acetylation. The PAS reaction was also partially destroyed by the action of neuraminidase, suggesting the presence of a ketosidically-linked sialic acid.

Bassot (1959) and Nicol (1960) have noted the presence, histochemically, of glycoproteins beneath the surface epithelium of the light organs situated in the siphon of the piddock *Pholas dactylus*. Two glandular layers could be distinguished, an outer one containing mucus cells and an inner layer of two types of granular cell with different staining properties. Both the mucus cells and one of the granular cell types secreted a glycoprotein-like protein-polysaccharide complex.

7. *Miscellaneous Molluscan Glycoproteins:* Blood group activity has been observed in aqueous extracts of whole oysters (Springer *et al.*, 1954). It was suggested that mucosubstances of the glycoprotein type might be responsible.

Nagy and Laszlo (1964) carried out a histochemical and biochemical study of mucosubstances in the tissues of the swan mussel *Anodonta cygnea*. The histochemical tests showed PAS positive substances to be present in the foot, mantle, gill, adductor muscle, syphon, heart and kidney. These substances had the staining properties of neutral mucopoly-saccharides, whereas the glandular tissues contained acid mucopoly-saccharides. Hexosamine determinations gave values of 674·6, 700·8, 563·5, 740·3, 632·9, 658·6 and 307·4 mg % for the foot, kidney, gill, heart, mantle, syphon and adductor muscle respectively.

The presence of neutral heteropolysaccharides associated with protein in the extra-pallial fluids of several molluscan species has been noted by Kobayashi (1964).

Two glycoproteins have been isolated from the digestive juice of the snail *H. pomatia* (Got and Marnay, 1968). These glycoproteins were enzymes hydrolysing o-nitrophenyl-β-D-galactopyranoside. The digestive juice was precipitated with four volumes of a 50% solution of acetone in pH 5·3, 0·1 M acetate buffer. The precipitate formed was redissolved in buffer and reprecipitated with acetone at 30% concentration. The precipitate was redissolved in pH 5·3 buffer, containing sodium chloride at 0·1 M concentration, and subjected to gel-filtration on a column of Sephadex G200. The material separated into at least four components, two of which had enzymic activity. These two peaks were however still impure, as the results of disc electrophoresis showed. Peak I contained some β-glucuronidase and N-acetylglucosaminidase activity in addition to β-galactosidase activity. Peaks I and II were precipitated with 2·8 M ammonium sulphate. The precipitate from peak I was redissolved in pH 5·3 acetate-sodium chloride buffer and chromatographed on a column of DEAE-Sephadex A-50 eluted in steps from 0·1 through 0·15 and 0·20, to 0·25 M sodium chloride. Three components separated, all with β-galactosidase activity, the principal peak (G2) eluting below 0·15 M NaCl. The ammonium sulphate precipitate of peak II was chromatographed on a column of CM-cellulose and yielded a component (G1) with β-galactosidase activity and with a trace of non-glycoprotein inactive impurity. The two β-hexosidases were physically characterized by ultracentrifugation and by electrophoresis over a range of pH values (Table VIII). The iso-electric points were closely similar but the mobilities differed on electrophoresis

TABLE VIII

Physical Parameters of the β-Hexosidase Glycoproteins
from *Helix pomatia* Digestive Juice[a]

	Galactosidase 1	Galactosidase 2
0·5% S_{20} in phosphate buffer 0·02 M-NaCl 0·08 M pH 7	6·4	10·4
Isoelectric point	4·5	4·6

[a] Got and Marnay (1968).

in acrylamide and starch gels; the assumption was that the molecular weights were different.

The contents of hexose, fucose, and hexosamine are shown in Table IX. Thin-layer chromatography of hydrolysates indicated that both enzymes contained galactose, mannose, glucose, fucose and 2-amino-2-deoxy-D-glucose. No sialic acid could be detected in either enzyme. The glycoproteins had no immunological identity.

TABLE IX

Carbohydrate Composition of the Two β-Hexosidase Glycoproteins from *Helix pomatia* Digestive Juice[a]

	Galactosidase 1	Galactosidase 2
Hexose (%)	8	9
Fucose (%)	1·8	1·6
Hexosamine (%)	6·7	4·4
Sialic acid (%)	0	0

[a] Got and Marnay (1968).

8. *Biosynthesis of Molluscan Glycoproteins:* When the tissues of the slug *Limax maximus* were extracted with 10% trichloracetic acid in order to obtain acid-soluble nucleotides (Wheat, 1960), a number of nucleotide sugar derivatives were identified. The acid extract was precipitated with an equal volume of ethanol, the precipitated material rejected and the extract desalted with charcoal. The extract was adsorbed onto Dowex-1-formate resin and all the nucleotides except the uridine diphosphate derivatives eluted with 3 M formic acid. The UDP sugars were eluted with 4 M ammonium formate, desalted by charcoal treatment and chromatographed on a Dowex-1-formate column with a gradient of 3 M formic acid to 4 M ammonium formate, pH 3. Five peaks were obtained, the major one containing over 80% of the 260 mμ absorbing material. This peak had 250/260 mμ and 280/260 mμ absorption ratios of 0·76 and 0·42 respectively, suggesting the presence of a uridine derivative. Paper chromatography on Whatman No. 1 paper, using ethanol-M sodium acetate, 7·5 : 3, first at pH 7·5 and then at pH 3·7, resolved this major peak into three ultra violet absorbing components. The slowest band contained 30% uridine diphosphate-glucose, identified by comparison with authentic samples, by degradation with UDP-glucose dehydrogenase and by acid hydrolysis and identification of D-glucose with glucose oxidase. The second band contained uridine diphosphate-2-acetamido-2-deoxy-D-glucose and 2-acetamido-2-deoxy-D-galactose. The amino sugars were identified, after hydrolysis of the nucleotide derivatives, by paper chromatography, ion-exchange

chromatography and by degradation to arabinose and xylose (also identified by paper chromatography). The third band contained an unidentified complex of nucleotide and reducing substance. The isolation of nucleotide amino sugars is of interest since these could be involved in glycoprotein biosynthesis, although UDP-2-acetamido-2-deoxy-D-glucose might also have been a chitin precursor; UDP-glucose can follow several pathways, including glycogen synthesis.

INSECTA

1. Plasma Glycoproteins: Two glycoproteins have been isolated from the plasma of the cockroach *Periplaneta americana* using electrophoretic techniques (Lipke, 1960; Lipke *et al.*, 1965a). Moulting in the cockroach had earlier been shown to be accompanied by changes in the bound carbohydrate of the plasma glycoproteins (Siakotos, 1960a). Plasma obtained by centrifugation, according to the method of Siakotos (1960b) was subjected to electrophoresis at 2° on Whatman No. 3MM paper at eight volts per cm for 20 hours in 0·05 M barbital buffer, pH 8·8. The electrophoretic components containing bound carbohydrate were identified by staining test strips with periodate-fuschin and eluted with a one per cent sodium chloride-0·05 N sodium bicarbonate solution. After dialysis, the protein was precipitated with ethanol, redissolved in 0·005 N sodium hydroxide, and reprecipitated with trichloracetic acid. The precipitates were washed with ethanol-ether (1 : 1) and dried. Both glycoproteins contained 2-amino-2-deoxy-D-glucose, 2-amino-2-deoxy-D-galactose, glucose, mannose, galactose, arabinose and xylose. On the day prior to moulting a third glycoprotein appeared in the plasma. The alternations in the composition of the bound carbohydrates are shown in Table X. Mannose was found to be the principal neutral sugar constituent of the glycoproteins; fucose and sialic acid were not detected. These workers also found bound mannose and hexosamine in the cockroach fat body, carbohydrate in an alkaline extract of the cuticle and glucose, galactose, mannose, xylose and arabinose in the cuticle and exuvium. These latter monosaccharides were considered to be constituents not of glycoprotein but of the chitin itself. The pattern of accumulation from ^{14}C-labelled glucose into the hexosamine moiety of the plasma and fat body glycoprotein suggested periods of little or no turnover and other periods when significant rates were manifest (Lipke *et al.*, 1965b). In the middle of the moult cycle, $t_{\frac{1}{2}}$ was 16 days for fat body hexosamine and 29 days for plasma hexosamine with no obvious precursor-product relationship between the two compartments.

Sissakian and Veinora (1958) have demonstrated the presence of an acid soluble glycopeptide, containing four per cent arabinose, in the body fluids of silkworm (*Bombyx mori*) pupae. The coelomic fluid was centrifuged

TABLE X

Amino Sugar and Neutral Sugar Compositions of the Glycoproteins
Separated from *Periplaneta americana* Plasma by Paper Electrophoresis[a]

Molt Stage	Total Protein mg/ml	Electro-phoretic Component	% Total Protein	Neutral Sugar		Hexosamine	
				μ moles /100 mg	Mole Ratio[b]	μ moles /100 mg	Mole Ratio[c]
−1	91	II	30	34	0·4 : 1·0 : 0·8	5·0	20 : 1
		IV	34	22	0·1 : 1·0 : 1·6	2·8	10 : 1
		V	25	26	0·3 : 1·0 : 2·5	2·3	20 : 1
M	90	II	33	39	0·1 : 1·0 : 2·4	8·0	20 : 1
		IV + V	65	30	0·3 : 1·0 : 1·9	5·3	0·6 : 1
+4	59	II	48	30	0·2 : 1·0 : 2·4	5·2	20 : 1
		IV	51	37	0·3 : 1·0 : 2·6	6·5	9 : 1
±	63	II	46	35	0·6 : 1·0 : 4·1	5·7	20 : 1
		IV	50	28	0·5 : 1·0 : 1·6	6·8	4 : 1

[a] Lipke, Grainger and Siakotos (1965a).
[b] Galactose to glucose to mannose.
[c] 2-amino-2-deoxy-D-glucose to 2-amino-2-deoxy-D-galactose.

at 3000 r.p.m. for 20 minutes and the sediment discarded. The supernatant was freeze-dried, redissolved in water and precipitated with 10% trichloracetic acid. The precipitate was removed by centrifugation and the supernatant precipitated with four volumes of ethanol. Following centrifugation the precipitate was discarded and the supernatant dialysed (Flow sheet, Fig. 6). The dialysate contained a complex which on examination by moving boundary electrophoresis gave one component only with a mobility of $0·4 \times 10^{-5}$ cm^2 v^{-1} sec^{-1} at pH 7·2 ionic strength 0·1 and a mobility of $3·5 \times 10^{-5}$ cm^2 v^{-1} sec^{-1} at pH 5·5. Paper chromatography however resolved the complex into two ninhydrin positive spots which gave positive reactions for phosphate. The complex contained 1·8–3·0% P, 3% N, 4·0–4·5% pentose and 2·8–3·0% hexose. Treatment of the dialysed-complex solution with two volumes of acetone yielded a precipitate separable into two fractions, one water soluble (I) and one insoluble (II). The acetone treatment appeared to split lipid from the complex. Fraction I was precipitable by two volumes of ethanol to give a gelatinous water-soluble material. Hydrolysis of Fraction I, followed by paper chromatography, indicated the presence of the amino acids alanine, phenylalanine, cystine or cysteine, valine, proline, leucine, lysine, tyrosine, arginine, and threonine or glutamic acid together with the sugars glucose, mannose and arabinose. Arabinose was the predominant monosaccharide. Fraction II contained only a little carbohydrate amounting to 1·4% of the material. This fraction dissolved in acids with the evolution of carbon dioxide. It was

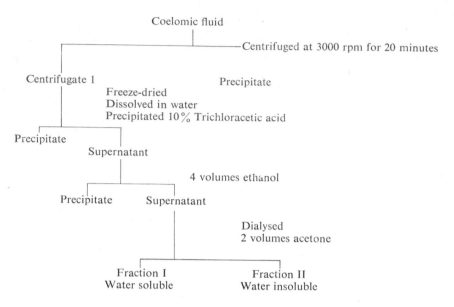

Fig. 6. Flow sheet for the preparation of acid soluble glycopeptides from *Bombyx mori* coelomic fluid (Sissakian and Veinora, 1958).

thought that the carbon dioxide might have been bound to amino groups in the complex. The isoionic point of Fraction II was 9·4. After hydrolysis, Fraction II yielded on chromatography the amino acids alanine, cystine or cysteine, tyrosine, valine, phenylalanine, leucine, lysine, arginine and threonine or glutamic acid. There was no tryptophane.

The intact complex had an ultraviolet absorption spectrum with a peak at 257 mμ and a minimum at 240 mμ. After acetone treatment and hydrolysis with five per cent perchloric acid the maximum shifted to 260 mμ. Fraction I, after perchloric acid hydrolysis, had an absorption maximum at 265 mμ with a 245 mμ minimum while Fraction II had maxima at 255 and 270 mμ and a 243 mμ minimum. These absorptions appeared to be due in part to the presence of bound nucleotide.

Hydrolysis and paper chromatography indicated the presence in the intact complex of adenine and guanine. The complex contained a pigment which was found to be riboflavine, at a concentration of 50 μg/g complex; the adenine was probably derived, in part at least, from this. Fraction I contained 62·5% of the total acid hydrolysable phosphate of the complex and appears to be a new class of nucleotide-glycopeptide derivative.

2. Salivary Gland Secretions: The secretion of the salivary gland of *Drosophila virilis* appears to contain glycoprotein with glucose, galactose, mannose and 2-amino-2-deoxy-D-glucose as constituent monosaccharides

(Perkowska, 1963). Immuno electrophoresis studies of the secretion gave a wide range of precipitin lines but it was shown that the most characteristic material present consisted mainly of glycoprotein.

2-amino-2-deoxy-D-glucose has also been demonstrated chromatographically in hydrolysates of *D. melanogaster* prepupal salivary gland secretion (Kodani, 1948). The secretion was precipitated from the excised glandular tissue by the addition of ethanol and was separated manually from the tissue debris. A variety of protein tests, including biuret, ninhydrin, and Molisch, indicated the presence of a protein low in aromatic residues. The material, which was water soluble, showed no specific absorption between 210 and 310 mμ, confirming the low content of aromatic amino acids. Tests indicated the presence of a small amount of carbohydrate. The secretion contained 10·8% nitrogen; a low value thought to be due to the presence of inorganic matter. Two-dimensional chromatography of acid hydrolysates indicated the presence of cysteic acid, aspartic acid, glutamic acid, serine, glycine, threonine, alanine, tyrosine, methionine sulphoxide, valine, lysine, arginine, proline, leucine, phenylalanine and 2-amino-2-deoxy-D-glucose. Glutamic acid, serine, and threonine were present in large amounts, a well known characteristic of vertebrate glycoproteins.

The salivary secretion of larval *Chironomus thumni* (*Tendipes thumni*) were separated, using electrophoretic and immunoelectrophoretic techniques, into at least 10 glycoprotein fractions (Kato *et al.*, 1963). Defretin (1951a) has similarly noted polysaccharide material in the saliva of *Chironomus plumosus* larvae.

Glycoproteins have been reported, on the basis of histological evidence, in the salivary secretions of the milkweed bug, *Oncopeltus fasciatus*, and have been located in the posterior and anterior lobes of the maxillary gland (Linder, 1956; Salkeld, 1960). Day (1949) histochemically detected glycogen, water-labile glycoprotein and acid mucopolysaccharide in the salivary glands of *Periplaneta*.

3. *Egg shells and Egg shell Adhesives: Drosophila melanogaster* egg shells contain an adhesive which appears to be secreted by the female accessory glands (Riley and Forgash, 1967). Hydrolysates of the adhesive substance yielded up to 10 amino acids which included lysine, arginine, serine, glycine, aspartic acid, threonine, alanine, valine, leucine and/or isoleucine. 2 amino-2-deoxy-D-glucose was also identified in the hydrolysate suggesting that the adhesive might be a glycoprotein. Wilson (1960a, b) has also identified amino acids and hexosamine in hydrolysates of *D. melanogaster* egg shell. Aspartic acid and alanine were the most abundant amino acids followed by serine, glutamic acid, glycine and tyrosine; lysine and arginine were present in equivalent amounts and also present were threonine, valine, proline, hydroxyproline, leucine and/or isoleucine and histidine. Traces

G*

of cystine or cysteine were present. The egg shells contained 0·6% 2-amino-2-deoxy-D-glucose, probably mostly in the adhesive.

McFarlane (1962) examined the action of pepsin and trypsin on the egg shell of the house-cricket *Acheta domesticus*. The maternal endocuticle of the swollen egg shells was digested by crystalline trypsin only after a preliminary treatment with pepsin; this latter enzyme appeared to digest the protein moiety of a glycoprotein. Presumably the glycoprotein was acting as a trypsin inhibitor. The role of the glycoprotein was thought to be to cover the eggs and lubricate their passage down the genital ducts.

Not all these egg coats are glycoprotein, however. The viscous hygroscopic material on the outside of *Nepa cinerea* (Hemiptera) egg shell seemed to be an acid mucopolysaccharide functioning as a cement and also protecting against humidity variations (Hinton, 1961).

Histochemical tests have indicated the presence of polysaccharide, in some form, in the capsule surrounding the eggs of *Mesoleius tenthredinis*, an ichneumonid parasitic in the larch sawfly *Pristiphora ericksonii* (Bronskill, 1960). The eggs are encapsulated by the larvae of the host, the capsule material appearing to inhibit embryonic development of the parasite by limiting its oxygen supply.

4. Miscellaneous Sources: Bonhag (1956) histologically detected protein-carbohydrate complexes in the oöcyte of the earwig *Anisolabis maritima* and in the tunica propria of the cockroach (*P. americana*) ovariole (Bonhag and Arnold, 1961).

The secretion product of the male accessory reproductive gland of the Japanese beetle *Popilla japonica* consists of a glycoprotein-like protein-polysaccharide complex (histochemical) associated with phospholipid (Anderson, 1950).

Several authors have reported the presence of glycoproteins in nervous tissues mainly on the basis of histochemical tests. The presence of a glycoprotein in a secretion from the neurosecretory cells of the adult silkworm has been suggested by Ganguly and Basu (1962). A glycoprotein gonadotrophic hormone has been observed in the corpora cardiaca and corpora allata of the female cockroach *P. americana* (Ganguly and Nanda, 1964). Ashhurst has found neutral mucopolysaccharides, with the histochemical properties of glycoprotein, to be present in the neural lamellae of the connective tissue sheaths of the nervous systems of *P. americana*, *Schistocerca gregaria* (locust) and the wax moth, *Galleiria mellonella* (Ashhurst, 1961, 1965; Ashhurst and Richards, 1964a) and also in the adipohemocyte blood cells of *Galleiria* (Ashhurst and Richards, 1964b).

Glycoproteins have been detected histochemically in the connective tissues of the sucking lice, *Linognathus vitali*, *Haemotopinus asini*, and *H. eurysternus* (Pipa and Cook, 1958).

Solid components accumulating in the gut of the aphid *Aphis fabae* have been tentatively identified by Edwards (1966) as neutral mucopolysaccharides associated with protein. Hydrolysis of this material yielded glucose and fructose as major components with perhaps xylose and arabinose.

5. *Biosynthesis:* Since glycoproteins containing neutral and amino sugars appear to be present in insects it would be expected that their nucleotide precursors might also be present.

Extracts of the pupal hemolymph of *Telea polyphemus* and the pupal fat body of *Platysamia cecrophia* have been found to contain high concentrations of nucleotide sugars (Carey and Wyatt, 1960). The 0·3 N perchloric acid extracts were subjected to chromatography on Dowex-1-formate and the uridine diphosphate derivatives were eluted with a gradient of ammonium formate in formic acid. The principal derivative found was UDP-2-acetamido-2-deoxy-D-galactose together with UDP-2-acetamido-2-deoxy-D-glucose, UDP-glucose and UDP-galactose. The level of the UDP-sugar in *Cecrophia* hemolymph changed with development. Thus in the larva the level of UDP-sugar was 0·01 μmole/ml of cell-free plasma rising to 0·2–0·9 μmole/ml in the pupa. These values largely represent the 2-acetamido-2-deoxy-D-glucose derivative. In the fat body, however, the level remained constant at 0·2 μmole/g with equal quantities of the derivatives. Wing epidermis contained 0·8 μmole/g of nucleotide sugar, in diapause, rising to the very high level of 4 μmole/g in early adult development. While the presence of UDP-glucose might be related to trehalose and glycogen synthesis and UDP-2-acetamido-2-deoxy-D-glucose to chitin synthesis it is difficult to account for the presence of the other derivatives except on the grounds of glycoprotein biosynthesis.

CRUSTACEA

Definite reports of glycoproteins in the crustacea are few. The presence of 2-amino-2-deoxy-D-glucose in a protein fraction obtained from the blood of the crab *Cancer pagurus* has been noted by Roche and Dumazert (1940), while gel-electrophoresis of *Carcinus maenas* serum has indicated the presence of at least 10 components, one of the major ones staining as a glycoprotein (Frentz, 1958).

ECHINODERMATA

1. *Nuclear Glycoprotein:* An unusual hexosamine-protein complex has been detected in the tissues of the sand dollar *Echinarachnius parma* (Rosenkranz, 1967; Rosenkranz and Carden, 1967). DNA isolated from the tissues of the adult sand dollar was found to contain a component with a buoyant density identical with that of polydeoxyadenylate-thymidylate.

This material was resistant to thermal denaturation even when the heating process was carried out in the presence of formaldehyde. It was not digested by either ribonuclease or deoxyribonuclease and could therefore be purified by enzymic digestion of the accompanying DNA followed by removal of low-molecular weight material on a column of Sephadex G100. The resultant purified material had an unusually high sedimentation coefficient of 907s comparable with values obtained for virus particles. It contained no deoxyribose or ribose but was strongly ninhydrin-positive after hydrolysis with acid. The hydrolysates gave positive reactions for neutral sugars as well as a slightly positive reaction for amino sugar. On electrophoretic analysis at pH 6·3 the hydrolysate separated into three ninhydrin-positive spots. One of these migrated to the negative electrode with a mobility identical to that of 2-amino-2-deoxy-D-galactose, the second migrated to the positive electrode and the third remained at the origin. The spot having a mobility identical to that of 2-amino-2-deoxy-D-galactose also gave a positive reaction for amino sugar. The whole material was not susceptible to attack by α amylase. Digestion with crude trypsin resulted in loss of the banding property in gradients of CsCl. Exposure to purified trypsin did not completely digest the material but caused an increase in the buoyant density. The material exhibited no specific absorption in the ultra violet. The results suggest that this material is glycoprotein, although the molecular weight must be abnormally high. No information is available with regard to the amino acid composition; the protein present was largely undegraded, one would assume, by the relatively mild acid hydrolysis conditions used, i.e. 2N HCl at 100° for two hours, hence the low number of components demonstrated on electrophoresis of the hydrolysate.

2. *Egg Glycoprotein:* When a trichloracetic acid extract of unfertilized *Paracentrotus lividus* eggs was extracted with ether a precipitate formed at the interface (Monroy and Vittorelli, 1960). This precipitate was found to contain 43·0–52·7% neutral carbohydrate, 1·6% 2-amino-2-deoxy-D-glucose and 5·6–7·4% N. No precipitate was formed from extracts of five-minute fertilized eggs. The neutral sugar was identified as glucose. After acid hydrolysis 16 amino acids were detected on paper chromatograms. These amino acids were cysteine, arginine, lysine, histidine, aspartic acid, glutamic acid, glycine serine, alanine, proline, methionine, valine, threonine, phenylalanine, leucine and isoleucine.

3. *Perivisceral Fluid:* The perivisceral fluid of *Arbacia lixula* has a glucose-containing glycoprotein constituent (Messina, 1957). On isolation from the animal the perivisceral fluid clots. Removal of the clotted material by centrifugation (20,000 g at −5°) yielded a clear liquid with a protein absorption spectrum having a maximum between 277 and 280 mμ. Precipitation of this liquid with three volumes of 95% ethanol yielded a material

containing 20·5 % carbohydrate. This carbohydrate was made up of glucose, mannose and small amounts of fructose. The glycoprotein was homogeneous by electrophoresis at pH 8·6 in barbital buffer.

4. *Connective Tissue Glycoprotein:* A neutral mucopolysaccharide, probably associated with protein, has been histochemically identified in the connective tissue-muscle layer of the gut regions in the purple sea urchin *Strongylocentrotus purpuratus* (Holland and Nimitz, 1964).

ACOELOMATA AND MINOR ACOELOMATA

The biochemistry of the tissues and secretions of the acoelomate and minor acoelomate phyla has attracted the attention largely of parasitologists rather than chemists or biological chemists; this is of course a result of the emphasis on a parasitic mode of life which occurs in these groups. In consequence little or nothing is known from the biochemical point of view of the large majority of the free-living members of the various phyla. The occurrence of glycoproteins and related substances in parasitic acoelomates has been comprehensively reviewed by Von Brand (1966).

1. *Glycoproteins in Trematode Tissues and Secretions:* Schistosome cercaria secrete a mucin which appears to serve a variety of functions, protective, adhesive, developmental and antigenic. The mucin is secreted from three pairs of postacetabular glands opening at the rim of the oral sucker. The contents of the glands of *Schistosoma mansoni* and the mucin itself stain with lithium carmine, mucicarmine, orange G, aldehyde fuchsin, orcein, aniline blue, methylene blue at pH4, Evans' blue, Trypan blue, thionin, toluidine blue and periodic acid—Schiff. Acid hydrolysates of the mucin contain 2-amino-2-deoxy-D-glucose and 2-amino-2-deoxy-D-galactose together with arginine, glycine, proline, hydroxyproline, glutamic acid, alanine and phenylalanine in relatively high proportions. Lipid is also present (Stirewalt, 1959, 1963, 1965; Stirewalt and Evans, 1960). It seems probable that this mucin contains both glycoprotein and acid mucopolysaccharide.

The antigenic fraction of *Schistosoma mansoni* has been separated by electrophoresis and shown to contain, in addition to five proteins and a lipoprotein, five glycoproteins (Biguet *et al.*, 1962).

Mehlis' gland, a gland entering the uterus of *Fasciola hepatica* proximal to the region where the egg shell is formed and thought to play a role in shell formation, has been examined histochemically by Clegg (1965). These studies revealed the presence of granules containing a glycolipoprotein. The secretion was extracted with 0·02 M phosphate buffer at pH8 and subjected to ion-exchange chromatography on DEAE-Sephadex. Three protein fractions were obtained, two eluting from the column with 0·02M phosphate at pH8 and a third being eluted by 0·8M sodium chloride in the same buffer.

While all three fractions had glycoprotein properties, i.e. gave positive Folin and PAS reactions, only the third fraction contained Sudanophilic material, i.e. lipid. Cellulose acetate strip electrophoresis at pH 8·6 in veronal buffer ($\mu = 0.1$) indicated that the first two fractions contained five and three components respectively while the lipid-containing fraction was apparently homogeneous. The lipid could be removed from the latter fraction by extraction with chloroform-methanol (2 : 1) and was found to account for 39% of the dry weight of the glycolipoprotein while carbohydrate accounted for 1·0%, of which 0·2% was hexosamine. No sialic acid could be detected. Specific staining reactions showed that the glycolipoprotein formed a thin film or membrane on the inner and outer surfaces of the egg shell and perhaps accounted for the semipermeable properties of the shell.

Maekawa et al (1954) have purified and crystallized an antigenic substance from F. hepatica which is apparently largely protein but with a small proportion of carbohydrate.

The presence of glycoprotein in the cuticle of F. hepatica has been suggested on histochemical grounds (Lal and Shrivastava, 1960). The cuticle was positive to biuret and xanthoproteic tests and to PAS before and after chloroform extraction. A negative reaction to PAS was obtained however after acetylation, the staining property being restorable by treatment with 0·1 N potassium hydroxide. The cuticle did not exhibit γ metachromasia with toluidine blue or a positive staining reaction below pH4 with methylene blue.

Histochemical evidence has also suggested the presence of glycoproteins in metacercarial walls (Singh and Lewert, 1959; Bogitsh, 1962). However, hydrolysates of Posthodiplostomium minimum and its cyst wall yielded uronic acids, glucose, galactose and ribose, while hexosamines were found only in the parasite body (Lynch and Bogitsh, 1962).

2. Glycoproteins in Cestode Tissues and Secretions: The occurrence of glycoproteins in hydatid acoelomates has been reviewed by Smyth (1964). The antigens obtained from the tapeworm Taenia saginata, as well as the Nematode Ascaris lumbricus, were shown at an early date to contain protein, neutral sugar, and hexosamine (Kuzin et al., 1949). The hexosamine content of 1·5% was found to be made up of 2-amino-2-deoxy-D-glucose, while the neutral sugar was made up of glucose, The glucose content was high (70–80%) and a proportion of it was likely to have been glycogen. Von Brand et al. (1964) were able to isolate an alkali-stable polysaccharide from larval and adult Taenia taeniaformis. This material yielded 2-amino-2-deoxy-D-glucose and 2-acetamido-2-deoxy-D-glucose but no galactose, on acid hydrolysis. It was suggested that this material was a neutral mucopolysaccharide, perhaps glycoprotein.

Cmelik (1952) extracted the cyst wall of *Echinococcus granulosus* with water and precipitated the extract with trichloracetic acid. The protein-aceous precipitate was hydrolyzed with pepsin and trypsin. The "core" material remaining after proteolysis was a polysaccharide which contained hexose and a large amount of hexosamine. The polysaccharide was toxic causing leucocytosis in rabbits. Molar sodium chloride extracts of the cyst membranes (Cmelik and Briski, 1953) yielded a glycoprotein, containing 12% N and 28% reducing sugar, together with a second glycoprotein con-taining 3·6% N and 54·2% reducing sugar consisting of glucose, galactose and 2-amino-2-deoxy-D-glucose.

Agosin *et al.* (1957) found small amounts of polysaccharide, in the pres-ence of large quantities of protein, occurring in the hydatid cyst scoleces of *E. granulosus.* This material yielded galactose and 2-amino-2-deoxy-D-glucose on acid hydrolysis. Uronic acid was not present. Agosin (1959) isolated a mixture of polysaccharides from *Echinococcus* protoscoleces and fractionated it electrophoretically into two components. One of these was glycogen; the other yielded, on hydrolysis, galactose and 2-amino-deoxy-D-glucose. Kilejian *et al.* (1962) also isolated a polysaccharide material con-taining galactose and 2-amino-2-deoxy-D-glucose from the same source. It is not clear whether these polysaccharides originate as glycoprotein prosthetic groups or not.

3. Glycoproteins in Nematode Tissues and Secretions: The perivisceral fluid of *Ascaris* has been reported to contain glycoproteins (Giraldo-Cardona, 1960) and has been shown by Savel (1954) and Yasuo (1961) to contain some 50% of the total protein as albumin and 31·7% as a γ-globulin. Small amounts of α and β globulins were also present.

A glycoprotein was obtained by Fukushima from *Ascaris* hemolymph by trichloracetic acid extraction and ethanol precipitation (Fukushima, 1966). Digestion of the ethanol precipitate with trypsin, chymotrypsin and pepsin yielded an ethanol precipitable glycoprotein containing about 40% hexo-samine. The glycoprotein appeared, on the basis of the anthrone reactivity, to contain little or no hexose. Paper and column chromatography identi-fied the hexosamine as 2-amino-2-deoxy-D-glucose and 2-amino-2-deoxy-D-galactose; the ratio of total hexosamine to N-acetyl groups in the glyco-protein was 1 : 1. Not all of the hexosamine residues had equal lability to acid hydrolysis. About 40% of the total hexosamine was released into the free state by six hours hydrolysis in 3 N hydrochloric acid. This labile hexosamine contained the two amino sugars identified in approximately 1 : 1 ratio. The remaining hexosamine was contained in a peak which eluted far in advance of the positions of the free amino sugars on the ion-exchange column (Dowex 50W, H$^+$ eluted with 0·3 N HCl). This material was re-covered and a further more drastic hydrolysis (6 N hydrochloric acid for

six hours at 100°) revealed the presence of 2-amino-2-deoxy-D-glucose together with an unidentified reducing and ninhydrin positive component.
Material with a similar composition to this acid resistant "oligosaccharide"
was obtained after alkaline treatment (conditions not given) and ethanolic
precipitation of hemolymph. Perhaps this latter material and the acid
resistant material represent a prosthetic group linked to the protein by an
alkaline labile O-glycosidic link and distinct from a second prosthetic
group containing both 2-amino-2-deoxy-D-glucose and 2-amino-2-deoxy-
D-galactose and linked by a different and alkaline resistant bond.

The larvae of *Trichinella spiralis* contain large amounts of protein bound
to carbohydrate (Tanner and Gregory, 1961). There are at least 11 electrophoretically distinct saline-soluble antigens; four behaving like γ globulins,
one each like β_1 and β_2 globulins and albumin, and two like α_1 and α_2
globulins. Oliver-Gonzalez (1953) has noted the inhibition of blood agglu-
tinins by substances isolated from *Trichinella* and *Ascaris*.

The egg envelope of the nematode *Cruzia americana* has an outer protein
layer which contains a lipoglycoprotein. This material gives positive
staining reactions for protein and carbohydrate (PAS); it is insoluble in
water but soluble in picric acid, 1–3% potassium hydroxide and 10%
trypsin; and it swells in dilute acetic acid (Crites, 1958).

ANNELIDA

The compositions of the tubiculous mucin secretions of a variety of
annelid worms have been investigated chromatographically by Defretin
(1951 b, c, d; Defretin *et al.*, 1949). Glycoproteins were detected in the tubes
of *Myxicola infundibulum*, *Sabella pavonina*, *Lanice conchilega*, and *Pectinaria koreni*. The wide range of monosaccharides found in these materials
is shown in Table XI. Amino acids identified in hydrolysates, by paper
chromatography, included aspartic acid, glutamic acid, serine, glycine,
alanine, valine, methionine, leucine and phenylalanine: proline and basic
amino acids were tentatively identified as being present in small amounts.
Similar results were obtained for *Hyalinoecia tubicola* and *Spirographis
spallanzanii* tubes (Defretin, 1955). Levulose was reported to be a constituent of some of these mucins. An unusual constituent of *Sabella* tube was
lanthionine.

UROCHORDA

A water soluble protein has been isolated from the test of the ascidian
Cynthia roretzi and purified by repeated precipitation with acetic acid and
ethanol (Tsuchiya and Suzuki, 1962). This material was electrophoretically
homogeneous (paper, pH 8·6 veronal and pH 5·6 phosphate buffers). The
mobility towards the cathode was six times as great at pH 8·6 than at

TABLE XI
Monosaccharides Identified in Hydrolysates of Mucins
from Tubicolous Annelid Worms[a]

	Myxicola infundibulum	Sabella pavonina	Lanice conchilega	Pectinaria koreni
2-amino-2-deoxy-D-glucose	+	−	−	−
Glucuronic acid	+	−	−	−
Galactose	+	+	+	+
Glucose	+	+	+	+
Mannose	+	?	−	−
Ribose	+	−	−	+
Fucose	+	+	+	+
Arabinose	+	+	−	+
Fructose	+	−	−	+

[a] Defretin (1951d).

pH 5·6. Positive reactions were given to a variety of tests for both protein and carbohydrate. Weight was added to the suggestion that this might be a glycoprotein by its response to the action of a number of denaturation methods. Thus it was not precipitable from aqueous solution by heat coagulation and only incompletely precipitated by trichloracetic, phosphotungstic and tannic acids. It was however precipitated by lead acetate, sulphosalicylic acid, acetic acid, dilute hydrochloric acid, saturated ammonium chloride and ethanol. The glycoprotein contained 10·39% N and 16·28% hexosamine as 2-amino-2-deoxy-D-glucose. Acid hydrolysis and paper chromatography in two different solvent systems indicated the presence of the monosaccharides rhamnose, xylose, fructose, mannose, galactose, glucose in the approximate ratios 1 : 1 : 1 : 3 : 2 : 1 respectively. Amino acids identified were cystine, lysine, histidine, arginine, glutamic acid, aspartic acid, glycine, alanine, serine, threonine, valine, leucine, phenylalanine, tyrosine, proline, methionine and tryptophane.

COELENTERATA

A PAS positive, non-metachromatic, neutral mucopolysaccharide was reported by Goreau (1956) to be present in the epidermis, mesoglea and gastroderm of 32 species of coral.

INVERTEBRATE ENZYMES WHICH DEGRADE
GLYCOPROTEIN PROSTHETIC GROUPS

Complexity of structure in the oligosaccharide prosthetic groups of vertebrate glycoproteins has prompted a number of workers to search for selective methods for the stepwise degradation of these units. Obviously enzymic methods would be likely to be most effective. Surprisingly perhaps, invertebrates have proved to be a rich source of suitable enzymes and even have yielded enzymes capable of breaking the protein-carbohydrate linkages of vertebrate glycoproteins. The presence of such enzymes in invertebrates naturally suggests either a dietary function or, in many cases more probably, a function in the turnover of the animal's own tissues and secretions. Thus these enzymes imply the existence of invertebrate glyco-proteins which have closely similar structures to vertebrate materials.

Invertebrate β glucosaminidases were first noted by Karrer and Hofmann (1929; Karrer and von Francois, 1929) as being present in the hepato-pancreas of *Helix pomatia*. This enzyme seems more likely to be associated with the degradation of chitin rather than glycoprotein. Freudenberg and Eichel (1935), however, found that an enzyme preparation from *Helix* inactivated a blood group A substance, obtained from human urine, con-comitant with the release of 2-acetamido-2-deoxy-D-glucose and galactose and with a 25% increase in reducing power. Neuberger and Pitt-Rivers (1939) were able to separate β glucosidase and β glucosaminidase activities in *Helix* extracts and noted their activity towards ovalbumin. Similar activities were noted by Zechmeister and Toth (1939). Enzymes from *Helix laeda* hydrolysed hog gastric blood group substance (Utusi *et al.*, 1949; Nagaoka, 1949; Yoshizawa, 1951). An α galactosaminidase, obtained from the hepatopancreas of *H. pomatia*, when allowed to act on blood group A substances from ovarian cyst fluid and hog gastric mucosa resulted in almost complete loss of serological A specificity and considerable enhance-ment of H activity (Tuppy and Staudenbauer, 1966). Terminal 2-acetamido-2-deoxy-D-galactose units were found to be released.

An enzyme, also decomposing blood group A substance, was isolated from liver extracts of the whelk *Busycon canaliculatum* (Howe and Kabat, 1953). This enzyme preparation liberated fucose from blood group A and (OH) substances and increased their cross reactivity with horse anti-SXIV. Blood group activity was reduced but (OH) activity was unaffected. 2-acetamido-2-deoxy-D-glucose was liberated from A substance but not from O(H) substances. The action of these enzymes seemed to be con-fined to oligosaccharide side chains.

Enzyme preparations from the marine gastropod *Charonia lampas* have been found to have α-D-mannosidase, β-D-mannosidase and N-acetyl-glucosaminidase activities, releasing mannose and 2-acetamide-2-deoxy-D-glucose from ovalbumin glycopeptides (Muramatsu and Egami, 1965; Muramatsu, 1966, 1967).

Perhaps the most important and significant advance was that made by Bhargava and Gottschalk (1966) when they found O-seryl-N-acetylgalacto-saminide glycohydrolase activity in the digestive juice of *H. pomatia*. This enzyme was separated from associated N-acetylhexosaminidase activity by gel-filtration on Sephadex G-150 and G-200 and its activity estimated using neuraminic acid-free ovine submaxillary gland glycoprotein as substrate. Schauer and Gottschalk (1967, 1968) were able to obtain a pure preparation of this same enzyme from the tissues of the earthworm *Lumbricus terrestris*. The enzyme was purified by three successive applications on Sephadex G-200 columns giving an increase in specific activity by a factor of 190. The specific activity was more than three times higher than that of the same enzyme purified from ox spleen. The final preparation was free from N-acetyl hexosaminidase and proteolytic activities. These authors noted that this enzyme is usually accompanied by β-N-acetyl glucosaminidase in both vertebrate and invertebrate tissues.

A proportion of the glycosidase and glycosaminidase activities of these mulluscan preparations is undoubtedly concerned with the degradation of chitin, perhaps acid mucopolysaccharides and also dietary polysaccharides. The presence of enzymes in the organs of the vegetarians *Helix* and *Lumbricus*, however, capable of breaking the O-glycosidic protein-carbohydrate bond, must surely indicate the presence in invertebrates of glycoproteins containing this type of bond and serves to underline our relative ignorance of the distribution and structure of invertebrate glycoproteins.

REFERENCES

Agosin, M. (1959). *Biológica, Santiago*, **27**, 3.
Agosin, M., Von Brand, T., Rivera, F. and McMahon, P. (1957). *Expl. Parasit.* **6**, 37.
Anderson, J. M. (1950). *Biol. Bull. mar. biol. lab.*, *Woods Hole*, **99**, 49.
Arcardi, J. A. (1963). *Ann. N.Y. Acad. Sci.* **106**, 451.
Arcardi, J. A. (1965). *Ann. N.Y. Acad. Sci.* **118**, 987.
Arvy, L. (1963). *Ann. N.Y. Acad. Sci.* **106**, 472.
Ashhurst, D. E. (1961). *Q. Jl. Microsc. Sci.* **102**, 455.
Ashhurst, D. E. (1965). *Q. Jl. Microsc. Sci.* **106**, 61.
Ashhurst, D. E. and Richards, A. G. (1964a). *J. Morph.* **114**, 237.
Ashhurst, D. E. and Richards, A. G. (1964b). *J. Morph.* **114**, 247.
Bailey, K. (1958). *Biochem. J.* **69**, 46P.
Bailey, K. and Worboys, B. D. (1960). *Biochem. J.* **76**, 487.
Barrois, T. (1889). *Rev. Biol. Nord. du France*, **5**, 351.
Bassot, J. M. (1959). *C.r. hebd. Séanc. Acad. Sci. Paris*, **249**, 1267.
Bayne, C. J. (1966). *Comp. Biochem. Physiol.* **19**, 317.
Berkeley, C. (1923). *J. exp. Zool.* **37**, 477.
Bhargava, A. S. and Gottschalk. A. (1966). *Biochem. biophys. Res. Commun.* **24**, 280.
Biguet, J., Capron, A. and TranVanky, P. (1962). *Annls Inst. Pasteur, Paris*, **103**, 763.
Bogitsh, B. J. (1962). *J. Parasit.* **48**, 55.
Bonhag, P. F. (1956). *J. Morph.* **99**, 433.

Bonhag, P. F. and Arnold, W. J. (1961). *J. Morph.* **108**, 107.
Bronskill, J. F. (1960). *Can. J. Zool.* **38**, 769.
Carey, F. G. and Wyatt, G. R. (1960). *Biochim. Biophys. Acta*, **41**, 178.
Clegg, J. A. (1965). *Ann. N.Y. Acad. Sci.* **118**, 969.
Cmelik, S. (1952). *Biochem. Z.* **322**, 456.
Cmelik, S. and Briski, B. (1953). *Biochem. Z.* **324**, 104.
Coupin, H. (1900). *C.r.hebd. Séanc. Acad. Sci., Paris*, **130**, 1214.
Crites, J. L. (1958). *Ohio St. J. Sci.* **58**, 343.
Day, M. F. (1949). *Aust. J. scient. Res.* **B2**, 421.
Defretin, R. (1951b). *C.r. hebd. Séanc. Acad. Sci., Paris*, **232**, 888.
Defretin, R. (1951a) *C.r. hebd. Séanc. Acad. Sci., Paris*, **233**, 103.
Defretin, R. (1951c). *C.r. Séanc. Soc. Biol.* **145**, 115.
Defretin, R. (1951d). *C.r. Séanc. Soc. Biol.* **145**, 117.
Defretin, R. (1955). *C.r. Congr. Socs Sav., Paris, Sect. Sci.* 167.
Defretin, R., Biserte, G. and Montreuil, J. (1949). *C.r. Séanc. Soc. Biol.* **143**, 1208.
Doyle, J. (1964). *Biochem. J.* **91**, 6P.
Doyle, J. (1965). *Biochem. J.* **96**, 73P.
Doyle, J. (1966). *In* "Some Contemporary Studies in Marine Science" (H. Barnes, ed.). Allen and Unwin, London.
Doyle, J. (1967). *Biochem. J.* **103**, 41P.
Edwards, J. S. (1966). *J. Insect Physiol.* **12**, 31.
Eichwald, E. (1865). *Justus Liebig's Annln chem.* **134**, 177.
Fish, G. R. (1955). *Nature, Lond.* **175**, 733.
Frentz, R. (1958). *C.r. hebd. Séanc. Acad. Sci. Paris*, **247**, 2470.
Freudenberg, K. and Eichel, H. (1935). *Justus Liebig's Annln Chem.* **518**, 97.
Fukushima, T. (1966). *Expl. Parasit.* **19**, 227.
Ganguly, D. N. and Basu, B. D. (1962). *Acta histochem.* **13**, 31.
Ganguly, D. N. and Nanda, D. K. (1964). *Acta histochem.* **18**, 251.
Geldmacher-Mallinkrodt, M. and May, F. (1957). *Hoppe-Seyler's Z. physiol. Chem.* **307**, 179.
Geldmacher-Mallinkrodt, M. and Horstmann, H. J. (1963). *Hoppe-Seyler's Z. physiol. Chem.* **333**, 226.
Geldmacher-Mallinkrodt, M. and Traexler, G. (1961). *Hoppe-Seyler's Z. Physiol. Chem.* **325**, 116.
Gennaro, J. F., Lorincz, E. A. and Brewster, H. B. (1965). *Ann. N.Y. Acad. Sci.* **118**, 1021.
George, J. C. (1952). *Biol. Bull. mar. biol. lab., Woods Hole*, **102**, 118.
George, J. C. and Jura, C. (1958). *Proc. K. ned. Akad. Wet.* **C61**, 598.
Ghirett, F. (1960). *Ann N.Y. Acad. Sci.* **90**, 726.
Giraldo-Cardona, A. J. (1960). *Rev a ibér. Parasit.* **20**, 425.
Goreau, T. (1956). *Nature, Lond.* **177**, 1029.
Got, R. and Marnay, A. (1968). *Eur. J. Biochem.* **4**, 240.
Gottschalk, A. and Graham, E. R. B. (1966). *In* "The Proteins" (H. Neurath, ed.). Vol. 4, p. 96. Academic Press, New York and London.
Hammarsten, O. (1885). *Pflüger's Archs ges. Physiol.* **36**, 373.
Hashimoto, Y. and Sato, T. (1955). *Bull. Jap. Soc. Scient. Fish.* **21**, 352; *Chem. Abstr.* **50**, 10934d.
Hashimoto, Y. and Sato, T. (1956). *Nippon Suisan Gakkaishi.* **22**, 417; *Chem. Abstr.* **52**, 8391c.

Hashimoto, Y., Sato, T. and Miyamoto, K. (1954). *Bull. Jap. Soc. Scient. Fish.* **20**, 406; *Chem. Abstr.* **49**, 7138c.

Hinton, H. E. (1961). *J. Insect Physiol.* **7**, 224.

Holland, N. D. and Nimitz, A. (1964). *Biol. Bull. mar. biol. lab., Woods Hole,* **127**, 280.

Horstmann, H. J. (1964). *Biochem. Z.* **340**, 458

Howe, C and Kabat, E. A. (1953). *J. Am. chem. Soc.* **75**, 5542.

Hozumi, M. (1959). *Sci. Rep. Tokyo, Bunrika Daig.* **9**, 37.

Hunt, S. and Jevons, F. R. (1963). *Biochem. biophys. Acta*, **78**, 376.

Hunt, S. and Jevons, F. R. (1965). *Biochem. J.* **97**, 701.

Iida, K. (1963). *J. Biochem.* **54**, 181.

Inoue, T. (1960). *Tokyo Gakugei Daigaku Kenkyu Hokoku*, 3-bu, *Shizen Kagaku.* **11**, 27; (1964) *Chem. Abstr.* **60**, 8380h.

Johansson, J. (1945). *Ark. Zool.* **36**, No. 13, 1.

Karrer, P. and Hofmann, A. (1929). *Helv. chim. Acta*, **12**, 66.

Karrer, P. and von Francois, G. (1929). *Helv. chim. Acta*, **12**, 986.

Kato, K. I., Perkowska, E. and Sirlin, J. L. (1963). *J. Histochem. Cytochem.* **11**, 485.

Kilejian, A., Sauer, K. and Schwabe, C. W. (1962). *Expl. parasit.* **12**, 377.

Kobayashi, S. (1964). *Nippon Suisan Gahkaishi*, **30**, 325; *Chem. Abstr.* **64**, 2467g.

Kodani, M. (1948). *Proc. natn Acad. Sci. U.S.A.* **34**, 131.

Kuzin, A. M., Babadzhanov, S. N. and Polyakora, O. I. (1949). *Biokhimiya* **14**, 65.

Lal, M. B. and Shrivastava, S. C. (1960). *Experientia*, **16**, 185.

Levene, P. A. (1925). *J. biol. Chem.* **65**, 683.

Linder, H. J. (1956). *J. Morph.* **99**, 575.

Lipke, H. (1960). *11th Int. Congr. Ent., Vienna.* (see Carey and Wyatt, 1960).

Lipke, H., Grainger, M. M. and Siakotos, A. N. (1965a). *J. biol. Chem.* **240**, 594.

Lipke, H., Graves, B. and Leto, S. (1965b). *J. biol. Chem.* **240**, 601.

Lynch, D. L. and Bogitsh, B. J. (1962). *J. Parasit.* **48**, 241.

Mackintosh, N. A. (1925). *Q. Jl. microsc. Sci.* **69**, 317.

Maekawa, K., Kitazawa, K. and Kushiba, M. (1954). *C.r. Séanc. Soc. Biol.* **148**, 763.

McFarlane, J. E. (1962). *Can. J. Zool.* **40**, 553.

McMahon, P., Von Brand, T. and Nolan, M. O. (1957). *J. cell. comp. Physiol.* **50**, 219.

Morrill, J. B., Norris, E. and Smith, S. D. (1964). *Acta Embyrol. Morph. exp.* **7**, 155.

Messina, L. (1957). *Publ. Staz. zool. Napoli*, **30**, 177.

Mitra, S. B. (1901). *Q. Jl. microsc. Sci.* **44**, 591.

Monroy, A. and Vittorelli, M. L. (1960). *Experientia*, **16**, 56.

Moret, V., Serafini-Fracassini, A. and Gotte, L. (1964). *J. Atheroscler, Res.* **4**, 184.

Muramatsu, T. (1965). *Archs Biochem. Biophys.* **115**, 427.

Muramatsu, T. (1967). *J. Biochem.* **62**, 487.

Muramatsu, T. and Egami, F. (1965). *Jap. J. exp. Med.* **35**, 171.

Nair, N. B. (1955). *Curr. Sci.* **24**, 126.

Nagaoka, T. (1949). *Tohoku J. exp. Med.* **51**, 137.

Nagy, I. Z. and Laszlo, M. B. (1964). *Acta biol. hung.* **15**, 39.

Neuberger, A. and Pitt-Rivers, R. V. (1939). *Biochem. J.* **33**, 1580.

Newell, B. S. (1953). *J. mar. biol. Ass. U.K.* **32**, 491.
Nicol, J. A. C. (1960). *J. mar. biol. Ass. U.K.* **39**, 109.
Oliver-Gonzalez, J. (1953). *Proc. Soc. exp. Biol. Med.* **82**, 559.
Perkowska, E. (1963). *Expl. Cell. Res.* **32**, 259.
Pipa, R. L. and Cook, E. F. (1958). *J. Morph.* **103**, 353.
Riley, R. C. and Forgash, A. J. (1967). *J. Insect. Physiol.* **13**, 509.
Roche, J. and Dumazert, C. (1940). *Annls Inst. océanogr., Paris*, **20**, 87.
Rosenkranz, H. S. (1967). *Can. J. Biochem.* **45**, 281.
Rosenkranz, H. S. and Carden, G. A. (1967). *Can. J. Biochem.* **45**, 267.
Salkeld, E. H. (1960). *Can. J. Zool.* **38**, 449.
Savel, J. (1954). Thesis, University of Paris, France.
Schauer, H. and Gottschalk, A. (1967). *Z. Naturf.* **226**, 1030.
Schauer, H. and Gottschalk, A. (1968). *Biochim. biophys. Acta*, **156**, 304.
Shashoua, V. E. and Kwart, H. (1959). *J. Am. chem. Soc.* **81**, 2899.
Siakotos, A. N. (1960a). *J. gen. Physiol.* **43**, 999.
Siakotos, A. N. (1960b). *J. gen. Physiol.* **43**, 1015.
Singh, K. S. and Lewert, R. M. (1959). *J. infect. Dis.* **104**, 138.
Sissakian, N. M. and Veinora, M. K. (1958). *Biokhimiya*, **23**, 52.
Smith, B. J. (1965). *Ann. N.Y. Acad. Sci.* **118**, 997.
Smyth, J. D. (1964). *In* "Advances in Parasitology". (B. Dawes, ed.). Vol. 2, p. 169. Academic Press, New York and London.
Springer, G. F., Rose, C. S. and György, P. (1954). *J. Lab. clin. Med.* **43**, 532.
Stirewalt, M. A. (1959). *Expl. Parasit.* **8**, 199.
Stirewalt, M. A. (1963). *Ann. N.Y. Acad. Sci.* **113**, 36.
Stirewalt, M. A. (1963). *Ann. N.Y. Acad. Sci.* **113**, 36.
Stirewalt, M. A. (1965). *Ann. N.Y. Acad. Sci.* **118**, 966.
Stirewalt, M. A. and Evans, A. S. (1960). *Expl. Parasit.* **10**, 75.
Tanner, C. E. and Gregory, J. (1961). *Can. J. Microbiol.* **7**, 473.
Tsuchiya, Y. and Suzuki, Y. (1962). *Bull. Jap. Soc. Science Fish.* **28**, 227.
Tuppy, H. and Staudenbauer, W. L. (1966). *Biochemistry, N.Y.* **5**, 1742.
Utusi, M., Huzi, S., Matumoto, S. and Nagoaka, T. (1949). *Tohoku J. exp. Med.* **50**, 175.
Von Brand, T. (1966). "Biochemistry of Parasites". Academic Press, New York and London.
Von Brand, T., McMahon, P., Gibbs, E. and Higgins, H. (1964). *Expl. Parasit.* **15**, 410.
Wheat, R. W. (1960). *Science, N.Y.* **132**, 1310.
Wilson, B. R. (1960a). *Ann. ent. Soc. Am.* **53**, 170.
Wilson, B. R. (1960b). *Ann. ent. Soc. Am.* **53**, 732.
Yamaguchi, M., Nagase, K., Yago, N. and Takatsuki, S. (1959). *Dobutsugaku Zasshi*, **68**, 390; (1961) *Chem. Abstr.* **55**, 22632i.
Yasuo, S. (1961). *Folio pharmac. Jap.* **57**, 393.
Yonge, C. M. (1926). *J. mar. biol. Ass. U.K.* **14**, 295.
Yoshizawa, Z. (1951). *Tohoku J. exp. Med.* **55**, 35.
Zechmeister, L. and Toth, G. (1939). *Naturwissenschaften*, **27**, 367,

CHAPTER 13

Sialic Acid

INTRODUCTION

The sialic acids or neuraminic acids occur widely in glycoproteins, glycopeptides and glycolipids throughout the vertebrate groups, but their distribution in the invertebrata is more sporadic and uncertain. This monosaccharide appears to be of immense, perhaps fundamental, importance to vertebrates. In view of this, its apparent total absence from the majority of invertebrate orders is surprising. Neuraminic acids are encountered here and there in certain invertebrate groups, however, and it is perhaps important that we should consider the situation in invertebrates in more detail.

Sialic acids were first isolated as crystalline substances by Blix (1936) from the mucin of the bovine submaxillary gland and later by Klenk (1941) from brain tissues (hence the name neuraminic acid). The basic structure, upon which a variety of substitutions may take place, is indicated in Fig. 1.

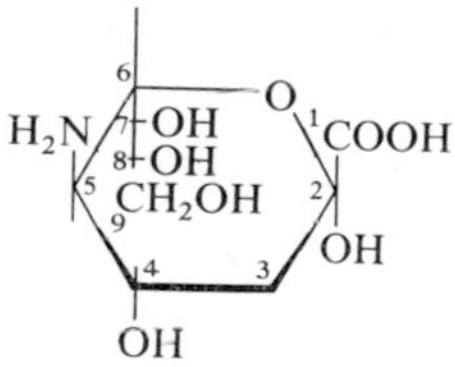

Fig. 1. Neuraminic acid, the backbone of the sialic acids.

This structure has been established largely as a result of the efforts of Alfred Gottschalk (see Gottschalk, 1960). Consideration of the structural formula will indicate that the monosaccharide, neuraminic acid, is in fact a condensation product of the uncommon amino sugar 2-amino-2-deoxy-D-mannose and pyruvic acid where carbon atoms 4 to 9 are derived from 2-amino-2-deoxy-D-mannose. It is curious perhaps that this is the only form in which 2-amino-2-deoxy-D-mannose or in the commonest form of sialic acid, 2-acetamido-2-deoxy-D-mannose occurs in the living animal, and in fact, as far as our present knowledge is concerned, in nature. It does not appear to form a constituent of glycoproteins or mucopolysaccharides in its own right as do 2-acetamido-2-deoxy-D-glucose and 2-acetamido-2-deoxy-D-galactose. An active nucleotide derivative of 2-acetamido-2-deoxy-D-mannose is not known; perhaps this therefore is the significant factor. Activation for inclusion of sialic acids in glycoproteins is via the CMP derivative. Mannose itself is usually activated for synthetic purposes as the GDP derivative. The formation of the nuceleotide monophosphate derivative is in itself singular; all the other known nucleoside sugars are diphosphate derivatives.

The vertebrates yield five of the six known naturally occurring neuraminic acids; N-acetyl, N-glycolyl, N-acetyl-4-O-acetyl, N-acetyl-7-O-acetyl, and N-acetyl-7,8(or 9)-di-O-acetyl (Fig. 2). Probably more occur. The

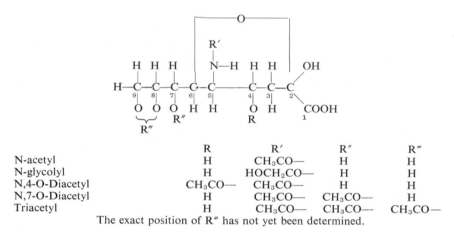

	R	R'	R''	R'''
N-acetyl	H	CH_3CO-	H	H
N-glycolyl	H	$HOCH_2CO-$	H	H
N,4-O-Diacetyl	CH_3CO-	CH_3CO-	H	H
N,7-O-Diacetyl	H	CH_3CO-	CH_3CO-	H
Triacetyl	H	CH_3CO-	CH_3CO-	CH_3CO-

The exact position of R''' has not yet been determined.

Fig. 2. The five natural neuraminic acid derivatives found in vertebrates.

parent acid, neuraminic acid or 5-amino-3:5-dideoxy-D-erythro-L-gulu-nonulosonic acid, is not known in the natural state. The term sialic acid is a general one relating to any of the derivatives of the parent molecule. Sialic or neuraminic are generic names which have found a more universal acceptance than the more accurate nonulosonic acid.

In the vertebrate glycoproteins sialic acid usually occurs ketosidically linked through position 2 to the 6 position of a 2-acetamido-2-deoxy-D-hexose residue (Fig. 3) and probably always occupies a terminal position at the non-reducing end of the carbohydrate chain. This latter deduction follows from the observation that sialic acid is usually the only monosaccharide to be released on the treatment of glycoproteins with neuraminidase or the first to be released on mild acid hydrolysis. Linkages of the $2 \rightarrow 4$ neuraminosyl-acetylhexosamine type have also been postulated on the basis of an unusually high resistance to acid hydrolysis of certain sialic

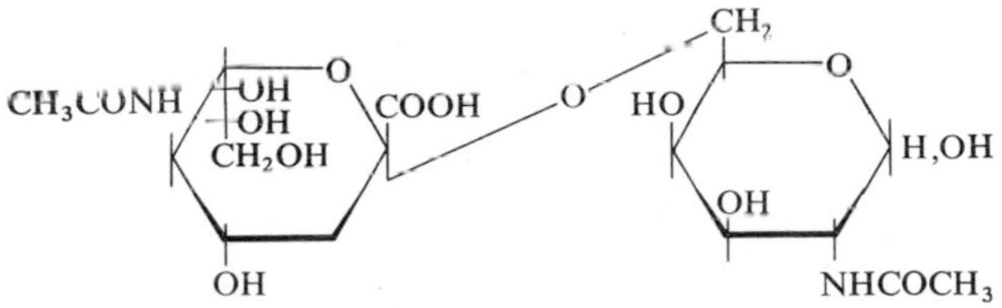

Fig. 3. Sialic acid-carbohydrate link of the type found in vertebrate glycoproteins.

acid residues in some glycoproteins. In the milk oligosaccharide, N-acetyl-neuraminlactose, the linkage between the two monosaccharides is 2 → 3, to the galactose.

Since the sialic acids are strongly acidic, with pK's in the region of 2·6, they give a pronounced polyanionic character to the glycoprotein molecules in which they occur. Where the prosthetic groups are numerously and rather evenly distributed along the protein chain they confer highly-viscous solution properties. This is a direct result of mutual electrostatic repulsion, by the neuraminic acid residues, imparting a highly-extended configuration to the glycoprotein molecule. In the vertebrates the sialoglycoproteins do frequently fulfil important physiological roles in which viscosity plays a key part, e.g. lubrication. It seems likely however that sialic acids have more fundamental functions to fulfil. The surface properties of vertebrate cell membranes, in particular their charge, is probably an important function of their sialic acid residues. This may have far-reaching implications. A role in calcification has been proposed for the peculiarly arranged sialic acid residues of bone glycoprotein. The whole subject is really rather open at the present time.

Apart from glycoproteins and cell wall glycopeptides, sialic acids occur as constituents of a wide variety of vertebrate glycolipids derived from nervous tissues and cell membranes. Here too the sialic acid occupies a terminal position on the carbohydrate moiety. Sialic acids are important constituents of blood group substances and milk oligosaccharides (Stacey and Barker, 1962).

INVERTEBRATE SIALIC ACIDS

The role of macromolecules in elevating the viscosity of physiological fluids or providing lubrication at epithelial surfaces in the invertebrates seems to be largely mediated by acid mucopolysaccharides; that is to say sulphated or carboxylated linear polysaccharides rather than the heavily sialic acid-substituted glycoproteins found ubiquitously in the vertebrates. With regard to the occurrence of sialic acids in cell membrane constituents, nervous system glycolipids and glycoproteins other than those of epithelial origin, little is known.

The presence of sialic acids in invertebrates was a fact not firmly established at all until 1959 when Perlmann and his associates (Perlmann et al., 1959) noted the occurrence of a sialic acid in the gametes of several species of Mediterranean sea urchin. Identification of the sialic acid as such was based on the production of positive reactions with the direct Ehrlich and Bial's orcinol reagents and by chromatography on columns of ion-exchange resin. Approximately half of the sialic acid found in *Paracentrotus lividus* gametes was localized in a particular fraction. The level of sialic acid detected in the eggs rose during embryological development.

In a later similar study Warren (Warren and Hathaway, 1960; Warren et al., 1960) was also able to identify sialic acids in both the eggs and semen of echinoids. *Arbacia punctulata* eggs each contained $3\cdot8 \times 10^{-4}$ μg bound sialic acid capable of being completely released within two hours by $0\cdot1$ N sulphuric acid at $80°$. None of this sialic acid gave a positive thiobarbituric acid reaction ("Warren's test") before hydrolysis and hence was all in a bound form. None of the sialic acid seemed to be localized in the jelly coat of the egg. At least two types of sialo-substance were present however. One complex was soluble in n-butanol leaving a remaining 75% of an insoluble form. This lipid-sialic acid complex was soluble in a chloroform methanol mixture (2 : 1v/v), and also in ether, suggesting in the latter case that the carboxyl of the sialic acid residue might conceivably be blocked. Free sialic acids are insoluble in ether. It seems more probable however that the solubility in ether would be a function of a more dominant lipid moiety. *Vibrio cholerae* neuraminidase released the sialic acid from the complex and hence it is likely that the link between the sialic acid and the rest of the complex is of the ketosidic type.

Chromatography, on Dowex-1-acetate, of the enzymatically released sialic acid yielded a pure product with all the properties, chemical and chromatographic, of authentic N-glycolylneuraminic acid. The remaining sialic acid of the egg was not characterized.

Sialic acid was present in the bound form in both the seminal plasma and sperm of *Arbacia*. The plasma contained two forms, one readily hydrolysable and a more firmly bound type. The former was released within 30 minutes by an $80°$ treatment with $0\cdot1$ N sulphuric acid and constituted 67% of the total sialic acid; the remaining sialic acid required four hours treatment under these conditions for complete release. Possibly the difference lies in the nature of the bond to other residues in the complex. Total release of sialic acid by *Vibrio cholerae* neuraminidase could not be achieved either; in the one experiment reported, 50% of the total sialic acid residues were cleaved from the complex. Acid-resistant sialic acid residues have been reported to occur in vertebrate glycoproteins. Single *Arbacia* sperm contained about $9 \cdot 5 \times 10^{-6}$ µg sialic acid. When sea water-washed sperm were treated with *Arbacia* fertilizin, a non-dialysable soluble sialic acid containing complex was released from them. The process takes under one minute. Addition of further fertilizin will cause the release of further sialic acid, but even so only about 20% of the total sialic acid seems to be available for release by this means. Heated fertilizin (80° for five minutes) is also active; this might be expected since the fertilization substances are largely acid mucopolysaccharide and hence not liable to denaturation. The actual basis of the process is unclear.

Sialic acid-containing substances were also released from sperm by treatment with detergents; both cationic (sodium lauryl sulphate) and anionic (cetyltrimethylammonium bromide) were effective. Eighty-nine per cent of the sperm sialic acid could be released by four successive treatments with $7 \cdot 5 \times 10^{-4}$ M sodium lauryl sulphate in sea water and surprisingly the sperm were still mobile even after the second extraction when $73 \cdot 8\%$ of the total sialic acid had been removed. Detergent extraction caused a 25% loss in weight of the sperm. It was also noticed that a gradual release of sialic acid from sperm takes place when they are suspended in distilled water or more rapidly when they are incubated at 45° or higher for five minutes.

The non-dialysable material released by fertilization treatment also seems to contain the acid stable sialic acid; four hours incubation with $0 \cdot 1$ N acid at 80° was required for release. Moreover, five per cent of the sialic acid in this material was available for release by neuraminidase. Chemical and paper chromatographic tests of the sialic acid released by acid and purified by ion-exchange chromatography confirmed its identity with N-glycolylneuraminic acid. Hence this derivative occurs in both the eggs and sperm of *Arbacia*.

A hitherto unknown sialic acid was isolated by Warren (1964) from the tissues of the asteroid (starfish) *Asterias forbesi*. Homogenized viscera were incubated at 80°, in water brought to pH 1 with sulphuric acid, for one hour. The solution was clarified by centrifugation (8000 g) and the residue re-extracted as before. The combined supernatants were then precipitated with ethanol (final concentration 20%) and the resultant residue discarded. The solution remaining was neutralized (Ba(OH)$_2$), filtered, and the filtrate plus filtrate washings concentrated *in vacuo*. At this stage an orcinol assay indicated that the viscera of 46 starfish had yielded $2 \cdot 27$ g sialic acid. The preparation was now further purified by chromatography on Dowex-1-acetate, eluting with 2 M acetic acid and then by preparative paper chromatography (70 : 30 ethanol-water, Whatman No. 17). This yielded a sialic acid preparation, which while chromatographically pure, could not be crystallized. The Rf values of this material were close to, but distinct from, N-acetyl and N-glycolyl neuraminic acids. Also the absorption spectra of the coloured chromogens produced in the orcinol, direct Ehrlich, resorcinol, diphenylamine, and thiobarbituric acid reactions, were identical with those produced by the two latter sialic acids.

An elemental analysis suggested an empirical formula $C_{12}H_{21}NO_{10} \cdot H_2O$ and a molecular weight of 357. The Zeizel method indicated the presence of one methyl group (0·94 of a molar equivalent of methoxyl groups). The C/O ratio was 0·87 and a yield of 1·04 moles of glycolic acid per mole of sialic acid was obtained. No phosphate or sulphate residues could be detected. The sialic acid was reducible at pH 7·0 by sodium borohydride. Substitution at hydroxyl 8 seemed to be indicated by the failure of the molecule to consume periodate. (N-acetyl and N-glycolyl neuraminic acids both consume two equivalents of periodic acid per mole). Ester substitution could not be detected. The most probable structure, therefore, is that shown in Fig. 4; N-glycolyl-8-O-methylneuraminic acid.

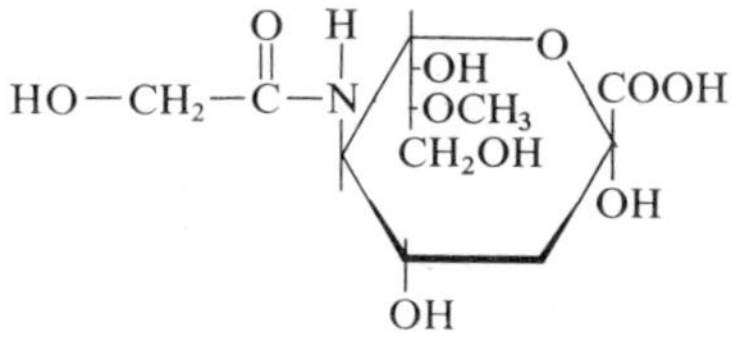

Fig. 4. N-glycolyl-8-O-methylneuraminic acid from *Asterias forbesi*.

The sialic acid which occurred *in vivo* in a bound form, i.e. non-dialysable was insusceptible to release either by neuraminidase from *V. cholerae* or *Clostridium perfringens*. Whether this insensitivity of the bond between the sialic acid and the rest of the complex (whatever this might be) reflects a link of different character to the more common ketosidic link, or whether the substitution by a methyl group at position 8 is instrumental in negating the enzymes specificity for sialic acid residues, is not clear. However, in view of the relative ease of acid cleavage, the latter possibility might be considered more attractive. The unbound form of the sialic acid was not degraded by treatment with *C. perfringens* N-acetyl-neuraminate-pyruvate-lyase (E.C.4.1.3.3.).

A comparative study of the distribution of sialic acids in nature has been made by Warren (1963). In general either whole organisms or tissues were homogenized, brought to pH 1 with sulphuric acid and incubated at 80° for one hour. Cell debris was removed by centrifugation, sialic acid adsorbed from the supernatants on Dowex-1-acetate and then eluted with 1 M ammonium acetate, pH 4·8 (Svennerholm, 1958). Sialic acids in the eluates were quantitated by the thiobarbituric acid, orcinol and direct Ehrlich assay.

Apart from the vertebrates, where sialic acids seem to occur universally from the fish through to the mammals, forms of sialic acid were detected in cephalochordates, hemichordates, echinoderms, crustaceans, molluscs and platyhelminths, although not in all the species examined in the latter three groups (Table I). In several instances where sialic acids were identified, larger quantities were isolated by preparative scale ion-exchange, chromatography and paper chromatography. These materials were examined for their sensitivity to N-acetyl-neuraminate-pyruvate-lyase (Table II). With the exception of a sialic acid isolated from the acoel

TABLE I
List of Invertebrate Animals in which Sialic Acid has been detected[a]

Phylum	Sub-group	Animal	Organ	Sialic acid content mg/g
Platyhelminth	Turbellaria	*Polychoerus carmelensis*	Whole animal	0·22
	Trematoda	*Fascioloides magna*	Whole animal	0·043
Mollusca	Gastropoda	*Charonia lampas*[b]	Digestive gland	—
	Cephalopoda	*Loligo pealii*	Digestive gland	0·38
Arthropoda	Crustacea	*Homarus americanus*	Digestive gland	0·046
Echinoderma	Asteroidea	*Asterias forbesi*	Viscera	0·05
		Asterias vulgaris	Viscera	0·59
		Henricia sanguinolenta	Viscera	0·08
	Ophiuroidea	*Ophioderma brevispinum*	Whole animal	0·02
		Unidentified	Whole animal	0·02
	Echinoidea	*Strongylocentrotus droebachiensis*	Viscera	not detected
			Testes	not detected
		Arbacia punctulata	Eggs, sperm, Viscera	not detected
		Echinarachnius parma	Viscera	0·04
	Holothuroidea	*Thyone briareus*	Viscera	0·01
		Synapta inhaerens	Viscera	0·14
	Crinoidea	*Nemaster* sp.	Viscera	0·03
Chordata	Hemichorda	*Dolichoglossus kowalevskii*	Whole animal	0·14
	Cephalochorda	*Amphioxus*	Whole animal	0·58

[a] Warren (1963). [b] Inoue (1965).

TABLE II
Susceptibility of Invertebrate Sialic Acids to
N-acetyl-neuraminate-pyruvate-lyase[a]

Phylum	Class	Animal	Presence of glycolyl group	Degradation by enzyme
Echinoderma	Asteroidea	*Asterias forbesi*	+	−
	Holothuroidea	*Thyone briareus*	+	±
	Ophiuroidea	*Ophioderma brevispinum*	+	+
	Echinoidea	*Arbacia punculata*	+	+
	Crinoidea	*Nemaster* sp.	+	+
Cephalochorda		*Amphioxus*	+	+
Hemichorda		*Dolichoglossus kowalevskii*	+	+
Mollusca	Cephalopoda	*Loligo pealii*	+	+
Arthropoda	Crustacea	*Homarus americanus*	+	+
Platyhelminthes	Turbellaria	*Polychoerus carmelensis*	−	+

[a] Warren (1963).

turbellarian *Polychoerus*, all of the invertebrate sialic acids seem to be of the N-glycolyl type.

Sialic acids were not detected in the phyla Prosopygia, Ctenophora, Coelenterata, Porifera, Annelida, Sipuncloidea, Nemertea and Chaetognatha nor in the sub-groups Cestoda, Nematoda, Amphineura, Gastropoda, Pelecypoda, Arachnoidea, Insecta and Urochorda.

One of the difficulties inherent in this type of study is that one cannot be absolutely certain in many cases that the source of the material isolated is actually the tissues or secretions of the organism itself; thus in several instances here the sialic acid might have had an exogenous origin. Caution is obviously required. Warren (1963) notes that the sialic acid found in *Fascioloides magna* might have originated in blood taken into the digestive system from the vertebrate host but that the sialic acid found in *P. carmelensis* was probably of endogenous origin since the food is plant material; he had failed to detect sialic acids in any of the plant phyla. However, sialic acids have since been detected in higher plants (Mayer *et al.*, 1964) and in green algae (Correll, 1964); the algal derivative was probably N-acetyl and that of the higher plant N-glycolyl.

In the case of both the cephalod and the crustacean sialic acids, the origins are also doubtful. Only the digestive glands contain sialic acids; no other tissue or secretion showed any trace of this. Moreover, it is rather disturbing to note that the sialic acids from both sources were already largely, if not entirely, in the free state. We must therefore be careful to ask ourselves whether or not the sialic acids could have originated in the food. In the crustacea it would be quite conceivable that the partially-digested food might be found in the digestive glands, and indeed if this were so, the free state of the material might be readily accounted for by the neuraminidase which Warren detected in the gland. In the case of the cephalopod the presence of food in the digestive gland seems to be a more doubtful circumstance; much depends upon what exactly Warren means by the term "digestive gland". Cephalopod digestive glands are in fact divided into a solid bi-lobed glandular liver, which is the principal storage organ for food reserves and which probably only receives these via the blood-stream, and a pancreas which secretes digestive enzymes via a duct into the caecum. In all probability all food is absorbed via the wall of the alimentary canal; the pancreas is however partially excretory. On balance one would consider that the sialic acid in the cephalopod gland might be endogenous were it not for the fact that 85% of it is in the free state.

Sialic acids are clearly present in the echinoderm phylum, however, and Warren has suggested that they are probably a general and primitive characteristic of the group. The importance (quantitatively) of sialic acid in the phyllum, in comparison to other invertebrate groups, is of particular interest when one considers that affinities between the echinoderms and chordates have long been suggested on morphological, embryological and other grounds. Of interest also is the detection of sialic acids in two of the protochordate subphyla and it is perhaps rather surprising that the third subphylum, the urochordates, where chordate affinities have always been considered to be particularly strong, should have failed to yield any detectable sialic acid.

While Warren's identification of an endogenous sialic acid in a cephalopod mollusc might be considered somewhat equivocal, the presence of N-glycolyl neuraminic acid in a gastropod would seem to have been firmly established by Inoue (1965). The sulphated sialic acid-containing heteropolysaccharides present in the digestive gland of *Charonia lampas* have already been described in the chapter on polysaccharide sulphates. Inoue has rejected the idea that these materials might have been of exogenous origin, mainly on the grounds that they could only be detected in the glands themselves and not in the digestive tract.

The sialic acid is released by mild acid hydrolysis to give a coloured chromophore in the thiobarbituric acid assay identical with that given by authentic sialic acids and co-chromatographs in two separate solvent systems with N-glycolylneuraminic acid.

The sialic acid is all released within one hour by 0·1 N sulphuric acid at 80° and 100% release seems to coincide with 20–25% appearance of the total reducing power of the polysaccharide. The molecule probably therefore is situated in a terminal position, ketosidically linked to another monosaccharide.

The N-glycolyl neuraminic acid is accompanied in the polysaccharide by 2-amino-2-deoxy-D-glucose and 2-amino-2-deoxy-D-galactose as well as fucose, mannose, galactose and glucose. This is the first firm indication we have that invertebrate sialic acids are linked to oligosaccharide chains of similar types to those found in the vertebrate sialoglycoproteins.

BIOSYNTHESIS

Vertebrates synthesize their sialic acids by a complex pathway (Fig. 5), culminating in the case of N-acetylneuraminic acid in the condensation of a molecule of 2-acetamido-2-deoxy-D-mannose-6-P and a molecule of phosphoenolpyruvate to give one molecule of N-acetylneuraminate-9-phosphate; the phosphate is subsequently removed by a specific phosphatase.

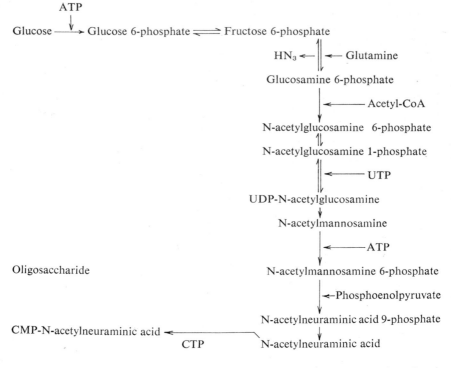

Fig. 5. Simplified form of the biosynthetic pathway leading to the formation of activated N-acetylneuraminic acid.

The neuraminic acids are peculiar in that they are activated in the verte-brates for polysaccharide synthesis not as nucleotide diphosphate deriva-tives but as the derivatives of cytidine monophosphate. Details of the path-ways leading to the biosynthesis of neuraminic acids and the prosthetic groups of glycoproteins will be found in Kent and Draper (1968) and in Kent and Allen (1968). The source of the N-glycolyl groups of mammalian N-glycolylneuraminic acids has not been established; a possible precursor is 2-(1,2-dihroxyethyl) thiamine pyrophosphate which could be formed from hydroxypyruvate or from fructose. Serine would yield hydroxypyruvate by deamination.

Relatively little is known of the mode of sialic acid formation in the invertebrates. Warren (1963) noted a net synthesis of sialic acid from 2-acetamido-2-deoxy-D-mannose-6-phosphate and phosphoenolpyruvate in cell-free extracts of the ovaries and viscera of *A. punctulata*. No synthesis could be detected in similar systems from *Dolichoglossus kowalevskii*, lobster, squid, *Thyone briarens* and *A. forbesi*. In further studies of the latter organism, Warren (1964) was unable to produce any synthesis of N-glycolyl-8-O-methylneuraminic acid by cell-free extracts of *A. forbesi* containing phosphoenolpyruvate, 2-glycolamido-2-deoxy-D-mannose and adenosine triphosphate, nor would replacement of the glycolyl derivative either by the acetyl amino sugar, or the 6-phosphate derivative of the latter, yield any sialic acid.

Cytidine-5'-monophospho-N-glycolyl and N-acetyl neuraminic acids were formed when extracts of *D. kowalevskii*, *A. punctulata* and *A. forbesi* were incubated with the appropriate neuraminic acids and cytidine triphosphate (Warren, 1963, 1964). Squid and lobster digestive gland extracts failed to produce nucleotide neuraminic acids.

EVOLUTION

Warren (1963) considers the sialic acids to have arisen late in evolution. Certainly they appear most prominently in the vertebrates and in some of the phyla lying on the chordate evolutionary line; those groups in fact which possess dipleurula larvae. Sialic acid is present in at least one mol-luscan species, however, and while this might be explained by invoking parallel evolution, origin from a common ancestor would seem more likely and more attractive. The detection of sialic acid in an acoel turbellarian would in fact place the ability to produce sialic acids back at the point of separation between the mollusc-arthropod and the chordate evolutionary lines. Admittedly, no sialic acids have been detected yet in the simpler metazoa, or in the porifera, but they do occur in plants, in microscopic green algae and also in the bacteria, where they form cell wall constituents.

H

On these grounds therefore, it would perhaps be justifiable to suggest that the biosynthetic pathway for the formation of sialic acids might have originated at a very early period in the evolution of life, in some primitive unicellular organism. Biosynthesis of sialic acid in bacteria does admittedly follow a different, rather less complex route, than that outlined above for the vertebrates. The phosphoenolpyruvate condenses with 2-acetamido-2-deoxy-D-mannose after, rather than before, removal of the 6-phosphate residue, while the formation of 2-acetamido-2-deoxy-D-mannosamine-6-phosphate proceeds directly from the 2-acetamido-2-deoxy-D-glucose-6-phosphate instead of via the 2-acetamido-2-deoxy-D-glucose-1-phosphate and the nucleotide derivative. We should not, however, consider this too serious an objection to the hypothesis of a common ancestral synthetic system. A direct interconversion, in mammals, of 2-acetamido-2-deoxy-D-glucose and 2-acetamido-2-deoxy-D-mannose, not involving nucleotides and requiring only traces of ATP, has been noted by Ghosh and Roseman (1962), while an aldolase catalysing the reversible synthesis of sialic acid from 2-acetamido-2-deoxy-D-mannose and pyruvate has been observed in vertebrate tissues (Comb and Roseman, 1960; Brunetti et al., 1962). Affinities with the bacterial system therefore seem possible. Moreover we might consider that the complexities of the main vertebrate pathway could reflect the evolution from a simpler system of a more specifically-controlled route. Thus the UDP-2-acetamido-2-deoxy-D-glucose-L-epimerase is the subject of end-product inhibition by CMP-N-acetyl-neuraminic acid.

Bacteria seem to have evolved a method, which other organisms lack, of linking sialic acid residues into chains; colominic acid produced by E. coli, K235 is a polysaccharide composed of N-acetylneuraminic acid alone. Animals and plants, as far as we know, can only link individual sialic acid residues to other classes of monosaccharide. Again, such differences need not argue against affinities of the synthetic systems; they probably arose at a later date.

In a sense there is little point in arguing for or against parallel or divergent evolutionary lines in the systems producing sialic acids until the enzymes have been isolated from the organisms concerned, and their sequences determined. Homologies, or lack of them, would then definitely indicate the presence or absence of a common ancestral synthetic route.

REFERENCES

Blix, G. (1936). *Hoppe-Seyler's Z. physiol. Chem.* **240**, 43.
Brunetti, P., Jourdain, G. W. and Roseman, S. (1962). *J. biol. Chem.* **237**, 2447.
Comb, D. G. and Roseman, S. (1958). *Biochim biophys. Acta* **29**, 653.
Correll, D. J. (1964). *Science, N.Y.* **145**, 588.
Ghosh, S. and Roseman, S. (1962). *Fed. Proc. Fedn Am. Socs exp. Biol.* **21**, 89.

Gottschalk, A. (1960). "The Chemistry and Biology of Sialic Acids". Cambridge University Press, London and New York.

Inoue, S. (1965). *Biochim. biophys. Acta* **101**, 16.

Kent, P. W. and Allen, A. (1968). *Biochem. J.* **106**, 645.

Kent, P. W. and Draper, P. (1968). *Biochem. J.* **106**, 293.

Klenk, E. (1941). *Hoppe-Seyler's Z. physiol. Chem.* **268**, 50.

Mayer, F. C., Dam, R. and Pazur, J. H. (1964). *Archs Biochem. Biophys.* **108**, 356.

Perlmann, P., Böstrom, H. and Vestermark, A. (1959). *Expl. Cell Res.* **17**, 439.

Stacey, M. and Barker, S. A. (1962). "Carbohydrates of Living Tissues". Van Nostrand, London

Svennerholm, L. (1958). *Acta. chem. scand.* **12**, 547.

Warren, L. (1963). *Comp. Biochem. Physiol.* **10**, 153.

Warren, L. (1964). *Biochim. biophys. Acta* **83**, 129.

Warren, L. and Hathaway, R. (1960). *Biol. Bull. mar. biol. lab., Woods Hole,* **119**, 354.

Warren, L., Hathaway, R. and Flaks, J. G. (1960). *Biol. Bull. mar. biol. lab., Woods Hole,* 355.

CHAPTER 14

Chromoglycoproteins and Chromoglycolipoproteins

The general class of conjugated proteins collectively called chromoproteins is well known. Five instances of chromoproteins which are also
glycoproteins or glycolipoproteins occur in the invertebrates. These are
found in the eggs and ovaries of certain molluscs and crustaceans and are
true carotenoproteins, the coloured prosthetic groups being constituted
either solely or partly by astaxanthin (3 : 3′-dihydroxy-4 : 4′-diketo-β-
carotene, Fig. 1), bound apparently by weak linkages to the protein.

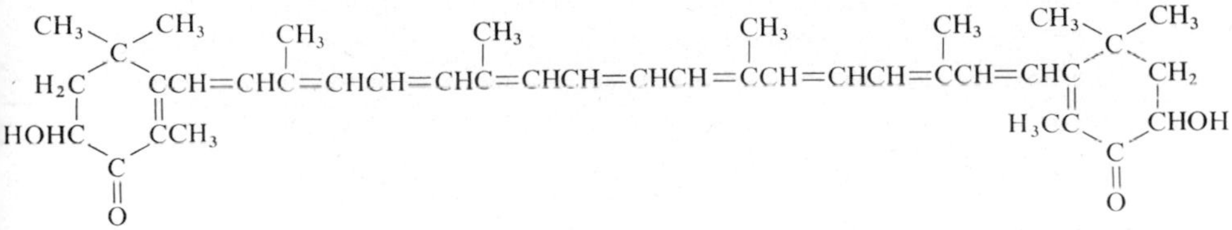

Fig. 1. Astaxanthin.

OVORUBIN

Among the prosobranch gastropod molluscs the genus Pomacea are
notable in that their eggs are in several species brightly coloured; red,
orange or green (Villela, 1956). Comfort (1947, 1949) has shown that the
rose-red colour of the eggs of *Pomacea canaliculata australis* (d'Orbigny)
is due to a heat stable carotenoid-protein complex, susceptible to alcohol,
acetone, and pyridine dissociation and readily adsorbed on aluminium
hydroxide gel. This complex has been further studied by Cheesman
(1954, 1956, 1958).

Homogenates of *Pomacea* eggs, in water, centrifuged at 10,000 g for
15 minutes, yielded a turbid suspension with absorption maxima at
427, 520 and 545 mμ (Cheesman, 1958). Treatment of this homogenate
with a suspension of aluminium hydroxide gel resulted in a quantitative
adsorption of the chromoprotein. The chromoprotein, ovorubin, could
be eluted from the adsorbent with 0·2 M Na_2HPO_4. Adsorption, after
adjustment to pH 6·0, and elution were repeated until $E_{280/510}$ reached a

constant value between 2·6 and 2·8. The ovorubin solution was dialysed, concentrated and lyophilized in the dark. Fresh mollusc eggs had to be used to achieve reproducible results.

Ovorubin has a nitrogen content of $12·6 \pm 0·2\%$ N and gives a strong Molisch reaction, suggesting that the low nitrogen content may be accounted for by the presence of a large carbohydrate component (Cheesman, 1958). The carotenoprotein is homogeneous by electrophoresis at pH 8·6 in 0·12 M veronal buffer and moves towards the anode. Thoroughly dialysed solutions have a pH of 5·5, suggesting an isoionic point in this range.

Ovorubin was found to be remarkably heat stable. Heating solutions of pH 4 to 7 at 100° for long periods produced only a slight turbidity, although a reversible partial dissociation of the carotenoid group, from the protein, seemed to be indicated by a change in colour to orange above 70°. Prolonged heating at pH values above 7 however seemed to produce an irreversible dissociation of the carotenoid; moreover, reduction of the pH to 5, with acetic acid, after such treatment, caused coagulation of the protein. Adjustment of ovorubin solutions to extremes of alkaline or acid pH (pH 14 and 1) in the cold produced at pH 1 an irreversible dissociation of the chromo group and precipitation of the protein, whereas exposure to pH 14 conditions resulted in a colour change to orange-pink, reversible on readjustment to pH 6. The chromoprotein was not readily precipitated by trichloracetic acid although addition of this reagent to five per cent concentration followed by addition of two volumes of acetone caused quantitative precipitation.

The properties of the carotenoid prosthetic group of ovorubin are largely outside the scope of this work; the carotenoid is astaxanthin (Fig. 1), not as was thought at first, a derivative of astaxanthin (Cheesman, 1958; Lee and Zagalsky, 1966).

The absorption spectrum of ovorubin in water (Fig. 2) is typical of aromatic amino acid-containing proteins. Above 300 mμ absorptions at 330, 480, 510 and 545 mμ are due to the carotenoid.

The apoprotein of ovorubin is readily isolated (Cheesman, 1958). Treatment of a one per cent solution of the chromoprotein at 0° with four volumes of ice cold acetone precipitated the protein which was spun down, redissolved in water and reprecipitated three times.

Apo-ovorubin is readily soluble in water and rapidly denatures when heated in solution above 70° at pH 5. Since the apo-protein still contains the carbohydrate moiety the heat stability of the intact chromoprotein must be a function of the carotenoid group and not of the carbohydrate, as might have been inferred from reference to the heat stable properties of such glycoproteins as seromucoid. The apo-protein and the carotenoid readily recombine in acetone-water mixtures and appear to show consider-

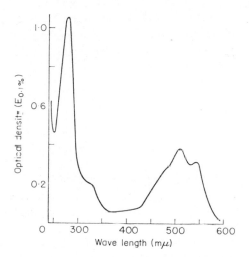

Fig. 2. Ultraviolet and visible absorption spectrum of ovorubin in water (Reproduced with kind permission from Cheesman, 1958).

able specificity for one another. The heat stable properties are restored on recombination.

The minimum molecular weight of ovorubin, based on an assumed molar extinction coefficient of 115000 for astaxanthin (Kuhn *et al.*, 1939; Karrer and Würgler, 1943) was $335,000 \pm 110,000$.

Ovorubin apo-protein is a glycoprotein. The carbohydrate moiety comprises $4 \cdot 9 \pm 0 \cdot 1\%$ amino sugar (as 2-amino-2-deoxy-D-glucose) and $14 \cdot 8 \pm 0 \cdot 2\%$ hexose (as glucose). A paper chromatographic analysis of acid hydrolysates of the protein has indicated that the main monosaccharide constituents are 2-amino-2-deoxy-D-glucose, mannose, and galactose, together with an unidentified reducing component (Cheesman, 1958).

About one third of the original ovorubin content of the egg (approximately 0·8 mg) is utilized in the development of the embryo, while the remainder, amounting to some five to six per cent of the snail's wet weight, is consumed during the first week of life (Cheesman, 1958). Ovorubin is quantitatively the most important nitrogenous constituent of the egg, the percentage of total nitrogen due to it being $72 \cdot 8\%$.

Ovorubin has low surface activity, with little tendency to froth or form unimolecular films, suggesting a considerable stability of the native configuration of the protein. This stability appears to be conferred by the carotenoid; the apo-protein has much greater surface activity (c.f. the heat stability).

A colourless glycoprotein analogous to ovorubin has been isolated from the eggs of the prosobranch mollusc *Pila orata* (Cheesman, 1958). Cheesman has suggested that the function of ovorubin is to protect the embryo from harmful radiation and assist water conservation at high temperatures, the snail being a tropical species.

OVOVERDINE

The green carotenoprotein ovoverdine is found in the eggs of the lobster *Homarus vulgaris* (L.) and was first noted and named by Stern and Salomon (1937, 1938). In spite of its difference in colour from ovorubin the prosthetic groups of the two chromoproteins are identical, being in both instances astaxanthin (Kuhn and Sörensen 1938a). The colour difference is presumably due to changes in conjugation within the astaxanthin residue caused by binding to protein.

Ovoverdine from *Homarus gammarus* (L.) has been isolated, purified and characterized by Ceccaldi, Cheesman and Zagalsky (1966). Lobster eggs were extracted with 0·05 M phosphate buffer, pH 7, and the extracts subjected to fractionation on DEAE and CM cellulose columns. The ovoverdine obtained was adsorbed onto calcium phosphate gel, at pH 5·5 and, after washing the gel four times with five per cent sodium chloride, eluted with two applications of 0·2 M phosphate buffer pH 7. Dialysis against 0·00125 M phosphate buffer, pH 7, resulted in precipitation of the ovoverdine. The precipitated material was soluble in water and proved to be homogeneous by electrophoresis on cellulose acetate at pH 5 and 8. Freeze-dried preparations of ovoverdine are red-orange coloured in contrast to the green of aqueous solutions (Ceccaldi, 1964).

Ovoverdine contains $11·3 \pm 0·2\%$ nitrogen increasing to $15·7 \pm 0·4\%$ nitrogen when the chromoprotein is extracted with 2 : 1 v/v chloroform-methanol (Ceccaldi *et al.*, 1966). The chloroform-methanol extract contains lipids amounting to $22 \pm 3\%$ of the original weight of the ovoverdine. Of these lipids chloesterol and cholesterol esters constitute $19 \pm 1·5\%$ and phospholipids $65 \pm 3\%$. The chromoprotein thus has lipoprotein character unlike ovorubin.

The de-lipidized proteins contains $3·2 \pm 0·2\%$ hexose and $1·6 \pm 0·1\%$ hexosamine, identified and estimated as mannose and 2-amino-2-deoxy-D-glucose, respectively (Ceccaldi *et al.*, 1966; Zagalsky *et al.*, 1967). Ovoverdine is therefore a chromoglycolipoprotein and obviously of a complex structure.

Gel-filtration of ovoverdine on Sephadex G200, in 0·05 M phosphate-0·2 M sodium chloride buffer, pH 7 gave a single component with an elution volume corresponding to a protein with a molecular weight of 380,000 (Ceccaldi *et al.*, 1966). This is in good agreement with the value of 300,000 obtained by Wyckoff (1937) from ultracentrifuge sedimentation data for *Homarus americanus* ovoverdin and with the minimum value of 170,000 (Zagalsky, 1964) obtained for *H. gammarus* ovoverdin on the basis of its astaxanthin content. The agreement of the gel-filtration data with the molecular weight obtained by hydrodynamic techniques suggests that the protein has a compact structure, probably roughly spherical. Comparison of the minimum molecular weight data with the physical data suggests a complex of two molecules of astaxanthin with one molecule of protein Kuhn and Sörensen (1938b) obtained a minimum molecular weight for *H. vulgaris* ovoverdine of 144,000, based on a nitrogen content of 16·4%.

OTHER CHROMOGLYCOLIPOPROTEINS

Using similar techniques to those applied to ovorubin and ovoverdin, carotenoid-protein complexes have been extracted from the ovaries of the edible crab *Cancer pagurus* and the scallop *Pecten maximus* and also from the eggs of the prawn *Plesionika edwardsi* (Zagalsky *et al.*, 1967). The compositions of these proteins were similar to that of ovoverdine. All three chromoproteins were also glycolipoproteins rich in phospholipids and cholesterol and in protein-bound phosphate. Whereas astaxanthin was the sole carotenoid component of the *P. edwardsi* complex, the other chromoproteins contained mixtures of carotenoids. In the case of the *C. pagurus* complex the principal carotenoid, of six carotenoids present, was astaxanthin, while in *P. maximus* chromoprotein the major component in a mixture of five carotenoids, including astaxanthin, was pectenoxanthin, a carotenoid of unknown structure. The absorption spectra of the three chromoproteins are shown in Figs. 3, 4 and 5. Minimum molecular weights, calculated from carotenoid and protein contents, of the *C. pagurus*, *P. edwardsi* and *P. maximus* chromoproteins were 420,000, 630,000 and

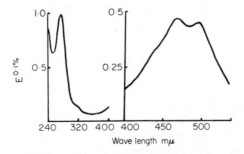

Fig. 3. Ultraviolet and visible absorption spectra of the carotenoid-glycoprotein complex from the ovary of *Pecten maximus*. Solvent 0·2 M phosphate buffer pH 7. (Redrawn from Zagalsky *et al.*, 1967).

H*

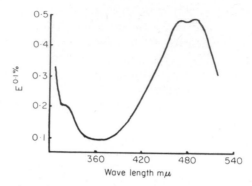

Fig. 4. Absorption spectrum of the carotenoid-glycoprotein complex from the ovary of *Cancer pagurus*. Solvent 0·2 M phosphate buffer pH 7. (Redrawn from Zagalsky, *et al.*, 1967).

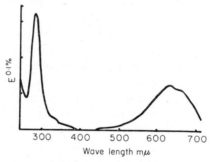

Fig. 5. Ultraviolet and visible absorption spectrum of the carotenoid-glycoprotein complex from the eggs of *Plesionika edwardsi*. Solvent 0·2 M phosphate buffer, pH 7. (Redrawn from Zagalsky *et al.*, 1967).

320,000, respectively, while molecular weights determined by gel-filtration on Sephadex G200 were 310,000, 530,000 and 700,000 respectively.

The carbohydrate compositions of all three apo-proteins were similar; all contained both mannose and 2-amino-2-deoxy-D-glucose. *C. pagurus* apoprotein contained $2·5 \pm 0·2\%$ mannose and $1·12 \pm 0·01\%$ 2-amino-2-deoxy-D-glucose; *P. edwardsi* apoprotein contained $1·8 \pm 0·1\%$ mannose and $0·75 \pm 0·01\%$ 2-amino-2-deoxy-D-glucose, and *P. maximus* apoprotein contained $2·6 \pm 0·1\%$ mannose and $0·75 \pm 0·02\%$ 2-amino-2-deoxy-D-glucose. Mannose and 2-amino-2-deoxy-D-glucose were the only sugar components detected. Sialic acid could not be detected in either the glycoprotein or the lipid components. Uronic acid could not be detected either. The lipid composition was complex in all cases. While the problem of the nature of the linkage of carotenoid to the glycolipoprotein was not resolved, it appeared that in *C. pagurus* and *P. maximus* chromoproteins the carotenoid might be dissolved in the lipid components; in the *P. edwardsi* chromoprotein the linkage seemed to have a more specific character comparable to that found in ovorubin and ovoverdine (Zagalsky *et al.*, 1967).

While the apoproteins of the chromoproteins described here are apparently true glycoproteins, the occurrence of carbohydrate prosthetic groups in other invertebrate chromoproteins does not appear to be a general rule. For example, crustacyanin, the blue chromoprotein of the lobster (*H. gummurus*) carapace, lacks carbohydrate although the carotenoid prothetic group is astaxanthin. Crustacyanin also lacks lipid (Cheesman *et al.*, 1966). The distribution and properties of carotenoproteins in invertebrates has been reviewed by Cheesman (Cheesman *et al.*, 1967).

To the author's knowledge chromoglyco- and chromoglycolipoproteins of this type do not occur elsewhere in the animal kingdom. Chromoglycoproteins have been reported to occur in certain algae (Fujiwara, 1963).

REFERENCES

Ceccaldi, H. J. (1964). *Recl Trav. stn mar. Endoume* **32**, 65.
Ceccaldi, H. J., Cheesman, D. F. and Zagalsky, P. F. (1966). *C.r. Séanc. Soc. Biol.* **160**, 587.
Cheesman, D. F. (1954). *Biochem. J.* **58**, xxxviii.
Cheesman, D. F. (1956). *J. Physiol.* **131**, 3P.
Cheesman, D. F. (1958). *Proc. R. Soc.* **B149**, 571.
Cheesman, D. F., Zagalsky, P. F. and Ceccaldi, H. J. (1966). *Proc. R. Soc.* **B164**, 130.
Cheesman, D. F., Lee, W. L. and Zagalsky, P. F. (1967). *Biol. Rev.* **42**, 131.
Comfort, A. (1947). *Nature, Lond.* **160**, 333.
Comfort, A. (1949). *Proc. malac. Soc. Lond.* **27**, 262.
Fujiwara, T. (1963). *Proc. 4th Int. Seaweed Symp.* 312. (Pergamon Press, Oxford).
Karrer, P. and Würgler, E. (1943). *Helv. chim. Acta* **26**, 116.
Kuhn, R. and Sörensen, N. A. (1938a). *Z. angew. Chem.* **51**, 365.
Kuhn, R. and Sörensen, N. A. (1938b). *Ber. dt. chem. Ges.* **71**, 1879.
Kuhn, R., Stene, J. and Sörensen, N. A. (1939). *Ber. dt. chem. Ges.* **72**, 1688.
Lee, W. L. and Zagalsky, P. F. (1966). *Biochem. J.* **101**, 9C.
Stern, K. G. and Salomon, K. (1937). *Science, N.Y.* **86**, 310.
Stern, K. G. and Salomon, K. (1938). *J. biol. Chem.* **122**, 461.
Villela, G. G. (1956). *Nature, Lond.* **178**, 93.
Wyckoff, R. W. G. (1937). *Science, N.Y.* **86**, 311.
Zagalsky, P. F. (1964). Thesis, University of London, England.
Zagalsky, P. F., Cheesman, D. F. and Ceccaldi, H. J. (1967). *Comp. Biochem. Physiol.* **22**, 851.

STRUCTURE PROTEINS

CHAPTER 15

The Fibrous and Structure Proteins

INTRODUCTION

For the purposes of this work the invertebrate fibrous and structure proteins will be considered only from the aspect of their association with carbohydrate. Fibrous proteins have been extensively reviewed on several occasions in recent years and much of the chemistry, biology, and biophysics of those derived from invertebrate sources has been well treated in these reviews. It would therefore be pointless to repeat the information already cited. For further details the reader is thus referred to the impressive account of structure proteins in general given by Seifter and Gallop (1966); that of silks by Rudall (1962), Lucas and Rudall (1968); and that of the comparative biochemistry of collagen by Gross (1963) and Rudall (1955, 1968). Two recent very detailed accounts of the chemistry and biology of collagen are an article by A. J. Bailey (1968) and "Treatise of Collagen," Vol. 1, edited by Ramachandran (1967) and Vol. 2 edited by Gould (1968).

Because of its almost universal distribution in the animal kingdom, in numerous guises, fulfilling a variety of structural functions, collagen occupies a position of major importance in this chapter, meriting inclusion in this book because of the as yet little understood, but probably profoundly significant role of carbohydrate in its structure. Discussed also are the structure proteins of the arthropod cuticle. Resilin, the remarkable elastic cuticle protein is considered, in spite of the absence of carbohydrate from the molecule itself; its association with cuticular chitin, as yet ill-defined, necessitates its inclusion. A number of other structure glycoproteins are discussed; those involved in calcification, tanned proteins, molluscan egg cases and the arthropod silks.

COLLAGEN

Collagen appears to have arisen early in the course of evolution, as it is found widely distributed throughout the whole of the animal kingdom, antedating even the appearance of the Metazoa. Fundamentally the role of this fibrous protein appears to be to act as an extracellular scaffolding, supporting and maintaining cells and organs in a fixed, ordered spatial arrangement. Over and above this a wide variety of secondary, but in a strict sense still structural functions have arisen through evolution: the tough byssus secreted by the feet of bivalved molluscs, serving to anchor the animal to the rocky substrate, is an extreme example of this type of collagen.

Physically, collagen may be defined in the general terms proposed by Astbury (1938), as a class of proteins of a fibrous nature, having a characteristic, unique wide-angle X-ray diffraction pattern with a major axial repeat at $\sim 2\cdot8$–$2\cdot9$ Å and an equatorial repeat of about 11 Å.

Phylogenetic studies of the collagen class of proteins, using X-ray diffraction techniques, have revealed a considerable uniformity of molecular organization over many phyla, both vertebrate and invertebrate (Bear, 1952).

$$NH_2-CH_2-COOH$$
Glycine (Aminoacetic Acid)

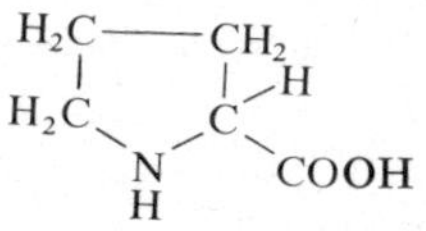

L-(—)-Proline (Pyrrolidine-2-carboxylic Acid).

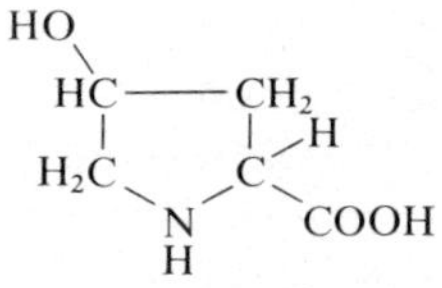

L-(—)-Hydroxyproline (4-Hydroxypyrrolidine-2-carboxylic Acid).

Fig. 1

Chemically, the collagens are characterized by a high glycine and proline content and by the presence of hydroxyproline (Fig. 1). Gross (1963) discussed the degree of variation of amino acid composition among the various known collagens, vertebrate and invertebrate. Apart from glycine, which had a variability never more than four per cent and a minimal value not less than 30% of the total residues, whatever the source, the invertebrate collagens show a far greater variability of amino acid composition than vertebrate collagens. Not only does the proline content vary widely but the proline-hydroxyproline ratios also show considerable variation even to the extent of the hydroxyproline content exceeding the proline content in the case of the earthworm cuticle collagen. The total amino acid content also shows a far greater range of variation in invertebrate collagens in comparison with vertebrate analyses. Although cysteine is absent from vertebrate collagens, it is present to a small extent in some, but not all, invertebrate collagens. Di-sulphide-linked collagenous proteins have been identified recently and isolated from the nematocyst capsules of the sea-anemone *Aiptasia pallida* (Blanquet and Lenhoff, 1966) and the cuticle of *Ascaris* (McBride and Harrington, 1967). While hydroxylysine (Fig. 2)

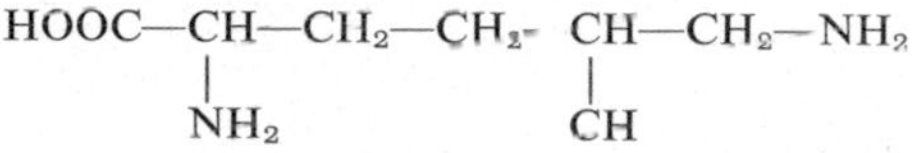

$$\text{HOOC—CH—CH}_2\text{—CH}_2\text{- CH—CH}_2\text{—NH}_2$$
$$\underset{\text{NH}_2}{|} \qquad \underset{\text{CH}}{|}$$

γ-Hydroxylysine (α, ε-Diamino-δ-hydroxycaproic Acid).

Fig. 2

appears to be a characteristic minor component of vertebrate collagens, it is absent from certain invertebrate collagens. Diodotyrosine and dibromo-tyrosine (Fig. 3) have been detected in the sponge collagen, spongin, and the coelenterate collagens, the gorgonins. Apart from the last two mentioned derivatives, it might be said that the composition of most invertebrate collagens is less atypical of proteins in general than the vertebrate collagens, where there is a highly distinctive and relatively invariable amino acid composition; thus the degree of molecular specialization is at a lower level. Watson (1958) and Gross (1963) have reviewed the amino acid analyses of invertebrate collagens.

An extensive discussion of the molecular organization of collagen is unnecessary for the purpose of this work and the reader is therefore referred

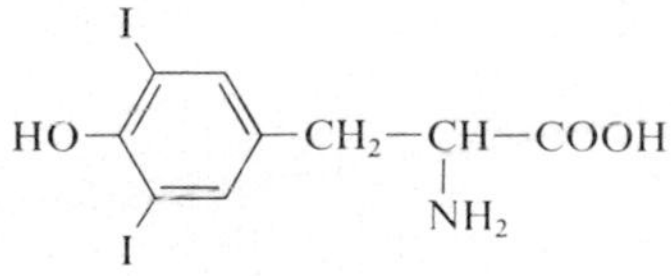

3,5-Diiodotyrosine.

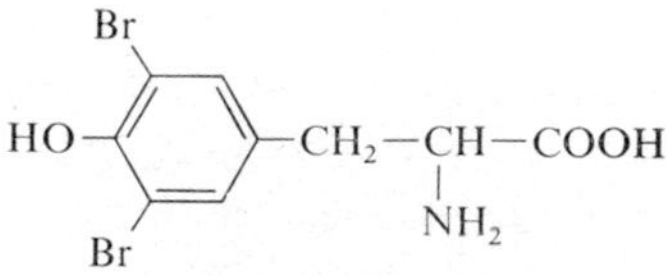

3,5-Dibromotyrosine.

Fig. 3

to the reviews already cited for details. Briefly, the structure may be summarized as follows: X-ray diffraction data have led to a helical structure being proposed for the fundamental sub-unit of collagen; the tropocollagen molecule. This sub-unit, about 3000 Å long by 14 Å wide, is composed of a rope of three polypeptide strands, differing in composition from one another and designated α_1, α_2 and α_3, each of molecular weight ~95,000. Each chain is twisted into a left-hand helix of the polyglycine II type and the three helices are wrapped round each other to form a right-handed superhelix (Fig. 4). The N-terminal ends are all together at one end where

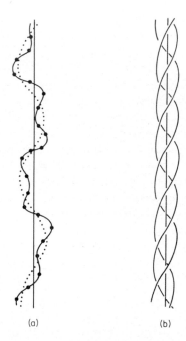

(a)   (b)

Fig. 4. Secondary and tertiary structure in the tropocollagen molecule. (a) An isolated α-chain showing the polyglycine II type coil structure upon which is superimposed a superhelical (dotted) structure. (b) The three chains (α_1, α_2, and α_3) superimposed into the entire superhelix.

there is a short non-helical region. The three chains are held together by inter-chain hydrogen bonds, probably linking the carbonyl oxygens and amino nitrogens of amino acids other than proline and hydroxyproline; other cross links may be present. Lysine appears to play an important role in cross linking. In the structure proposed by the Madras group (Ramachandran, 1963) there is a twist in the three chain structure of nearly 30°

for a height of three residues (residues other than the proline, hydroxy-proline, glycine triplets) and a residue height of 2·95 Å; these values refer-ring to unstretched collagen. The Madras model contains two hydrogen bonds per triplet as opposed to the single hydrogen bond structure proposed by Rich and Crick (1961). Much of the physical data obtained from X-ray diffraction, infrared dichroism and deuterium exchange can be satis-factorily explained by the Madras model, which is that most widely accepted at the present time. It should be noted that the tropocollagen molecule is not now thought of as assuming a single uninterrupted type of conformation for the entire length of its structure, but that the presence of less ordered regions are thought to be a requirement for flexibility and slight elasticity. There is evidence to suggest that the α-chains are not continuous poly-peptide chains but consist of sub-units joined by non-peptide, possibly ester, links.

The tropocollagen molecules are aggregated in a complex manner to give the quaternary structure of the collagen fibrils observed in the electron microscope. Characteristically, in most instances this linear, side by side aggregation gives rise to the well-known 640–700 Å axial periodicity, ex-plainable in part at least by postulating a structure in which tropocollagen units overlap (Fig. 5). While this type of fibril structure is that most fre-

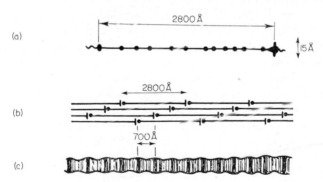

Fig. 5. Quaternary organization in collagen. (a) Tropocollagen molecule. (b) Over-lapping, linearly arrayed tropocollagen subunits producing the characteristic striated collagen fibril (c).

quently observed in vertebrates and invertebrates, absence of axial periodi-city is also known, suggesting a quite different arrangement of the tropo-collagen molecules. Collagen fibrils of this type are found typically in the collagen of the earthworm cuticle. While branching of the fibrils is unknown in the vertebrates and most invertebrates, it is found in the Keratozoan collagen, Spongin B.

The collagen fibrils themselves may be associated into fibres forming discrete parallel or interlacing bundles or sheets of laminae. With regard to the integumentary collagen fibres of both invertebrates and vertebrates, organization either into a geometrical laminated dermis, or into a more randomly-arranged pattern of bundles may take place. Fauré-Fremiet and Garrault (1944) have described a plywood-like arrangement of collagen fibrils in the cuticular collagen of the acoelomate worm *Ascaris*, while in the cuticle of the earthworm *Lumbricus* the arrangement appears as a highly-oriented regular quiltwork of non-striated collagen fibrils (Reed and Rudall, 1948). An oriented pattern of fibrous collagen has been observed in the coelenterata by Chapman (1953). Studies of the structure of invertebrate collagen fibres have been well reviewed by Bear (1952).

OCCURRENCE IN THE INVERTEBRATES

As already mentioned, collagens yield characteristic wide-angle X-ray diffraction patterns and the recognition of these probably remains the most reliable criterion by which a structural protein can be identified and assigned to the class of collagen-like proteins. This consideration applies particularly to invertebrate collagens where amino acid analyses, or molecular organizations visualized in the electron microscope, may not provide unequivocal grounds for identification as a collagen.

The distribution of collagen in the invertebrates has been reviewed in detail by Rudall (1955). Collagens have been identified in at least eight invertebrate phyla. In the molluscs collagens occur in the connective tissues of the body wall (Williams, 1960; Melnick, 1958; Schmitt *et al.*, 1942) and as a so-called "secreted" collagen in the form of the byssus of the lamelibranch molluscs (Jackson *et al.*, 1953; Brown, 1947; Singleton, 1957). Both body wall and secreted collagens occur also in the echinoderms. In the case of the secreted collagen this takes the form of filaments ejected by holothurian (sea cucumber) cuvierian tubules (Marks *et al.*, 1949; Watson and Silvester, 1959; Gross *et al.*, 1958; Piez and Gross, 1959).

In the annelid worms and in the parasitic nematode worms the cuticular collagens have received particular attention (Astbury, 1944; Fauré-Fremiet and Garrault, 1944; Picken *et al.*, 1947; Fraenkel and Rudall, 1947; Reed and Rudall, 1948; Singleton, 1957; Watson, 1958; Watson and Silvester, 1959; Maser and Rice, 1962, 1963a, b; Fujimoto and Adams, 1964; Anya, 1966; McBride and Harrington, 1967). Coelenterate collagens occur in the body walls and floats of sessile and free-living hydrozoans and anthozoans, as the corneins and gorgonins of the axial stalks of anthozoans, and in the nematocyst capsules of hydrozoans (Marks *et al.*, 1949; Astbury, 1940; Champetier and Fauré-Fremiet, 1942; Lenhoff *et al.*, 1957; Lenhoff and

Kline, 1958; Blanquet and Lenhoff, 1966; Lane and Dodge, 1958; Piez and Gross, 1959; Gross *et al.*, 1958; Roche and Michel, 1951).

While collagen plays a lesser role than chitin in the formation of arthropodan connective tissues, it has been identified in insect and arachnid tissues (Pipa and Cook, 1958; Ashhurst, 1961, 1965; Piez and Person, 1963). A silk-like secreted collagen has been identified by Rudall (1962) in the cocoons of gooseberry sawfly pupae. A partial characterization of blattarian collagen has been carried out by Harper *et al.* (1967). Connective tissues and the occurrence of collagen in insects have been reviewed by Ashhurst (1968).

The poriferan (sponge) collagens, the spongins, have been particularly closely studied (Piez and Gross, 1959; Gross *et al.*, 1958; Stary *et al.*, 1956; Tekman and Öner, 1963; Roche and Eysseric-Lafon, 1951).

In addition, collagen occurs in the body walls of the Onychopera, the peduncles of brachiopods, the proboscis and branchial skeletons of hemichordates and the tunics of urochordates (Rudall, 1955).

ASSOCIATION OF COLLAGEN WITH CARBOHYDRATE

Tissue collagen is almost always found in association with large amounts of carbohydrate. Much of this material is mucopolysaccharide and glycoprotein, and while the association is not fortuitous, since the presence of these other polymers is necessary to the architecture and functioning of the tissue, neither is it strongly covalent. Removal of these polysaccharides and purification of the collagen does, however, leave a residue of carbohydrate associated by a firm bond with the constituent tropocollagen molecules.

In vertebrate collagens this carbohydrate does not usually account for more than 0·6–2% of the molecule. In contrast, invertebrate collagens have been found containing as much as 15·5% carbohydrate. On the whole, collagens, vertebrate and invertebrate, of ectodermal origin have higher carbohydrate contents than those of mesodermal origin.

The monosaccharides associated with collagen by covalent linkage in vertebrates are limited in character; moreover, unlike most glycoproteins, hexosamine appears to be absent from the carbohydrate prosthetic group. Repeated reprecipitation of vertebrate acid-soluble collagen (Kühn *et al.*, 1959) has yielded a product containing hexose but no hexosamine. Subsequent studies have shown this hexose to be composed of D-glucose and D-galactose residues which, in ichthycol, are in the ratio of seven galactose residues to five glucose and which are bound to the protein as monosaccharides (Blumenfeld *et al.*, 1963). In the soluble collagen of guinea pig skin the hexose is again D-glucose and D-galactose, bound O-glycosidically to the hydroxyl groups of certain hydroxylysine residues of the collagen protein chain, as disaccharides of the two sugars (Butler and Cunningham, 1966). Butler and Cunningham (1966) have suggested that the glycopeptide region of the tropocollagen chain may be important in the formation of the intra and intermolecular cross-links which stabilize the network of mature collagen fibres. This cross-linking need not necessarily be via hexose bridges but may perhaps be mediated through a process of hydroxylysine activation by the bound carbohydrate, bringing about ϵ deamination of the glyco-sylated hydroxylysine, condensation of the resultant enol with a lysine on an adjacent chain and final deglycosylation.

A considerable number of carbohydrate analyses, qualitative and quanti-tative, have been carried out on collagens from a range of invertebrate species. These studies have indicated that as is the case with the amino acids of invertebrate collagens, invertebrate carbohydrate compositions are subject to a far greater variability than their vertebrate counterparts. This may perhaps represent an earlier stage in the evolution of the molecule with a corresponding lower order of specialization. The carbohydrate composi-tions of those invertebrate collagens examined to date are summarized in Table I.

At least one invertebrate collagen has its carbohydrate linked, in part at least, to the protein by bonds not totally dissimilar to those found in verte-brate collagen. Lee and Lang (1968) found earthworm cuticle collagen to contain some 12% D-galactose which could be released by alkaline treat-ment, although not entirely; the alkali labile carbohydrate comprised about 70% of the total carbohydrate. Concomitant with the release of the galac-tose, a fall in the serine and threonine content of the collagen was observed, corresponding to 20·5 residues of threonine and 5·8 residues of serine per 1000 total amino acid residues. This would suggest that the carbohydrate is linked O-glycosidically to the hydroxyl groups of the serine and threo-nine residues in analogy to the O-glycosidic link to hydroxylysine found in vertebrate collagen. When reduction by borohydride was allowed to accompany the alkaline splitting, reduced saccharides were produced which could be separated by gel-filtration and shown to be disaccharides and trisaccharides of galactose. Optical rotatory dispersion and nuclear mag-netic resonance observations indicated that the glycodisic links in these oligosaccharides had the α configuration.

In spite of there being a wider range of monosaccharides in the inverte-brate collagens, galactose and glucose seem to play the most important role, if quantity is any criterion. In this respect there is a direct analogy with the vertebrate collagens. Gross *et al.* (1958) have suggested that in view of the wide variety of sugars present in the invertebrate collagens, more than one molecular species of polysaccharide may be present. Thus the pros-thetic groups may differ in structure and composition, just as they do in certain vertebrate glycoproteins. If this were the case then it might be possible that groups like those linked to the vertebrate collagens could largely account for the galactose-glucose content; a second type, linked through hexosamine to the protein chain, might account for the 2-amino-2-deoxy-D-glucose, 2-amino-2-deoxy-D-galactose and other monosaccharides present. It would, however, be dangerous to speculate too far merely on the basis of the established structures of vertebrate materials. Hexose in all cases accounts for the largest fraction of carbohydrate, and the total monosaccharide content is in general higher in invertebrate than vertebrate collagens.

While, as stated earlier, purification or gelatinization always drastically reduces the hexosamine content of vertebrate collagens, this phenomenon is less noticeable where invertebrate collagens are involved; indeed in the case of *Physalia* (coelenterata) float gelatin (Gross *et al.* 1958) and *Ascaris* gelatin (Watson and Silvester, 1959) there is an enrichment in hexosamine content. Josse and Harrington (1964), however, found no hexosamine in the *Ascaris* collagen which they examined. Again *Thyone* (echinodermata), *Physalia* and *Ascaris* gelatins are enriched in hexose content over the parent material and the hexose contents in other cases in general are not greatly reduced by purification (Gross *et al.*, 1958; Watson and Silvester, 1959). Gross *et al.* (1958) report a remarkably large amount of methyl pentose, identified as fucose ($1 \cdot 2\%$) in *Thyone* skin. A suggestion by Gross that the carbohydrate might become bonded to the fibril as an "innocent by-stander" during fibrogenesis seems unlikely. Some of the other mono-saccharides found in certain of the invertebrate collagens may seem unusual in comparison with vertebrate tissue constituents, for example, rhamnose and arabinose. However, such sugars seem to occur quite generally in invertebrate glycoproteins and may simply reflect again a lower degree of molecular specialization in less highly-evolved life forms.

Rudall (1968) has noted that collagens with high carbohydrate contents are never like standard vertebrate collagens. The relationship between carbohydrate content and collagen structure, as observed at the electron microscope level, appears not to be a straightforward one. High carbo-hydrate content tends to render the 650 Å striation less distinct; often there is no banding at all or there may be a new type of banding which is associa-ted with a narrow core to the fibre. Anomalies such as the bandless, low carbohydrate, *Ascaris* cuticle collagen also exist. Often a high carbohydrate content is accompanied by an elevated acidic amino acid composition, but interpretation of this as indicating the presence of links between these amino acids and the carbohydrate must be treated with caution in the absence of more definite evidence. Narrow fibres often accompany high carbohydrate content, but again exceptions occur in nematode and annelid

TABLE I

Carbohydrate Compositions of Invertebrate Collagens[a]

Species and tissue	Hexosamine			Neutral sugar (%)			Sugars m moles/g			Sugars identified
	GlAm	GaAm residues/ 1000 total	Total %	Total anthrone positive	Hexose	Pentose	HexAm	Hexose	Pentose	
Annelida										
Lumbricus sp[b]	0·6	1·1	2·0	12						
Rhinodrilus fafneri[b]	12·0	18·0								
Allolobophora longa[c]	—	—	0·37	4·9						
Lumbricus sp[d]										
cuticle			1·76		14	2·4	0·098	0·78	0·16	Galactose
gelatin			0·88		10·6	—	0·049	0·59	—	
Lumbricus sp[e]										
cuticle	0	0	—	8·8						
Echinodermata										
Thyone sp[f,j]										
corium			1·1		3·3	0·6				Glucose, Galactose, Mannose, Arabinose, Fucose, GlAm, GaAm, Uronic acid and lactone.
gelatin	2·4	0	0·6[f] 0·38[j]		4·4	0·5				as above less Mannose.
Holothuria forskali[a]										
Cuvierian tubules			1·34		8·2	0·46				GlAm,
			1·34		7·8	0·40				GaAm.
Tubule collagen fibres			0·95		4·5	0·29				
Nematoda										
Ascaris lumbricoides[g]										
cuticle			0·79		0·19	0·10				
			0·51		0·19	0·19				
gelatin			1·29		0·50	0·15				

TABLE I (contd)

Species and tissue	Hexosamine GlAm residues/1000 total	GaAm residues/1000 total	Total %	Neutral sugar (%) Total anthrone positive	Hex-ose	Pent-ose	Sugars m moles/g HexAm	Hex-ose	Pent-ose	Sugars identified
Ascaris cuticle	0	0	0	0·4						
Porifera										
Euspongia offici-nalis[h]			0·4		5·4					Galactose, Xylose.
Spongia grammea[f,j]										
Spongin A	10		1·8[f] 2·01[j]		11·0	2·7				Glucose, Galactose, Mannose, Arabinose, Fucose, GlAm, GaAm, Uronic acid and lactone.
Spongin B	1·6	0	0·2[f] 0·22[j]		4·9	0·5				Glucose, Galactose.
Coelenterata										
Metridium[f,j] corium			1·0		8·9	0·9				
corium limed			0·6		5·1	0·8				
corium gelatin	4·0	0	0·7[f] 0·6[j]		6·7	0·8				⎧ Glucose, Galactose, Mannose, Arabinose, Ribose?, Fucose, Rhamnose?, GlAm, GaAm, Uronic acid and lactone
Physalia[f,j] float			0·5		5·7	0·8				
float limed			0·2		4·8	0·9				
float gelatin	2·5	3·2	0·9[f] 0·83[j]		6·0	0·9				
Mollusca										
Helix aspersa[i] body wall	0·17%	0·48%			12·1					Galactose, GlAm, GaAm.

[a] GlAm: 2-amino-2-deoxy-D-glucose; GaAm: 2-amino-2-deoxy-D-galactose; HexAm: 2-amino-2-deoxy-D-hexose. [b] Maser and Rice (1962). [c] Singleton (1957). [d] Watson (1958). [e] Josse and Harrington (1964). [f] Gross *et al.* (1958). [g] Watson and Silvester (1959). [h] Stary *et al.* (1956). Tekman and Öner (1963). [i] Williams (1960). [j] Piez and Gross (1959).

cuticle collagens. Such variability may suggest that there is more than one type of association of carbohydrate with the collagen molecule (Rudall, 1968a).

SKELETAL   PROTEINS

Any structural protein might in the strictest sense be considered to have a skeletal function; however, the term "skeletal" is usually less broadly interpreted.

All animals may be said to possess in some form or other a skeleton. This statement may perhaps occasion some mild surprise, for while internal skeletons in vertebrates, and to a lesser extent external or exo-skeletons in arthropods, may be familiar facts, other forms of skeleton may be less obvious. For example, vertebrates too possess an exo-skeleton in their horny keratinized skin and many invertebrates, at first sight completely soft-bodied, also possess analogous toughened ectodermal coverings acting as skeleta. Just as calcified skeletons occur internally in vertebrates, so too do they occur in the echinoderms and sponges, while external calcified skeletons are found in certain coelenterates, in the molluscs, the ectoprocta, the brachiopods and of course the crustacea.

As already stated, uncalcified exo-skeletons occur widely, the most obvious being the insect and arachnid cuticles, consisting of protein-chitin complexes (Chapter 8) to which rigidity is given not by embedding in calcium carbonate but by quinone cross-linking of the protein chains. In other invertebrate cases the ectodermal skeleton may possess little or no chitin, the important skeletal material being a protein or proteins analogous to those forming the vertebrate epidermis.

PROTEINS ASSOCIATED WITH CALCIFICATION

In the vertebrates collagen appears to be associated with the deposition of apatite in the process of bone formation. Considerable attention has therefore been directed towards establishing its role in the mineralization of cartilage tissues. Our knowledge of the chemistry of the proteins found in the calcified tissues of invertebrates on the other hand is rather more limited. Mineralization of cartilage apparently does not take place in the invertebrates. The calcified tissues which have been principally studied are the shells of molluscs and brachiopods and the crustacean integument. Collections of papers relating to calcification in biological systems have been edited by Sognnaes (1960) and Moss (1963) while the subject has been reviewed by Glimcher (1960), Moss (1964), Wilbur (1964), Lavine and Isenberg (1964), Florkin (1966) and Wilbur and Simkiss (1968).

THE MOLLUSCAN SHELL

In the molluscs the shell is secreted by the ectodermal epithelium of the mantle. Initially the shell is laid down by the mantle of the veliger larva and further extension takes place by secretion at the shell edge (Fig. 6).

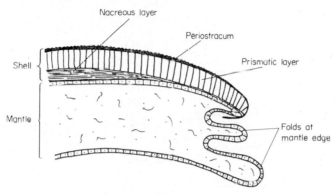

Fig. 6. A stylized representation of the relation between the mantle edge and the developing shell in a generalized mollusc.

The outer shell layer consists of a horny proteinaceous periostracum beneath which lies a "prismatic" layer of calcite or alternatively arragonite secreted by the thickened mantle edge (Fig. 7). The lowest layer, immediately above the mantle, constitutes the nacreous layer and is secreted by the cells of the whole mantle surface. The calcified regions of the shell, i.e. the prismatic and nacreous layers, are composed of organic material impregnated with calcium carbonate. This organic material is frequently referred to as "matrix conchiolin". In the nacre the organic matrix is made up of fibres of a mucopolysaccharide, nacroin, associated with a protein. Nacroin is essentially composed of alanine and glycine in conjunction with some 0·5–6 % chitin (Florkin, 1966).

Protein matrices (conchiolins) have been isolated from various molluscan shells, following decalcification. A number of amino acid analyses of these materials have been published (Stegemann, 1961; Tanaka et al., 1960; Florkin, 1966; Simkiss, 1965).

Early analyses of shell proteins by Japanese workers (Tanaka and Hatano, 1953; see also Gross and Piez, 1960) using oyster shell and pearl, indicated the presence of hydroxyproline, although weighed against this, the apparent absence of large amounts of glycine and proline prevented the classification of the proteins as collagens. Piez (1961, 1963) found both hydroxyproline and hydroxylysine in the shell of the snail *Australorbis glabratus*. The hydroxyproline content was low, although a high glycine

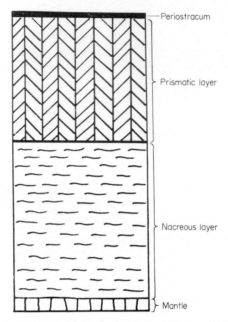

Fig. 7. Layers in the generalized molluscan shell.

content was found. Hare (1962, 1963, 1964, 1965) working with the shells of *Mytilus* failed to detect either hydroxyproline or hydroxylysine although the glycine content was high. The amino acid composition of the protein associated with sulphated acid mucopolysaccharide (Chapter 5) in the matrix of the shell of the oyster *Crassostrea virginica* (Simkiss, 1965) resembled more that of the outer prismatic calcite, or the inner nacreous aragonite, conchiolin of *Mytilus* shell (Hare, 1963) than the matrix of *Australorbis* shell (Piez, 1961). In this case also there was no evidence of hydroxyproline or hydroxylysine; only traces of cysteine, methonine and tyrosine were found (Table II).

The type of protein found appears to be related to the crystal form laid down. Hare (1962, 1963, 1964, 1965) noted that protein associated with calcite had a significantly more acidic composition than that associated with aragonite.

Wilbur and Watabe (1963) examining the X-ray diffraction patterns given by the matrix derived from molluscan shells noted that either an α or β-keratin pattern was obtained, depending upon the species studied.

Decalcification of the nacre of *Nautilus* (cephalopod) shell yields a protein which can be fractionated into two components: soluble and insoluble nacrine (Grégoire *et al.*, 1955; Florkin, 1966). Soluble nacrine

TABLE II
Amino Acids of Molluscan Shell Matrix Proteins[a]

Amino acid	Crassostrea[b,c]	Mytilus[d] (outer prismatic[c])	Mytilus[d] (inner nacreous[e])	Australorbis[f]
Cystine/2	trace	11·2	16·5	10·0
Hydroxyproline	—	—	—	3·2
Aspartic acid	104·6	98·6	102·0	183·0
Threonine	18·3	11·3	19·2	22·0
Serine	180·0	102·0	101·0	91·0
Glutamic acid	13·1	24·4	28·6	81·0
Proline	4·2	11·4	14·7	47·0
Glycine	432·0	298·0	269·0	265·0
Alanine	60·8	274·0	246·0	83·0
Valine	25·2	24·6	29·4	32·0
Methionine	trace	5·0	7·6	1·1
Isoleucine	14·7	13·5	17·1	15·0
Leucine	18·6	47·0	49·5	34·0
Tyrosine	trace	16·3	16·9	35·0
Phenylalanine	14·7	15·7	18·7	21·0
Hydroxylysine	—	—	—	6·2
Lysine	28·7	16·7	26·7	52·0
Histidine	8·2	2·5	5·5	3·8
Arginine	29·4	26·6	30·9	15·0

[a] Residues per 1000 total residues. [b] Simkiss (1965). [c] Calcite. [d] Hare (1963). [e] Aragonite. [f] Piez (1961).

dissolves in Trim's borate buffer, pH 9·2, while insoluble nacrine is extracted by five per cent sodium hydroxide at 53°. Insoluble nacrine is two to three times more abundant than soluble nacrine. The ultra structure of the conchiolin of nacre has been well reviewed by Florkin (1966) and appears to consist of a core of polypeptide or mucopolysaccharide fibrils surrounded by a double coat of insoluble and soluble protein. In the prismatic layers a prism conchiolin, built on a fibrous web, surrounds each prismatic mineral sub-structure.

The amino acid compositions of nacre and prism conchiolins differ markedly (Florkin, 1966), nacre conchiolin containing more arginine, serine and glutamic acid than prism conchiolin, and less glycine, tyrosine and valine. Later studies have shown that characteristically nacre conchiolin contains predominantly the amino acids glycine, alanine, serine and aspartic acid, while prism conchiolin contains predominantly glycine and aspartic acid (Bricteux-Grégoire et al., 1968) (Table III).

Carbohydrate: Bevelander and Benzer (1948) suggested, on the basis of histochemical properties, that the conchiolin of lamellibranch mollusc

TABLE III

Amino Acid Compositions of Nacre and Prism Conchiolins[a,b]

Amino acid	Atrina (Pinna) nigra		Pinna nobilis	
	Nacre	Prisms	Nacre	Prisms
Lysine	1·8	0·8	2·3	0·6
Histidine	trace	0·4	trace	0·5
Arginine	4·1	1·1	4·3	1·7
Aspartic acid	12·6	10·4	13·5	23·3
Threonine	1·4	1·1	1·8	1·5
Serine	9·5	4·0	9·0	4·6
Glutamic acid	4·1	1·8	4·1	1·9
Proline	2·3	2·2	2·0	2·2
Glycine	19·2	46·2	21·0	36·9
Alanine	30·8	4·7	28·9	5·7
Valine	2·5	6·3	2·4	5·6
Isoleucine	2·6	2·0	2·6	3·4
Leucine	4·2	7·4	3·8	4·5
Tyrosine	1·7	9·3	1·9	5·3
Phenylalanine	2·9	2·3	2·4	2·4

[a] Bricteux-Grégoire et al. (1968). [b] Residues per 100 total residues.

shells might contain one or more reducing sugars, while Horiguchi (1959) obtained definite evidence that carbohydrate in some form might be associated with conchiolin. Conchiolins of *Pteria martensii* and *Hyriopsis sclhegelii* had reducing sugar contents of 1·25–8·00%, composed mostly of hexosamine.

More recently the matrix proteins from the calcified tissues of a wide variety of molluscan species have been analysed for their amino acid and amino sugar contents (Degens and Spencer, 1966; Degens et al., 1967). Thirty three different species of gastropod were examined in these studies together with 23 species of pelecypod and four species of cephalopod (Table IV). This work may be said to be classic; it represents a significant milestone in comparative molecular biology. Using the statistical technique of factor analysis, Degens and his co-workers were able to describe the amino acid compositions of calcified and uncalcified molluscan tissues by a limited number of independent factors which were related to both phylogeny and environment. Apart from the important evolutionary implications of this study, several more generally interesting conclusions emerged. Shell-matrix proteins appear to vary considerably, both in quantity and composition, from species to species, although there is good agreement within a species. It would appear too that the matrix proteins are not of the collagen type, although an attempt to fractionate the proteins was made in only one case. The range and variability of amino acids was

TABLE IV
Amino Sugars in the Shell Organic Matrix of Mollusca[a]

Species	2-amino-2-deoxy-D-glucose (μg/g)	2-amino-2-deoxy-D-galactose (μg/g)	Protein: amino sugars
Gastropoda			
Achatinella lorata	58	26	82
Acmaea pustulata	11	4	68
Akera soluta	176	55	9
Aplysia willcoxi	164	5	15
Architectonica nobilis	3	1	22
Astraea caelata	113	6	37
Bulla striata	1	—	60
Cavolina tridentata	34	—	14
Colus trophius	—	—	—
Crepidula fornicata	16	5	50
Crepidula plana	—	—	—
Cypraea zebra	5	27	10
Dolabella scapula	1	—	273
Epitonium angulatum	16	3	16
Fissurella barbadensis[b]	3	2	280
Fissurella barbadensis[c]	4	1	101
Gastropod sp.	29,329	—	18
Haliotis crachfordi	765	102	13
Helisoma trivalis	420	11	11
Janthina janthina	23	8	22
Littorina littorea	89	49	19
Lunatia trisseriata	8	—	160
Melanella martini	0·2	—	295
Murex brevifrons	0·7	0·3	125
Nerita plexa	8	4	66
Nassarius trivittatus	5	—	124
Oxynoe viridis	46	6	65
Polinices duplicatus[d]	2	—	118
Polinices duplicatus[e]	2	—	121
Polinices duplicatus[e]	2	—	117
Polinices duplicatus[f]	1	—	162
Polinices duplicatus[g]	0·7	0·8	268
Siphonaria alternata	30	5	23
Succinea ovalis	186	64	77
Turitella terebra	0·7	0·2	58
Umbraculum indicum[h]	15	2	17
Umbraculum indicum[i]	7	1	16
Urosalpinx cinera	9	9	116
Viviparus georgianus	0·3	—	1060
Pelecypoda			
Aequipecten irradians	4	4	175
Aequipecten irradians	4	3	425
Anadara transversa[j]	—	—	—

TABLE IV (contd)

Species	2-amino-2-deoxy-D-glucose ($\mu g/g$)	2-amino-2-deoxy-D-galactose ($\mu g/g$)	Protein: amino sugars
Anadara transversa[k]	—	—	—
Anadara transversa[l]	11	4	91
Anadara transversa[k]	—	—	—
Anadara transversa[m]	14	14	55
Arctica islandica	64	88	15
Corbicula consobrina	4	1	62
Crassostrea virginica	11	2	301
Laevicardium mortoni	14	15	191
Limopsis compressus	20	2	62
Lyonsia hyalina	232	21	82
Macoma tenta	11	11	121
Macoma tenta	37	—	108
Malletia sp.	15	24	73
Mercenaria mercenaria	41	30	25
Mulinia lateralis[n]	4	5	206
Mulinia lateralis[j]	7	9	166
Mulinia lateralis[o]	4	5	176
Mulinia lateralis[p]	20	25	57
Mulinia lateralis[p]	14	16	54
Mulinia lateralis[l]	25	30	34
Mulinia lateralis[l]	5	11	252
Mytilus edulis	25	16	219
Mytilus edulis	75	47	74
Neotrigonia margaritacea	89	19	80
Nucula proxima	22	4	274
Nucula proxima	10	—	723
Nucula proxima	139	12	47
Nucula truncula	73	15	119
Periploma leanum	215	14	21
Petricola pholadiformis	12	—	431
Pitar cordata	15	4	48
Pitar morrhuana	—	79	47
Pitar morrhuana	—	53	64
Saxidomus nuttali	28	16	28
Tagelus divisus[q]	6	—	88
Tagelus divisus[r]	5	—	94
Yoldia limatula	5	13	180
Yoldia limatula	5	14	194
Cephalopoda			
Loligo sp. (pen)	245,678	3	2
Nautilus pompilius	664	24	30
Sepia sp. (cuttle bone)	5,678	57	2
Spirula spirula	10,046	12	2

[a] Adapted from Degens et al. (1967). [b-r] Indicates that specimens of indentical species were obtained from different geographical sources.

wide, but hydroxyproline was present only in a very few cases. The incidence of hydroxylysine was also low; cysteine and methionine were consistently present. Three, four-dihydroxyphenylalanine was present in a few instances. Wide angle X-ray diffraction and electron microscope data were also inconsistent with the presence of collagen and it was concluded that the proteins had more affinities with the keratin group of scleroproteins. A high percentage of the tissues examined contained 2-amino-2-deoxy-D-galactose. The ratio of protein to amino sugar showed a wide variation. Obviously not all of this amino sugar can be involved in chitin. Determination on five species revealed the presence of glucose and galactose in the matrix proteins.

Voss-Foucart (1968) has solubilized and fractionated the carbohydrate-containing fraction of *Nautilus pompilus* shell matrix which remains after removal of soluble nacrine. This material was dissolved in saturated lithium thiocyanate solution by holding at 130° for four hours. Dialysis of this solution against 8 M urea, in pH 5·6 acetate buffer, gave two fractions, one of which precipitated while the other remained in solution. The second, urea-soluble fraction precipitated on dialysis against distilled water. The urea-insoluble fraction had a much higher 2-amino-2-deoxy-D-glucose content than the urea-soluble fraction, suggesting that this material was a chitin-protein mucopolysaccharide. The soluble fraction had a high protein, low hexosamine content and may therefore be a glycoprotein. Both fractions had rather similar amino acid compositions with high glycine and alanine contents.

The 2-amino-2-deoxy-D-glucose contents of the insoluble organic parts of the shell have been determined for several groups of molluscs (Hare and Abelson, 1965). The highest concentration occurs in the cross-lamellar amphineuran shell followed by Archeogastropoda, Cephalopoda, and Anisomyaria; much lower hexosamine contents are found in the Heterodonta, Mesogastropoda and Neogastropoda.

Fossil Shell Proteins: The process of fossilization appears to bring far fewer drastic changes in already calcified tissues, such as the molluscan shell, than might at first be expected; in consequence, the paleobiochemistry of proteins surviving in fossil calcified tissues has evinced considerable interest. Amino acid analyses of the protein matrices from fossil molluscan shells several million years old have been published by several authors (Degens and Love, 1965; Degens and Parker, 1965; Degens and Schmidt, 1966; Florkin et al., 1961; Bricteux-Grégoire et al., 1968; Grégoire, 1959). The field has been reviewed by Florkin (1966) together with a presentation of much new important data. Strikingly the typical chemical composition of the conchiolin of mother of pearl has been found in fossil shells from the Holocene period (Table V).

I

TABLE V

Amino Acid Compositions of Fossil and Modern Conchiolins of
Mother of Pearl[a,b]

Amino acid	*Nautilus* (modern)	*Iridina* (holocene)
Aspartic acid	9·0	10·1
Threonine	1·5	1·6
Serine	10·9	7·8
Glutamic acid	5·5	5·0
Proline	1·8	2·0
Glycine	35·7	29·8
Alanine	27·2	28·2
Valine	2·2	4·0
Isoleucine	1·8	2·4
Leucine	1·9	4·0
Tyrosine ⎫ Phenylalanine ⎭	2·4	0
Histidine	0	0
Lysine	0	2·3
Arginine	0	1·9

[a] Florkin *et al.* (1961). [b] Residues per 100 total residues.

THE BRACHIOPOD SHELL

Like that of the mollusc, the brachiopod shell can be decalcified, leaving a web-like organic matrix consisting largely of protein produced by the outer epithelial cells of the mantle prior to calcification. The protein matrices from the shells of 11 species of brachiopod have been examined by Jope (1965, 1967a). Of these, eight species were of the class Articulata, a group having a non-chitinous carbonate shell; the remaining three species were from the group classified as the Inarticulata. One of these three latter species was a member of the superfamily, the Craniacea, which like the Articulata have a calcium carbonate shell. The other two species, representative of the Inarticulata as a whole (the Craniacea are anomalous), had chitinous shells impregnated not with carbonate but with calcium phosphate, in analogy to vertebrate bone. The carbonate shells of the articulate and craniate brachiopods contain only about 0·5% of their weight as organic matrix; in the regular inarticulates this rises to 15–25%. Amino acid analyses (Table VI) of these materials confirm the division between articulate and inarticulate classes and emphasize the anomalous position of the Craniacea. A basis for a classification of orders and families seemed to be indicated.

The shell proteins all had high glycine, aspartic acid, and proline contents. Those of the phosphatic inarticulates had high alanine values also and, unlike the articulate proteins, contained relatively high concentrations of hydroxyproline and small amounts of hydroxylysine. Cysteine was present in varying amounts in all but one (articulate) species. None of the articulate protein matrices were accompanied by 2-amino-2-deoxy-D-glucose, while all the inarticulate species, including the carbonaceous *Crania anomola*, contained this amino sugar. It is impossible to say whether this 2-amino-2-deoxy-D-glucose originates entirely or in part from chitin. Certainly the matrix of each species of inarticulate gave a positive chitosan reaction.

While the high glycine and glycine plus alanine contents, coupled with moderately elevated levels of serine and proline, are characteristic of scleroproteins, Jope considered that even in the case of the hydroxyproline-containing inarticulate matrices, the amino acid compositions did not satisfy the criteria required for their classification as collagens.

Significantly, the matrix proteins were susceptible to solution in reagents normally used to bring disulphide bridged scleroproteins into solution. Thus 8 M urea, 7 M urea buffered to pH 8·9, and anhydrous dichloracetic acid, coupled with performic acid or mercaptoethanol, all solubilized the proteins. In the case of the material dissolved by the buffered 7 M urea and subjected to thiol reduction, disc electrophoresis indicated the presence of only one component.

The association of hydroxyproline-containing proteins with the formation of phosphatic calcified tissues, as with the collagens of bone, is a noteworthy feature of the inarticulate protein matrix. Of particular significance is the hydroxyproline in *Lingula*, a genus with a demonstrably long ancestry from the Ordovician: $4·5 \times 10^8$ years, the time of emergence of bony structures generally. The order Lingulidae could carry phosphatic calcification and proline hydroxylases back to the Lower Cambrian (Jope, 1967a; Hare and Abelson, 1965). Intact protein matrices have been isolated from fossil brachiopods $4·2 \times 10^8$ years old and amino acid compositions determined (Jope, 1967b).

CRUSTACEA

While the chitinous nature of the calcified crustacean integument has been well studied, surprisingly little is known of the proteins involved in the tissue matrix. The process of calcification in the crustacean gastrolith and carapace and its association with protein, mucopolysaccharide and chitin has been expounded by Travis (1963).

The presence of protein in the calcified integument has been recognized for some time (see Dennel, 1960), with early qualitative amino acid analyses being carried out by Trim (1941) and Lafon (1943). Travis (1963) found the protein in the gastrolith matrix to be accompanied by some seven per cent hexose, as well as chitin. Hunt (unpublished) has isolated a protein, in

TABLE
Amino acid and Amino Sugar Composition of proteins

Phylum												
Class												Articulata
Order										Terebratulida		
Genus	*Laqueus*					*Neothyris*				*Waltonia*	*Terebratalia*	*Macandrevia*
Protein source in shell	Periostracum	Outer & Inner (O & I) calcite	Outer & Inner (O & I) calcite	Outer & Inner (O & I) calcite	Inner calc.	Periostracum	O & I calc.	O calc.	I calc	O & I calc.	O & I calc.	O & I calc.
Gly	32·4	20·9	18·2	19·9	28·8	39·2	26·8	24·7	31·7	22·6	23·4	19·6
Ala	3·7	4·9	5·4	4·4	5·5	4·2	12·6	11·7	14·3	11·0	7·4	7·1
Val	2·6	6·9	6·5	6·5	7·0	3·2	5·6	6·0	4·2	6·0	7·8	7·8
Leu	2·0	4·3	5·2	4·6	3·4	3·0	4·2	5·0	3·0	4·8	4·3	5·6
Ileu	1·2	3·8	4·6	4·5	3·6	2·3	3·8	4·2	2·6	4·7	3·7	4·3
Ser	7·5	8·5	7·5	8·1	6·0	8·5	6·6	7·1	6·4	7·0	9·0	9·1
Thr	1·6	4·8	5·1	5·1	4·2	1·5	3·7	3·9	2·7	4·0	5·1	5·3
Pro	5·2	4·0	5·8	5·1	2·3	1·9	5·1	5·0	4·6	7·3	5·8	6·0
Hypro	—	—	—	—	—	—	—	—	—	—	—	—
Phe	6·1	4·0	3·4	3·9	3·4	3·1	3·5	4·0	3·0	3·5	3·4	3·0
Tyr	5·9	6·2	5·3	6·3	6·8	0·3	—	—	0·1	—	—	—
Try	†	†	†	†	†	†	†	†	†	†	†	†
CyS	1·3	3·3	3·7	5·3	3·9	0·6	2·4	1·1	4·7	1·2	Tr	—
Met	0·1	1·6	1·7	1·8	1·0	Tr	1·3	1·3	0·9	1·1	1·4	1·6
Asp	13·5	10·1	9·2	8·2	9·2	15·2	10·0	10·1	9·6	10·5	10·5	12·3
Glu	4·7	6·9	7·2	6·1	5·6	6·6	5·7	6·5	4·6	6·4	6·5	8·6
Hylys	—	—	—	—	—	—	—	—	—	—	—	—
Lys	2·8	3·2	4·0	3·6	2·7	3·4	2·6	3·2	2·0	3·8	4·1	3·9
His	3·2	1·0	0·8	0·7	0·5	2·0	0·1	0·2	0·2	0·3	0·6	3·0
Arg	6·4	5·5	6·3	5·9	5·8	5·3	6·1	6·0	5·6	5·8	7·1	5·7
Glucosamine	—	—	—	—	—	—	—	—	—	—	—	—
Gly+ala	36·1	25·9	23·6	24·3	34·3	41·0	39·4	36·4	46·0	33·6	30·8	26·7
Gly/ala	8·8	4·2	3·3	4·5	5·2	9·4	2·1	2·1	2·2	2·1	3·2	2·7
Dicarb/basic	1·5	1·8	1·5	1·4	1·6	2·0	1·8	1·8	1·8	1·7	1·5	1·7

a Residues per 100 total residues.
b Jope (1967a).
* Without periostracum.
† Not estimated.

0·5 % yield, from the carapace of *Cancer pagurus*, following EDTA de-mineralization and 50 % saturation of the resulting solution with ammonium sulphate. The protein precipitated in this manner had an infrared spectrum containing characteristic carbohydrate bands. Degens *et al.* (1967) published amino acid analyses of calcified and uncalcified tissues from portunid crabs and noted that the degree of calcification varied inversely with the protein content and directly with hexosamine content. Hydroxyproline was not detected. The relation of these proteins to those found in the un-calcified cuticles of the other arthropods discussed below is not clear.

VI

from the Various Layers of Brachiopod Shells[a,b]

		Brachiopoda										
							Inarticulata					
	Rhynchonellida					Acrotretida			Lingulida			
Terebratulina	Hemithiris		Notosaria			Crania		Discina	Lingula			
O & I calc.		Soft protein layer	Hard protein layer	Whole calcite protein		Whole calcite protein		Whole shell*	Outer layer	Mid-layer	Inner layer	
25·8	25·0	40·5	26·0	33·2	35·1	14·1	14·4	22·2	15·3	15·6	13·0	13·7
5·1	3·5	9·8	5·2	8·6	9·6	6·1	7·1	18·8	21·5	15·6	16·6	17·9
9·9	9·1	3·2	7·1	3·1	3·3	5·0	4·6	1·9	5·3	3·7	7·1	6·0
5·2	5·1	2·6	5·8	2·3	2·5	4·5	4·6	1·5	3·5	4·8	4·6	3·7
5·5	3·6	1·1	3·7	1·4	1·5	3·0	3·8	1·0	2·1	2·8	2·8	2·1
8·6	7·9	5·6	6·3	6·9	6·5	14·0	12·9	3·5	5·3	6·3	5·6	4·9
4·8	4·2	2·2	3·7	2·7	2·7	5·8	5·3	2·7	3·4	4·0	3·9	3·2
2·9	5·5	6·1	6·0	8·8	7·2	4·9	6·6	5·5	6·7	7·1	7·7	7·7
—	—	—	—	—	—	—	—	13·7	4·3	6·3	5·5	6·1
2·4	2·5	6·2	2·5	8·3	6·2	2·2	2·9	0·7	2·1	2·3	2·6	1·8
—	—	—	—	—	0·2	—	—	0·1	2·5	2·2	2·7	1·8
†	†	†	†	†	†	0·1	†	†	0·5	†	†	†
0·1	1·5	1·6	0·7	0·2	1·3	1·5	Tr	0·7	0·1	0·6	0·3	0·1
1·1	1·0	0·3	0·7	0·9	0·7	1·0	0·9	0·1	0·5	0·5	0·9	3·7
12·1	11·8	10·5	10·6	11·9	10·6	18·5	18·4	7·0	10·9	11·3	10·3	9·9
7·9	7·1	4·2	5·2	4·6	4·4	8·8	9·4	4·9	5·8	6·1	5·9	5·2
—	—	—	—	—	—	—	—	0·1	—	0·1	—	—
3·7	3·8	1·4	2·7	1·3	1·6	4·5	4·8	1·9	2·0	2·5	2·7	5·0
0·6	0·4	1·4	2·8	1·5	0·9	1·9	1·7	4·5	0·8	0·9	0·9	0·9
8·3	6·0	3·3	11·6	3·9	4·0	4·2	5·2	8·3	7·6	7·0	6·8	6·1
—	—	—	—	—	—	3·4	5·4	1·7	5·6	3·8	7·4	6·3
30·9	28·5	50·2	31·2	41·7	44·7	20·2	21·5	41·0	36·8	31·2	29·6	31·6
5·0	7·1	4·1	5·0	3·9	3·7	2·3	2·0	1·5	0·7	1·0	0·8	0·8
1·6	1·9	2·5	0·9	2·5	2·3	2·6	2·4	0·8	1·6	1·3	1·6	1·7

This brief consideration of the proteins in the crustacean integument in fact anticipates discussion of that peculiar complex skeletal structure which we shall examine next: the arthropod cuticle.

So far in this chapter we have considered only collagen and certain as yet incompletely characterized proteins whose situation and function suggest analogy with collagen. There occur in the invertebrates, however, numerous other structure proteins, many of which fulfil roles analogous to counterparts among vertebrate tissues, and many which have functions peculiarly invertebrate.

Chitin as a polymeric molecule, linked to protein, has been discussed in Chapter 8. The proteins involved in linkage with or accompanying chitin in invertebrate skeletal tissues were not discussed in detail, however, as they fall more conveniently within the scope of this chapter. We shall therefore now consider the structural proteins of the arthropod cuticle and of other invertebrate chitinous tissue to the extent that they fall within the terms of reference of this volume, i.e. their association with carbohydrate. We shall examine principally the insect exo-skeleton.

THE INSECT CUTICLE

Much of the classic account of the insect cuticle by Richards (1951) still stands; a later account is that of Wigglesworth (1957), while very recent accounts of structure and chemistry have been given by Locke (1964) and by Hackman (1964). Rudall (1965) has briefly reviewed insect skeletal structures.

For the purposes of discussion, the insect cuticle may be divided into two main layers (Fig. 8). The main body of the cuticle is the thick inner procuticle, while overlying this is a thin layer composed of tanned protein, phenols and waxes, constituting the epicuticle. The layer of procuticle immediately adjacent to the epicuticle may also contain tanned proteins; in this case the epicuticle may be subdivided into tanned exocuticle and inner soft endocuticle. The procuticle contains mainly chitin and protein, some but not all of which is a part of the chitin mucopolysaccharide complex.

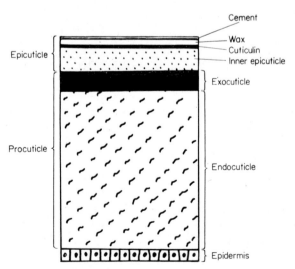

Fig. 8. Generalized insect cuticle.

It must not be assumed that this description represents the state of the whole insect integument; indeed the insect cuticle is far from uniform in its structure. Depending upon mechanical function and other less well understood physiological factors, the various regions of the cuticle differ in appearance, and in mechanical and chemical properties. The cuticle may be hard or soft, rigid or flexible, dark or light coloured, high or low in chitin content. Rigid cuticle may fulfil the diverse functions of protection (armour plating), maintenance of body shape, and provision of attachment and leverage points for the muscles. Soft cuticle permits movement, acting as a flexible hinge for the displacement of hard cuticular regions relative to one another. In this respect Weis-Fogh (1960, 1961a, b) has noted the rubber-like behaviour of certain parts of the insect (and crustacean) cuticle, a property arising from the presence of a protein called resilin, similar in its properties to the vertebrate elastic protein elastin.

Molecular order in cuticle may not be high, in the sense that the chitin chains do not form fibres by intermolecular interactions (in a manner analogous to plant cellulose), although the crystallites do lie parallel to the cuticle surface (Seifter and Gallop, 1966). This imparts to the cuticle moderate rigidity but low tensile strength. The degree of order is only slightly greater in rubber-like cuticle and its strength is comparable to solid cuticle. The process of tanning imparts to the cuticle considerable compressive strength and flexion rigidity without appreciable alteration of the properties of tensile strength and elastic coefficient found in such soft cuticles as old arthropod tibia.

RESILIN

Rubber-like cuticle with a high content of the elastic protein resilin occurs in patches in the cuticles of many insects (Weis-Fogh, 1960), but principally in the exo-skeleta of winged insects. It also occurs in the crustacea (Andersen and Weis-Fogh, 1964). Principal sources of resilin for study have been the main wing hinges and prealar arms of the desert locust *Schistocerca gregaria* and the elastic tendons of the dragonfly *Aeshna cyanea*. The function of the main wing hinge is implicit in its name, the resilin providing for the elastic recoil of the wings. The prealar arm mediates the anterior suspensions of the moveable plate between the forewings, while the elastic tendon is a swelling of the tendon for the pleuro-subalar muscle, also concerned with the flight process.

Jensen and Weis-Fogh (1962) believe the rubber-like cuticle of the locust to consist of parallel chitin plates, some $0.2\,\mu$ thick, cemented together by a "glue" of resilin $3\,\mu$ thick. Swollen with water this structure has the properties of an almost ideal isotopic rubber.

The chemistry and properties of resilin have been reviewed by its discoverer Weis-Fogh (Andersen and Weis-Fogh, 1964) and by Seifter and Gallop (1966).

Resilin is colourless, but under an ultraviolet light it emits a strong bluish-white fluorescence. Although highly solvated by water and other aqueous solvents, resilin is strikingly insoluble in all of the usual and unusual protein solvents. It does however yield readily to proteolytic enzymes.

Resilin shows a complete lack of flow under tension, a property unique among both natural and synthetic rubbers, and it exhibits no trace of secondary structure, existing simply in its native form as a three-dimensional network of randomly-coiled polypeptide chains stabilized by cross-links. The peculiar character of the cross-linking accounts for the insoluble nature of the protein.

Amino acid composition: Resilin contains no methionine, cystine or hydroxy-proline and less than 0.3% tryptophane. The amino acid composition (Table VII) as a whole, if compared with other proteins of the fibrous group, most closely resembles collagen; the relationship is however distant. Extensive, detailed studies have shown the cross-linking residues to be di- and tri-tyrosines (Fig. 9) (Andersen, 1963, 1964). An elastic protein, constituting the hinge ligament of the mollusc *Mytilus edulis*, has been shown by Andersen (1967) to contain yet another tyrosine derivative: 3,3'-methylene-bistyrosine (Fig. 10), which again probably has a cross-linking function. This amino acid was also present in the inner ligament proteins of *Spisula solidissima* and *Pecten maximus*. A second amino acid with a rather similar but as yet incompletely characterized structure accompanies 3,3'-methylene-bistyrosine in the *Mytilus* and *Spisula* proteins.

Association of Resilin with Polysaccharide or Carbohydrate: Resilin, which has been separated from chitin by partial hydrolysis with dilute acid, and following the removal of the insoluble chitin, subjected to further and complete hydrolysis, contains no hexosamine. Whether neutral mono-saccharides are present is not known.

Resilin has been dealt with here only scantily since its association with carbohydrate is uncertain. It is by no means clear as yet whether or not there is any specific link with the accompanying chitin molecules. The problem is a difficult one in view of the insolubility of the two macromole-cules involved.

Resilin and its Vertebrate Counterpart Elastin

The elastic fibres of vertebrate connective tissue intercellular space are composed of the protein elastin. This protein is in many ways the counter-part of invertebrate resilin. Like resilin, elastin is a rubber-like elastic

TABLE VII
Amino Acid Compositions of Resilin and Elastin[a]

Amino acid	Resilin[b]	Elastin[c] (bovine ligamentum nuchae)
4-Hydroxyproline	—	11·4
Aspartic acid	102·0	7·7
Threonine	29·6	8·6
Serine	78·6	8·0
Glutamic acid	50·4	15·2
Proline	79·4	109
Glycine	376·0	331
Alanine	111·0	223
Valine	25·6	141
Cystine/2	—	—
Methionine	0	trace
Isoleucine	20·4	27·0
Leucine	25·6	63·9
Tyrosine	29·2	7·7
Phenylalanine	27·4	35·0
Desmosine/4[d]	—	13·6
Lysine	4·9	3·2
Histidine	6·5	0·6
Arginine	33·6	7·0
Tryptophane	—	0·5
(NH_3)	69·2	—

[a] Residues per 1000 total residues.　[b] Bailey and Weis-Fogh (1961).
[c] Gotte et al. (1963).　[d] Includes isodesmosine.

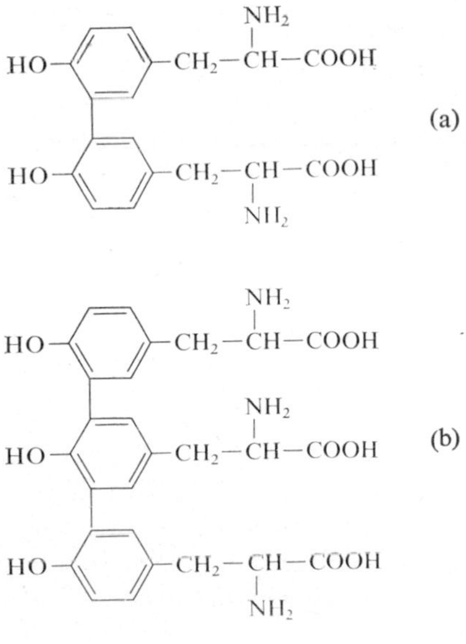

(a)

(b)

Fig. 9. (a) Dityrosine;
(b) Trityrosine.

1*

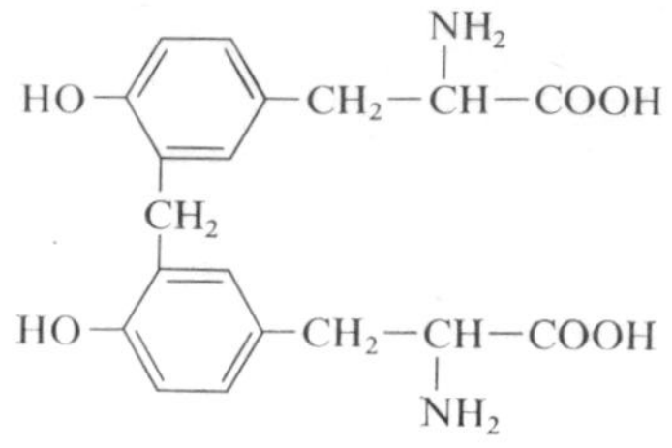

Fig. 10. 3,3′-methylene-bistyrosine.

protein without secondary structure but having a three-dimensional net-work structure of randomly-coiled polypeptide chains stabilized by cross-linkages of an unusual nature. Elastin shares the insolubility and swelling properties of resilin. There are, however, remarkably few polar side chains in elastin, whereas there are a large number in resilin. Elastin contains hydroxyproline (Table VII). The major cross-linking amino acids are not in this case di- and tri-tyrosines but are instead two related pyridinium derivatives and N-ε-(5-amino 5-carboxypentanyl)-lysine. The pyridinium derivatives contain amino acid side chains suitable for the cross-linking of peptide chains. These two compounds, desmosine and isodesmosine (Fig. 11), both contain four potential cross-linking side chains and hence four protein chains might be feasibly joined by a single unit. The desmosines impart a characteristic ultraviolet fluorescence to elastin.

It has been suggested that the elastic fibres of vertebrate connective tissue are a composite of filaments of fibrous elastin and an amorphous glycoprotein occurring in conjunction with mucopolysaccharide (cf. resilin and chitin), collagen fibres and other proteins.

While it has been suggested that resilin and elastin present a particularly interesting example of parallel evolution, it is worth noting that dityrosine has been detected in foetal elastin (LaBella *et al.*, 1967). Dityrosine has also been detected in small amounts in vertebrate collagen (LaBella *et al.*, 1968). Elastin-like materials have been reported in several invertebrates (Van Gansen-Semal, 1960; Elder and Owen, 1967).

TANNED PROTEINS

The sclerotins, skeletal proteins which owe their stability to tanning by orthoquinones, are characteristic of the arthropod exo-skeleton; they have been reviewed by Pryor (1962) and Brunet (1965).

Hackman (1953a), in a study of the water-insoluble proteins in the hard cuticle of the dark-brown elytra of the beetle *Aphodius howitti*, obtained in 9·4% yield (after removal of lipids and water-soluble proteins) a dark-coloured protein, by extraction with N-sodium hydroxide solution at 50°

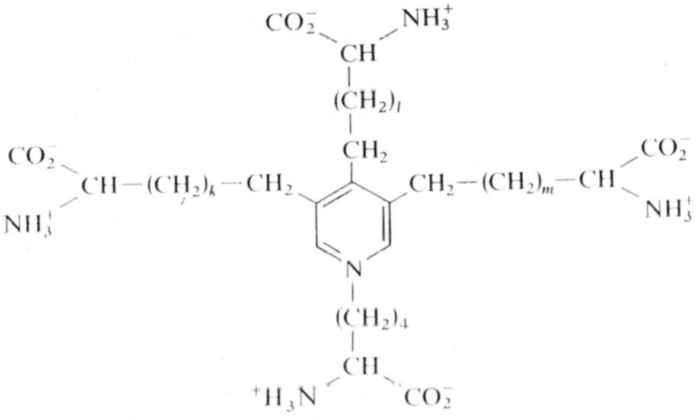

Desmosine ($k = m = 1, l = 2$)

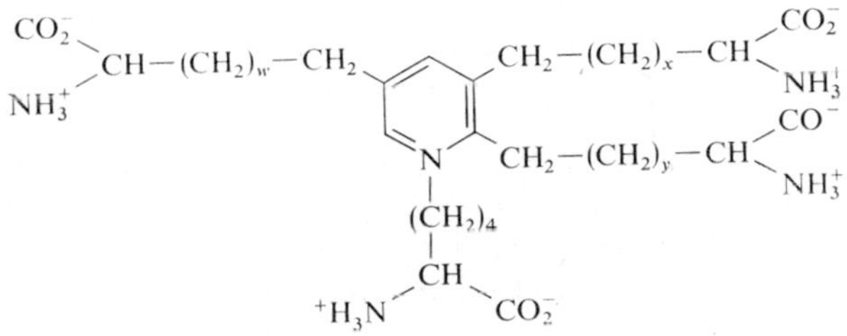

Isodesmosine ($w = x = 1, y = 2$)

$$HOOC-CH-(CH_2)_3-CH_2-NH-CH_2-(CH_2)_3-CH-COOH$$

with NH_2 groups below the first and second CH centres.

$N-\epsilon-$(5-amino-5-carboxypentanyl)-L-lysine. (Lysinonorleucine).

Fig. 11

for five hours, followed by precipitation with 50% saturated ammonium sulphate. This material was a glycoprotein containing some 3·3% total carbohydrate, estimated as glucose by the orcinol-sulphuric acid reaction. The nature of the carbohydrate was not determined but appeared not to be chitin and to be an integral part of the protein. The glycoprotein contained only traces of threonine and neither proline nor hydroxyproline. The total nitrogen content was 12·17% (on an ash free basis), the sulphur content 0·3% approximately, and the ash content 9·26%. Thus the glycoprotein must contain 18·9% of some other non-nitrogenous material. In view of the dark colour of the material, phenolic and/or quinonoid substances seem to be likely candidates for this missing percentage. The glycoprotein may not be homogeneous; dinitrophenylation followed by acid hydrolysis yielded DNP derivatives of glycine, serine, alanine, glutamic acid, and lysine.

Hackman's glycoprotein appears to be similar to a water-insoluble fraction obtained from the soft, larval cuticle of *Sphinx ligustri* by Trim (1941b), using warm five per cent sodium hydroxide. This protein had 11% ash, 12·4% total nitrogen, 0·57% sulphur, and 3·2% carbohydrate. A similar protein was found to occur in the larval cuticle of *Sarcophaga falculata*.

The residue remaining after removal of the water- and alkali-soluble proteins from *A. howitti* elytra (Hackman, 1953a) contained "hardened" protein (sclerotin) and chitin (55·1%). Carbohydrate other than chitin was not detected. Although hydroxyproline was present in the protein, proline, serine, threonine, and lysine were absent. No free amino groups could be detected.

ARTHROPODINS

Water-soluble proteins from the cuticles of several insect species have been isolated and studied by Trim (1941a, b) and Fraenkel and Rudall (1940, 1947). Trim, using either water or a buffer solution containing 1·9% w/v aqueous borax, water, ethanol, and ether in the proportions 5 : 5 : 4 : 1 respectively, extracted proteins from the soft cuticles of the larvae of *S. falculata* and *S. ligustri*. The proteins appeared to be quite distinct from collagen but bore some resemblance to silk sericin.

Fraenkel and Rudall gave the name "arthropodin" to the water soluble protein of insect cuticle, believing it at that time to be a fundamental protein common to all arthropod cuticles. Arthropodin, extracted from the larvel cuticles of blowflies and moths and also from the intersegmental cuticles of lobsters, was not denatured by hot water and was soluble in hot (but not cold) trichloracetic acid. It was precipitated from aqueous solutions by ethanol in concentrations above 45%, by one-third saturated ammonium sulphate and by saturated sodium chloride. After cold trichloracetic acid or ethanol precipitation the proteins readily redissolved in water, remaining in solution even after heating in boiling water. These properties, the authors commented, are reminiscent of collagen; they are also typical in many respects of glycoproteins. X-ray diffraction patterns did not in fact resemble those typically given by collagens but were of the β type (Fraenkel and Rudall, 1940, 1947). Birefringent fibres may be drawn from concentrated solutions of cuticle protein (Richards and Pipa, 1958).

Hackman (1953b), using Trim's buffer, isolated water-soluble proteins from the cuticles of seven insect species: *Diaphora dorsalis, Dasygnathus, Mastochilus, Tenebrio molitor* (all cuticles from larval Coleoptera), *Chortoicetes terminifera* (Orthoptera; final nymphal instar), *Cyclochila australasia*

(Hemiptera; final nymphal instar, cast skin) and *A. howitti* (Coleoptera; adult elytra). All seven proteins gave negative responses to tests for sulphur and carbohydrate. The amino acid compositions were remarkably similar and all contained hydroxyproline. It appears however, contrary to earlier opinions, that arthropodin is in fact a heterogeneous mixture. Moving boundary electrophoresis of the cuticle extract of *D. dorsalis* indicated the presence of at least four components, the fastest, anodic-migrating one of which contained almost no serine, threonine, phenylalanine, proline or hydroxyproline.

Hackman and Goldberg (1958) have subdivided the insect cuticular proteins into five classes: cold-water-soluble, soluble in 0·16 M sodium sulphite solution, soluble in 7 M aqueous urea, soluble in cold 0·01 N sodium hydroxide, and a final residue of insoluble protein equal in quantity to the chitin present. All of the soluble fractions are reported to be electrophoretically heterogeneous. Thus the soft larval cuticle of the beetle *Agrianome spinacollis* contained eight per cent water-soluble protein, one per cent sodium sulphite-soluble protein, 16% urea-soluble protein, two per cent alkali-soluble protein, and 35% insoluble protein accompanied by 37% chitin.

Moorefield has reported the isolation of electrophoretically heterogeneous water-soluble proteins from the cuticle of *Sarcophaga crassipalpis*. This material was monodisperse in the ultracentrifuge with an apparent molecular weight of 7000–8000. Such a small protein (or proteins) would be likely to be readily amenable to mild extraction procedures.

The failure to detect carbohydrate in some of these preparations must be interpreted with caution. Water-soluble, salt-soluble, and urea-soluble preparations are unlikely to be suspect; alkali-soluble materials however may have been severed from carbohydrate moieties by the extraction procedure, perhaps even from the chitin itself.

Proteins which appear to be quinone-tanned are not confined to the arthropod cuticle; they occur in the Nematodes, an example being the cysts of the eelworm, in the egg shells of Platyhelminths, in the radula teeth of Gastropod mollusca and in the byssus and shell hinges of Lamellibranch molluscs. They are probably present also in Oligochaete and Polychaete chaetae and in the tubes of Pterobranchs.

INVERTEBRATE SKELETAL AND STRUCTURAL PROTEINS WITH AFFINITIES TO VERTEBRATE KERATINS

Keratins have been defined as "natural, cellular systems of fibrous proteins cross-linked by cystine sulphur" (Lundgren and Ward, 1963).

"They have evolved primarily as a barrier to the environment, serving to protect the higher vertebrates—amphibians, reptiles, birds and mammals—from the stresses of life. Keratins occur as the principal constituents of the horny outermost layer of the epidermis and of related appendages, such as horns, hooves, scales, hair, and feathers, that are derived from the skin." Low sulphur contents are allowed by these authors in the case of the "soft" keratins.

Keratins are generalized as being hydrogen-bonded within (α) or among (β) individual peptide chains. The α component typically forms helical strands with a degree of crystallinity. Histologically, keratins are classified as "soft" or "hard". "Soft" keratin arises from epithelial cells which go through almost discrete granular and glassy stages and is observed in structures which are becoming horny, for example corns and callouses. Soft keratin has a low sulphur content. "Hard" keratin arises as the result of less demarcated cellular transformations so that granular or glassy stages may not be observed. Hard keratin is rich in sulphur and in disulphide bridges.

As Seifter and Gallop (1966) observe, "the molecular properties of keratins may endow tissues with toughness, elasticity, and insolubility. Thus the keratins are found in structures adapted for mechanical protection, warmth, abrasion resistance, grasping, tearing, feeling, digging, climbing, fighting, eating, flight and ornamentation." Many if not all of these requirements and activities are common to invertebrates as well as vertebrates. Whether keratins are peculiar to vertebrates is uncertain, however. Certainly many analogous but clearly distinct macromolecular systems are found in invertebrates occupying situations filled by keratins in vertebrates. We have already discussed some of these in this chapter, for example the tanned proteins. In many cases also chitin-protein systems replace keratin. As an extreme but attractive example of the latter case we may recall the pelt of hair-like chaetae on the dorsal surface of *Aphrodite* the sea mouse, an errant-polychaete annelid worm.

Proteins having the appearance, function and some of the physical and chemical characteristics of vertebrate keratins have however been detected in certain invertebrate situations. Those which are associated with carbohydrate will be considered here; vertebrate keratin would incidentally appear to be carbohydrate-free.

MOLLUSCA

Periostracum: The outside surface of the molluscan shell is covered by a protective skin, the periostracum, composed of a flexible, tough, horny, semi-transparent, brownish material. This material is composed principally of protein with a low percentage of associated oligosaccharide and some lipid, the latter possibly having a water-proofing function.

The periostracum of the whelk *Buccinum undatum* has been examined by Hunt (unpublished) and the amino acid composition determined (Table VIII). The cysteine content is low and the aspartic acid content far higher than any of the other amino acid residues. Proline, hydroxyproline, and, rather surprisingly, allohydroxyproline are present. The tyrosine content is also low and the ratio of acidic (aspartic and glutamic acids) to basic (lysine, histidine, and arginine) amino acids is approximately three to one. There is evidence to suggest the presence of small amounts of 3,4-dihydroxyphenylalanine in the protein.

The amino acid composition of this gastropod periostracum contrasts sharply with that of a typical bivalve mollusc *Mytilus californianus*, in which the glycine rather than aspartic acid content is high (Hare, 1963) (Table VIII).

Buccinum periostracum contains some five per cent amino sugar consisting of 2-amino-2-deoxy-D-glucose and 2-amino-2-deoxy-D-galactose in

TABLE VIII

Amino Acid Composition of *Buccinum undatum* and *Mytilus californianus* Periostraca[a]

Amino acid	B. undatum	M. californianus
Cysteic acid	17·9	14·4 (Half-cystine)
Hydroxyproline	16·4	—
Aspartic acid	218·6	26·0
allo Hydroxyproline	9·0	—
Threonine	94·4	8·7
Serine	87·3	69·0
Glutamic acid	83·7	25·0
Proline	47·3	23·7
Glycine	69·9	508
Alanine	50·1	8·0
Valine	53·2	32·0
Isoleucine	28·9	13·0
Leucine	66·9	9·0
3,4 Dihydroxyphenylalanine	trace	—
Tyrosine	15·0	155
Phenylalanine	36·1	10·5
Lysine	46·8	21·7
Histidine	17·0	16·7
Arginine	40·5	55·0

[a] Residues per 1000 total residues.

the ratio three to one. The hexose content, estimated by the orcinol-sulphuric acid reaction, is $3 \cdot 3 \%$ and the monosaccharides galactose, mannose, xylose, fucose and rhamnose have been identified in hydrolysates by paper and gas-liquid chromatography.

Possibly the low tyrosine content of this material may be only apparent, reflecting some involvement of aromatic residues in cross-linking. Periostracum is brown in colour; this may indicate a tanned structure. The protein seems to be insoluble in most of the more usual protein solvents and intractable to keratin solvents although it does swell and become almost colourless in thioglycollic acid. Periostracum does not fluoresce in ultraviolet light.

K. M. Rudall, in a personal communication to the author, has suggested that the X-ray diffraction pattern of *Buccinum* periostracum resembles the patterns given by the keratin-myosin-fibrinogen group of proteins, i.e. the α pattern.

Amino acid analyses for the periostraca of four gastropod, eight pelecypod, and one cephalopod species have been published by Degens *et al.* (1967) (Table IX). All of these contained hexosamine to a greater or lesser extent and the analyses indicated a fair degree of variability in amino acid composition between species. The gastropods constituted the most homogeneous group. Pelecypod periostracum was characterized by a high glycine content; in one instance glycine contributed over 66% of the total number of amino acid residues and in five out of eight species studied the glycine content was over 50%. Such a high proportion of this amino acid must significantly affect the form of the secondary and tertiary structure. Gastropod and cephalopod materials both had high acidic amino acid levels; those from pelecypods were low in these amino acids. None of the proteins from the pelecypod group contained hydroxyproline although three of the species yielded proteins containing hydroxylysine. Two of the gastropod proteins contained hydroxyproline while in the cephalopod periostracum the proportion of this amino acid was quite high (65 residues/ 1000 total). In one of the gastropods and four of the pelecypods 3,4-dihydroxyphenylalanine was detected in low concentration (highest, 9 residues/1000 total; lowest, 1 residue/1000 total in the gastropod). Degens and his associates suggest that the proteins of the periostracum may be stabilized by quinone tanning. This hypothesis is one which is likely to prove difficult to substantiate or disprove.

Meenakshi, Hare and Wilbur (Wilbur and Simkiss, 1968) have also given amino acid analyses for a range of molluscan periostraca (Table X).

The overall view suggested by Tables VIII to X would seem to be that, with the exception of the opisthobranch and certain marine prosobranch molluscs, glycine accounts for frequently more than half the total residues

TABLE IX

Amino Acid Composition of the Periostracum of Various Mollusca[a]

	OH-Pro	Asp	Thr	Ser	Glu	Pro	Gly	Ala	Cys	Val	Met	Iso	Leu	DOPA	Tyr	Phe	OH-Lys	Lys	His	Arg	Hexosamines
							Amino acid residues per 1000 total														Proteins
Gastropoda																					
Akera soluta	19	148	33	112	130	45	74	101	2	73	7	57	90	—	12	40	—	19	—	39	15
Aplysia willcoxii	—	163	52	120	141	37	91	89	2	71	5	48	93	1	12	31	—	23	1	21	17
Dolabella																					
scapula	—	135	55	105	115	31	73	79	1	58	5	68	104	—	8	49	—	35	42	30	18
Hydatina physis	12	132	72	73	137	38	72	89	7	66	10	57	100	—	14	33	—	38	8	45	27
Pelecypoda																					
Arctica islandica	—	16	5	23	13	25	669	21	14	11	20	21	12	9	70	13	—	19	6	33	390
Corbicula																					
consobrina	—	23	8	17	12	23	626	28	5	12	34	10	28	—	93	29	—	13	21	17	18
Mulinia lateralis	—	74	22	40	36	51	435	53	22	32	64	34	34	2	9	42	1	16	1	46	457
Mytilus edulis	—	52	19	60	15	24	510	38	2	43	3	16	27	—	97	21	—	15	12	46	775
Pitar morrhuana	—	91	47	54	58	47	333	61	10	49	3	29	42	2	30	34	5	37	9	59	351
Pitar morrhuana	—	105	54	65	68	48	286	79	8	50	3	29	43	1	12	34	5	38	10	61	235
Solemya velum	—	29	11	89	19	54	494	41	4	42	72	27	21	2	12	54	—	13	3	13	37
Tagelus divisus	—	23	23	20	24	37	550	42	0·4	34	60	8	16	1	69	41	—	24	3	25	20
Tagelus divisus	—	10	25	26	15	54	571	49	7	53	19	27	12	7	44	82	—	14	1	6	417
Tagelus divisus	—	7	24	28	13	54	547	54	7	57	19	4	10	—	39	84	—	24	4	18	123
Yoldia limatula	—	32	9	29	9	34	648	17	4	18	17	5	6	—	13	85	12	20	4	38	502
Yoldia limatula	—	35	9	26	9	32	661	18	4	16	17	5	6	—	13	86	4	22	4	34	523
Cephalopoda																					
Nautilus																					
pompilius	65	224	48	67	101	74	132	46	38	49	3	28	41	—	12	19	—	17	—	35	16

[a] Degens et al. (1967).

TABLE X
Amino Acid Compositions of Molluscan Periostraca[a,b]

Amino acid	Bivalvia	Marine prosobranch	Freshwater prosobranch	Terrestrial pulmonate
Half-cysteine	4–37	6–46	0–39	—
Hydroxyproline	—	—	—	—
Aspartic acid	26–60	90–164	16–60	8–109
Threonine	8–14	16–64	5–24	5–50
Serine	18–69	26–116	9–29	10–120
Glutamic acid	10–38	30–146	14–69	9–101
Proline	24–38	46–92	16–43	15–38
Glycine	400–546	71–398	460–628	250–632
Alanine	8–120	23–115	10–60	56–80
Valine	17–42	22–67	12–105	50–103
Isoleucine	13–40	13–52	7–53	33–46
Leucine	9–60	14–94	12–89	32–52
Tyrosine	91–135	6–184	18–139	3–59
Phenylalanine	10–70	12–99	9–91	17–35
Lysine	3–24	7–136	7–25	7–19
Histidine	10–20	0–12	1–20	0–2
Arginine	13–90	14–105	7–26	4–38

[a] Adapted from Wilbur and Simkiss (1968); work by Meenakshi, Hare and Wilbur (unpublished). [b] Residues per 1000 total residues.

of the periostracum. Glycine is particularly high in the fresh water species while acidic residues are low.

Operculum: A protein, rather similar in its appearance to periostracum, forms the operculum, the horny plate carried on the prosobranch gastropod foot and acting as a door to the shell when the body is withdrawn.

Degens *et al.* (1967) have given amino acid analyses for the opercula of *Lunatia trisseriata* and *Nerita plexa* (Table XI). The materials have a low cysteine content (1 and 0·7 residues/1000 total respectively) while hydroxyproline is absent and the glycine contents high (348 and 335 residues/1000 total respectively). *Lunatia* operculum contained 0·962 mg/g 2-amino-2-deoxy-D-glucose and 25 mg/g 2-amino-2-deoxy-D-galactose, while that of *Nerita* contained 0·081 and 0·072 mg/g of these sugars respectively. The operculum of *B. undatum* contains only traces of a single amino sugar, 2-amino-deoxy-D-glucose together with galactose, mannose, xylose, fucose and rhamnose (Hunt, unpublished). The amino acid compositions of these three opercula (Table XI) are all rather similar.

Buccinum opercula are dark brown in appearance and may therefore be tanned; like certain insect cuticles they are rather brittle. Much of their brown colour can be bleached by the action of sodium hydrosulphite ($Na_2S_2O_4$).

TABLE XI
Amino Acid Compositions of Gastropod Opercula[a]

Amino acid	Buccinum[b] undatum	Lunatia[c] trisseriata	Nerita[c] plexa
4-Hydroxyproline	—	—	—
Aspartic acid	146·2	104	80
Threonine	19·8	10	12
Serine	59·9	25	47
Glutamic acid	41·0	33	42
Proline	28·9	55	55
Glycine	276·7	348	335
Alanine	136·6	187	142
Cysteine	—	1	0·7
Valine	40·6	64	40
Methionine	—	0·4	1
Isoleucine	13·2	5	6
Leucine	38·0	35	52
3,4-Dihydroxyphenylalanine	—	—	—
Tyrosine	52·5	62	30
Phenylalanine	15·1	34	72
Hydroxylysine	—	—	7
Lysine	51·6	22	15
Histidine	48·4	2	7
Arginine	31·4	12	57

[a] Residues per 1000 total residues. [b] Hunt. [c] Degens et al. (1967).

Egg cases: The nature and mode of secretion of the molluscan egg case has from time to time attracted the interest of zoologists (Hancock, 1956; Fretter, 1941; Ankel, 1937; Drew, 1901); in spite of this however, interest has crystallized largely only in the study of the histology and histochemistry of these materials. The actual chemical character of the egg case remains conjectural, apart from the qualitative studies of Bayne (1966) and the more detailed analytical work of Hunt (1966). While Bayne's studies related to a land-living pulmonate gastropod, Hunt examined the egg case of a marine gastropod *B. undatum.* Degens et al. (1967) gave an amino acid analysis for a cephalopod egg case (see Table XII).

Buccinum egg case is composed of a horny semi-transparent, brownish material, insoluble in water, 0·5 M sodium chloride and 4 N sodium hydroxide. Sodium sulphide (five per cent w/v in water) causes the material to swell but fails to solubilize it, as do performic acid at 4° and thioglycolic acid. The brown colour does however tend to disappear in the latter solvent leaving the egg capsule with a pale, almost transparent, white appearance. If the sulphide bonds are present they cannot contribute significantly to the stabilization of the tertiary structure. It would therefore seem likely that

TABLE XII

Amino Acid Compositions of *B. undatum* and *Argonauta hians* Egg Cases[a]

Amino acid	*B. undatum*[b] (Gastropoda)	*A. hians*[c] (Cephalopoda)
Hydroxyproline	—	—
Cysteic acid	7·5	—
Aspartic acid	134·9	125
Threonine	62·3	66
Serine	79·0	167
Glutamic acid	175·8	70
Proline	14·9	19
Glycine	46·6	191
Alanine	75·6	32
Cysteine	—	76
Valine	67·4	46
Methionine	2·7	5
Isoleucine	62·4	50
Leucine	92·4	14
3,4-Dihydroxyphenylalanine	trace	—
Tyrosine	trace	29
Phenylalanine	57·2	39
Hydroxylysine	—	3
Lysine	73·2	39
Histidine	—	11
Arginine	38·1	19

[a] Residues per 1000 total residues. [b] Hunt (1966). [c] Degens *et al.* (1967).

there is some other factor accounting for the stability and insolubility of the material.

Buccinum egg cases exhibit considerable resistance to proteolytic digestion. Crystalline pepsin, trypsin, chymotrypsin and bacterial proteinase are almost completely without action and the powerful general unspecific protease pronase brings about only a low degree of degradation even after five days. More prolonged pronase treatment brings a yellowish fluorescent substance into solution; the egg cases themselves however seem to remain largely intact.

Egg cases fluoresce with a strong blue-white emission, when illuminated with an ultraviolet light source producing radiation of about 3500 Å wave length. This fluorescence persists even after the extraction of fluorescent lipid materials from the egg cases by prolonged extraction with a range of lipid solvents. This fluorescence may perhaps suggest the presence of aromatic cross-linking amino acids analogous to those found in resilin or elastin. The data at present available however seem to suggest that neither dityrosines nor desmosines are present.

The amino acid composition of *Buccinum* egg case is given in Table XII. Amino acids account for approximately 77·5% of the total dry weight of the egg case. Tyrosine and histidine are present in trace amounts only, while cysteine (as cysteic acid) occurs in low concentration. In some respects the amino acid analysis resembles that of certain α-keratins; Rudall (1968b) has noted a resemblance to the trypsin resistant proteins of hair medulla and inner root sheath.

The X-ray diffraction patterns of *Buccinum*, and other marine gastropod egg cases, also suggest a structure similar to that of α-keratin (Rudall, personal communication). Infrared spectra are consistent with this view, the characteristic amide bands being located at 1550, 1660, and 3300 cm^{-1}.

Rudall (1968b) has observed that the capsule, at the electron microscope level, is composed of very thin ribbons which exhibit an axial periodicity of about 960 Å, the banding pattern being symmetrical, unlike vertebrate collagen. There is a general resemblance of the pattern to light meromyosin but on a larger scale. The band period of 960 Å, if taken as an indication of molecular length, compares closely with the length of 1050 Å for prekeratin.

The infrared spectra also indicate the presence of carbohydrate, with bands in the 950–1100 cm^{-1} region and a characteristic hydroxyl band at around 3400 cm^{-1}. Paper and gas-liquid chromatography of acid hydrolysates of egg case reveal the presence of galactose, mannose, xylose, fucose, and rhamnose. Significantly perhaps these are the same sugars which are found in *Buccinum* periostracum and also in small amounts in *Buccinum* operculum (Hunt, unpublished). These sugars account for some five to six per cent of the egg case dry weight, while a further two per cent is composed of hexosamine comprising 2-amino-2-deoxy-D-glucose and 2-amino-2-deoxy-D-galactose in the molar ratio three to two (Hunt, unpublished). No detectable sialic acid is present. *A. hians* (Cephalopod) egg case contains 80 μg/g 2-amino-2-deoxy-D-glucose and 265 μg/g 2-amino-2-deoxy-D-galactose (Degens *et al.*, 1967).

While it has been suggested (Drew, 1901; Gersch, 1936) that the egg cases of gastropods (in this case *Nuculla delphinodonta* and *Gibbula tumida*) are derived from secretions of the hypobranchial gland, this seems unlikely for *B. undatum*, where the compositions of egg case and the normal hypobranchial mucin are totally different. The egg case is most probably secreted by the lower part of the female genital duct. It has been suggested that the secretion is kneaded into its final shape by a temporary groove along the base of the foot (Fretter, 1941; Cunningham, 1899); possibly it is at this time that hardening and insolubilization take place, perhaps involving the action of an enzyme.

The egg case of *Nucella lapilus* (Muricidae) has a similarly inert character to that of *Buccinum* (Ankel, 1937). The plug cementing the capsule does not dissolve in potassium hydroxide, acids, vertebrate enzymes or sea water. The material is however softened by extracts of crushed pre-hatching juveniles.

Murex truncus egg cases also resist the action of proteolytic enzymes while giving positive reactions to tests for protein (Dulzetto and Labruto, 1951). These egg cases were soluble in warm concentrated mineral acids and in five per cent sodium or potassium hydroxide at 100° (few proteins would resist these conditions). However, alkaline solutions gave a precipitate on neutralization and the supernatant remaining after filtration yielded a water-soluble gelatinous precipitate when treated with ammonium sulphate. This soluble material appeared to be a protein. Elementary analyses of the egg cases gave C, 57·02; H, 7·35; N, 14·58 and S, 4·7%. The nitrogen content was divided among ammoniacal N, 0·71; melamine N, 0·04; basic N, 10·19; amine N, 1·96; and immine N, 1·68%. The authors concluded that the egg case was composed of a keratin-like scleroprotein.

Thin layer chromatography of the acid-hydrolysed egg shell of the pulmonate gastropod *Agriolimax reticulatus* revealed the presence of "abundant" 2-amino-2-deoxy-D-glucose, galactose, fucose, and 11 amino acids: alanine, glycine, aspartic acid, glutamic acid, serine, threonine, leucine/isoleucine, proline, phenylalanine, valine and cystine (Bayne, 1966). There appeared to be more galactose than fucose.

Rangarao (1963a, b) found the egg cases of the pulmonate land snail *Ariophanta lingulata* to contain protein, acid mucopolysaccharide and calcium salts. The tests were however only histochemical.

PLATYHELMINTHS

The skeletal structures of the platyhelminths are thought to be composed of scleroproteins rather than chitin (Hyman, 1951). Lyons (1966) has demonstrated the affinities of monogenean attachment sclerites with vertebrate keratin by histochemical, chemical and physical tests. The association of these proteins with carbohydrate was however only tenuously established and the positive reactions which were obtained may have been due to the presence of aldehydes.

PROTOCHORDATES

In view of the possible evolutionary significance of the affinities of the protochordate groups with the vertebrate line it would be of considerable interest to establish whether keratins occur in this phylum. However, little seems to be known of the nature of the scleroproteins of hemichordates, urochordates and cephalochordates. A so-called pseudokeratin has been isolated from the defatted test of the urochordate *Cynthia roretzi* (Tsuchiya and Suzuki, 1962). The only carbohydrate reported associated with this protein was tunicin (cellulose) of the test.

SILK PROTEINS

Silk is the fibrous macromolecular material secreted by most species of insects and Arachnida at some stage of their lives for a variety of essentially structural functions. Silk is usually proteinaceous in nature but chitinous silks secreted by larvae of the praying mantis are known. Even in this latter case it is likely that there is an associated protein component. A silk of collagen is known (the cocoon of the gooseberry sawfly, *Nematus ribesii*) but on the whole, arthropod silk is composed of a unique class of protein, fibroin.

The occurrence and nature of arthropod silks has been reviewed by Lucas *et al.* (1958, 1960), by Rudall (1962) and by Seifter and Gallop (1966). By accepted convention, fibroins are recognized as those protein elements of silk which exhibit, either before or after stretching, a parallel β-type of wide-angle X-ray diffraction pattern. The fibrous components of silk, the fibroins, may exist naturally in various conformations depending upon the source of the silk; these have been discussed by Rudall (1962). For example, egg stalks of *Chrysopa flava* contain fibroin in the cross β-conformation, i.e. the polypeptide chains lie across or at right angles to the fibre axis; this protein on extension takes up a parallel β-conformation with the polypeptide chains lying parallel to the fibre axis. The most commonly-occurring silks are of the parallel or anti-parallel β-type. The family Aculeata of the Hymenoptera produce cocoon silks showing an α-conformation, i.e. a helical rather than pleated configuration.

While the fibroin component of silk gives rise to the characteristic X-ray diffraction pattern, arthropod silks usually contain two proteins, fibroin and sericin, both formed in the secreting glands as soluble proteins. Sericin appears to act as a gum covering the fibrous elements of the silk and uniting them. Silk leaving the arthropod silk gland consists of an insoluble double fibroin filament embedded in the still water-soluble sericin. It appears that the mechanical process of secretion causes the semicrystallization of the fibroin yielding the fibrous, solid end-product. Sericin is characterized by a high serine content (37% in *Bombyx mori* cocoon silk). It is possible that some of this serine is involved in cross-linking to the fibroin; perhaps by ester links to aspartic acid residues. This is substantiated by the need for dilute alkali to separate sericin from fibroin and by the formation of hydroxamic acid on treatment of lepidopteran silks with hydroxylamine.

The fibroins have unusually high contents of the three small amino acids glycine, alanine and serine. These may total 50% or more of the entire amino acid content. There is, however, a fair degree of variability of amino acid composition between orders, families and species and any hope of basing on it a system of molecular taxonomy for the insects and arachnids seems remote. While glycine, alanine and serine contents are in general high, one or more of these amino acids may be present only in low concentrations. Thus *Anaphe* (Lepidoptera; Thaeumetopoeidae) fibroins have low serine contents and are essentially copolymers of alanine and glycine. As a rule proline is either present in low concentrations or is completely absent.

This brief account of the silks would be completely unjustified were it not for the fact that certain of them have been demonstrated to be associated with small quantities of firmly-bound carbohydrate and to have mono-saccharide compositions which place them in the same class as vertebrate glycoproteins.

Thus Abderhalden and Heyns (1933) demonstrated the presence of 2-amino-2-deoxy-D-glucose in hydrolysates of tussah silk fibroin, while Ito and Komori (1939a) reported a neutral sugar content in silk fibroin of 0·37% but failed to detect amino sugar. These latter authors (1939b), however, did find hexosamine to be present in the sericin fraction.

Smith and Clark (1965) detected hexosamine in the raw silk of *B. mori*. The hexosamine, amounting to 1·52 μM/100 mg dry weight, was separated by column chromatography and found to be composed of 0·82 and 0·65 μM/100 mg dry silk 2-amino-2-deoxy-D-glucose and 2-amino-2-deoxy-D-galactose, respectively. It is not clear though whether the amino sugar detected here was associated with fibroin or sericin or with both compo-nents.

Sinohara and Asano (1967) fractionated the silk of *B. mori* into three components; fibroin, sericin A, and sericin B. The carbohydrate contents of each of the three fractions are shown in Table XIII. Each sericin fraction contains 2-amino-2-deoxy-D-glucose, 2-amino-2-deoxy-D-galactose, man-nose, and galactose while the fibroin contains only 2-amino-2-deoxy-D-glucose and mannose. No sialic acid or methyl pentose could be detected.

TABLE XIII
Carbohydrates of *B. mori* silk proteins[a,b]

Monosaccharide	Fibroin	Sericin A	Sericin B
2-amino-2-deoxy-D-glucose	0·16	0·18	0·65
2-amino-2-deoxy-D-galactose	0·00	0·62	1·61
Mannose	0·17	0·24	0·66
Galactose	0·00	0·06	0·16

[a] Sinohara and Asano (1967). [b] Dry weight (%).

Sericin B contained three times more carbohydrate than sericin A and the qualitative differences between the two sericins and fibroin leave one in no doubt that the carbohydrate content of the fibroin would be unlikely to have arisen from contamination by sericin. A minimum molecular weight for fibroin, estimated by assuming the presence of one residue each of 2-amino-2-deoxy-D-glucose and mannose per molecule, yielded a value of 100,000. This figure compares well with the value obtained by analysis of the cystine content.

It would be of interest to know whether these results are typical of other silks. Of particular interest would be data relating to the collagen-like silk of *Nematus ribesii* (gooseberry sawfly) for the purposes of comparison with the carbohydrate compositions of other invertebrate and vertebrate collagens. *Nematus* silk is in fact in its amino acid composition intermediate between collagen and fibroin.

The presence of carbohydrate, and in particular hexosamine, in protein rich in serine, may suggest carbohydrate-amino acid linkages of the type now known to frequently occur in vertebrate glycoproteins and mucopolysaccharides. To assume this, however, would be dangerous at this stage; the preparation and characterization of silk fibroin and sericin glycopeptides can be awaited with interest. Certainly such studies will have to avoid materials which have been subjected to alkaline treatment and the accompanying possibility of the loss of carbohydrate by β elimination from serine. This may in fact account for the surprising failure of so many workers to detect hexosamines and neutral sugars in silk, in spite of a vast number of amino acid analyses. Many of these silks may have been the subjects of alkaline cleaning and degumming procedures.

REFERENCES

Abderhalden, E. and Heyns, K. (1933). *Biochem. Z.* **259**, 320.
Andersen, S. O. (1963). *Biochim. biophys. Acta*, **69**, 249.
Andersen, S. O. (1964). *Biochim. biophys. Acta*, **93**, 213.
Andersen, S. O. (1967). *Nature, Lond.* **216**, 1029.
Andersen, S. O. and Weis-Fogh, T. (1962). *Phil. Trans. R. Soc.* **B245**, 137.
Ankel, W. E. (1937). *Verh. dt. zool. Ges.* **39**, 77.
Anya, A. O. (1966). *Parasitology*, **56**, 179.
Ashhurst, D. E. (1961). *Q. Jl. microsc. Sci.* **102**, 455.
Ashhurst, D. E. (1965). *Q. Jl. microsc. Sci.* **106**, 61.
Ashhurst, D. E. (1968). *A. Rev. Ent.* **13**, 45.
Astbury, W. T. (1938). *Trans. Faraday Soc.* **34**, 377.
Astbury, W. T. (1940). *J. int. Soc. Leath. Trades Chem.* **24**, 69.
Astbury, W. T. (1944). *Proc. R. Instn Gt Br.* **33**, 140.
Bailey, A. J. (1968). *In* "Comprehensive Biochemistry" (M. Florkin and E. H. Stotz, eds.). Vol. 26B, p. 297. Elsevier, Amsterdam.
Bailey, K. and Weis-Fogh, T. (1961). *Biochim. biophys. Acta*, **48**, 452.

Bayne, C. J. (1966). *Comp. Biochem. Physiol.* **19**, 317.

Bevelander, G. and Benzer, P. (1948). *Biol. Bull. mar. biol. lab, Woods Hole*, **94**, 176.

Bear, R. S. (1952). *Adv. Protein Chem.* **7**, 69.

Blanquet, R. and Lenhoff, H. M. (1966). *Science, N.Y.* **154**, 152.

Blumenfeld, O. O., Paz, M. A., Gallop, P. M. and Seifter, S. (1963). *J. biol. Chem.* **238**, 3835.

Bricteux-Grégoire, S., Florkin, M. and Grégoire, Ch. (1968). *Comp. Biochem. Physiol.* **24**, 567.

Brown, C. H. (1947). *Proc. 6th Int. Congr. exp. Cytol., Exp. Cell. Res., Suppl.* **1**, 351.

Brunet, P. C. J. (1965). *In* "Aspects of Insect Biochemistry" (Biochemical Society Symposium). No. 25, p. 49. Academic Press, London and New York.

Butler, W. T. and Cunningham, L. W. (1966). *J. biol. Chem.* **241**, 3882.

Champetier, G. and Fauré-Fremiet, E. (1942). *C.r. hebd. Séanc. Acad. Sci., Paris*, **215**, 94.

Chapman, G. (1953). *Q. Jl. microsc. Sci.* **94**, 155.

Cunningham, J. T. (1899). *Nature, Lond.* **59**, 557.

Degens, E. T. and Love, S. (1965). *Nature, Lond.* **205**, 876.

Degens, E. T. and Parker, R. H. (1965). *Bull. geol. Soc. Am.* 43.

Degens, E. T. and Schmidt, H. (1966). *Paläont. Z.* **40**, 218.

Degens, E. T. and Spencer, D. W. (1966). *Tech. Rep. Woods Hole oceanogr. Instn*, No. 66-27.

Degens, E. T., Spencer, D. W. and Parker, R. H. (1967). *Comp. Biochem. Physiol.* **20**, 553.

Dennel, R. (1960). *In* "The Physiology of Crustacea" (T. H. Waterman ed.). Vol. 1, p. 449. Academic Press, London and New York.

Drew, G. A. (1901). *Q. Jl. microsc. Sci.* **44**, 313.

Dulzetto, F. and Labruto, G. (1951). *Atti Accad. gioenia. Sci. nat.* **7**, 211.

Elder, H. Y. and Owen, G. (1967). *J. Zool. Lond.* **152**, 1.

Fauré-Fremiet, E. and Garrault, M. (1964). *Bull. biol. mar. biol. lab., Woods Hole*, **78**, 206.

Fretter, V. (1941). *J. mar. biol. Ass. U.K.* **25**, 173.

Florkin, M. (1966). "A Molecular Approach to Phylogeny," p. 133. Elsevier, Amsterdam, London and New York.

Florkin, M., Grégoire, Ch., Bricteux-Grégoire, S. and Schoffeniels, E. (1961). *C.r. hebd. Séanc. Acad. Sci., Paris*, **252**, 440.

Fraenkel, G. and Rudall, K. M. (1940). *Proc. R. Soc.* **B129**, 1.

Fraenkel, G. and Rudall, K. M. (1947). *Proc. R. Soc* **B134**, 111.

Fujimoto, D. and Adams, E. (1964). *Biochem. biophys. Res. Commun.* **17**, 437.

Gersch, M. (1936). *Zool. Anz. Suppl.* **23**, 40.

Glimcher, M. J. (1960). *In* "Calcification in Biological Systems" (R. F. Sognnaes, ed.). No. 64, p. 421. American Association for the Advancement of Science, Washington, D.C.

Gotte, L., Stern, P., Elsden, D. F. and Partridge, S. M. (1963). *Biochem. J.* **87**, 344.

Gould, B. S., ed. (1968). "Treatise on Collagen". Vol. 2A. Academic Press, New York and London.

Grégoire, Ch. (1959). *Bull. Inst. r. Sci. nat. Belg.* **35**, 1.

Grégoire, Ch., Duchâteau, Ch. and Florkin, M. (1955). *Annls Inst. océanogr., Paris*, **31**, 1.

Gross, J. (1963). *In* "Comparative Biochemistry" (M. Florkin and H. S. Mason, eds). Vol. 4, p. 397. Academic Press, New York and London.

Gross, J. and Piez, K. A. (1960). *In* "Calcification in Biological Systems" (R. F. Sognnaes, ed.). No. 64, p. 395. American Association for the Advancement of Science, Washington, D.C.

Gross, J., Dumsha, B. and Glazer, N. (1958). *Biochim. biophys. Acta*, **30**, 294.

Hackman, R. H. (1953a). *Biochem. J.* **54**, 362.

Hackman, R. H. (1953b). *Biochem. J.* **54**, 367.

Hackman, R. H. (1964). *In* "The Physiology of Insecta (M. Rockstein, ed.). Vol. 3, p. 471. Academic Press, New York and London.

Hackman, R. H. and Goldberg, M. (1958). *J. Insect Physiol.* **2**, 221.

Hancock, D. A. (1956). *Proc. zool. Soc. Lond.* **127**, 565.

Hare, P. E. (1962). Thesis, California Institute of Technology, Pasadena, California.

Hare, P. E. (1963). *Science, N.Y.* **139**, 216.

Hare, P. E. and Abelson, P. H. (1964). *Rep. Dir. geophys. Lab, Carnegie Instn.* **63**, 267.

Hare, P. E. and Abelson, P. H. (1965). *Rep. Dir. geophys. Lab, Carnegie Instn.* **64**, 223.

Harper, E., Seifter, S. and Scharrer, B. (1967). *J. Cell. Biol.* **33**, 385.

Horiguchi, Y. (1959). *Bull. Jap. Soc. scient. Fish.* **25**, 397.

Hunt, S. (1966). *Nature, Lond.* **210**, 436.

Hyman, L. J. (1951). "The Invertebrates". Vol. 2. McGraw-Hill, New York.

Ito, T. and Komori, Y. (1939a). *J. agric. Chem. Soc. Japan*, **15**, 53.

Ito, T. and Komori, Y. (1939b). *J. agric. Chem. Soc. Japan*, **15**, 300.

Jackson, S. F., Kelly, S. C., North, A. C. T., Randall, J. T., Seeds, W. E., Watson, M. and Wilkinson, J. R. (1953). *In* "The Nature and Structure of Collagen" (J. T. Randall, ed.), p. 106. Butterworth, London.

Jensen, M. and Weis-Fogh, T. (1962). *Phil. Trans. R. Soc.* **B245**, 137.

Josse, J. and Harrington, W. F. (1964). *J. molec. Biol.* **9**, 269.

Jope, M. (1965). *In* "Treatise on Invertebrate Palaeontology", Vol. (H), p. 156. Ed. by Moore, R. C. Kansas: University of Kansas Press.

Jope, M. (1967a). *Comp. Biochem. Physiol.* **20**, 593.

Jope, M. (1967b). *Comp. Biochem. Physiol.* **20**, 601.

Kühn, K., Grassmann, W. and Hofmann, V. (1959). *Z. Naturf.* **14b**, 436.

LaBella, F., Keeley, F., Vivian, S. and Thornhill, D. (1967). *Biochem. biophys. Res. Commun.* **26**, 748.

LaBella, F., Waykole, P. and Queen, G. (1968). *Biochem. biophys. Res. Commun.* **30**, 333.

Lafon, M. (1943). *Annls Sci. nat.* Zool. **5**, 113.

Lane, C. E. and Dodge, E. (1958). *Biol. Bull. mar. biol. Lab.*, *Woods Hole*, **115**, 219.

Lavine, L. S. and Isenberg, H. D. (1964). *J. Bone Jt. Surg.* **46A**, 1563.

Lee, Y. C. and Lang, D. (1968). *J. biol. Chem.* **243**, 677.

Lenhoff, H. M. and Kline, E. S. (1958). *Anat. Rec.* **130**, 425.

Lenhoff, H. M., Kline, E. S. and Hurley, R. (1957). *Biochim. biophys. Acta*, **26**, 204.

Locke, M. (1964). *In* "The Physiology of Insecta" (M. Rockstein, ed.). Vol. 3, p. 471. Academic Press, New York and London.

Lucas, F. and Rudall, K. M. (1968). *In* "Comprehensive Biochemistry" (M. Florkin and E. H. Stotz, eds). Vol. 26B, p. 475. Elsevier, Amsterdam.

Lucas, F., Shaw, J. T. B. and Smith, S. G. (1958). *Adv. Protein Chem.* **13**, 107.

Lucas, F., Shaw, J. T. B. and Smith, S. G. (1960). *J. molec. Biol.* **2**, 339.

Lundgren, H. P. and Ward, W. H. (1963). *In* "Ultrastructure of Protein Fibres" (R. Borasky, ed.), p. 39. Academic Press, New York and London.

Lyons, K. M. (1966). *Parasitology*, **56**, 63.

Marks, M. H., Bear, R. S. and Blake, C. H. (1949). *J. exp. Zool.* **111**, 55.

Maser, M. D. and Rice, R. V. (1962). *Biochim. biophys. Acta*, **63**, 255.

Maser, M. D. and Rice, R. V. (1963a). *Biochim. biophys. Acta*, **74**, 283.

Maser, M. D. and Rice, R. V. (1963b). *J. Cell Biol.* **18**, 569.

McBride, O. W. and Harrington, W. F. (1967). *Biochemistry, N.Y.* **6**, 1484.

Melnick, S. C. (1958). *Nature, Lond.* **181**, 148.

Moorefield, H. H. (1953). "Studies on Some Integumental Proteins of Insects." Thesis, University of Illinois, Urbana, Illinois.

Moss, M. L. (1963). *Ann. N.Y. Acad. Sci.* **109**, 410.

Moss, M. L. (1964). *Int. Rev. gen. exp. Zool.* **1**, 297.

Picken, L. E. R., Pryor, M. G. M. and Swann, N. M. (1947). *Nature, Lond.* **159**, 434.

Piez, K. A. (1961). *Science, N.Y.*, **134**, 841.

Piez, K. A. (1963). *Ann. N.Y. Acad. Sci.* **109**, 256.

Piez, K. A. and Gross, J. (1959). *Biochim. biophys. Acta*, **34**, 24.

Piez, K. A. and Person, P. (1963). *Ann. N.Y. Acad. Sci.* **109**, 113.

Pipa, R. L. and Cook, E. F. (1958). *J. Morph.* **103**, 353.

Pryor, M. G. M. (1962). *In* "Comparative Biochemistry" (M. Florkin and H. S. Mason, eds). Vol. 4, Academic Press, New York and London.

Ramachandran, G. N., ed. (1963). "Aspects of Protein Structure". Academic Press, New York and London.

Ramachandran, G. N., ed. (1967). "Treatise on Collagen". Vol. 1. Academic Press, New York and London.

Rangarao, K. (1963a). *J. Anim. Morph. Physiol.* **10**, 158.

Rangarao, K. (1963b). *Curr. Sci.* **32**, 213.

Reed, R. and Rudall, K. M. (1948). *Biochim. biophys. Acta*, **2**, 7.

Rich, A. and Crick, F. H. C. (1961). *J. molec. Biol.* **3**, 483.

Richards, A. G. (1951). "The Integument of Arthropods". University of Minnesota Press, Minneapolis, Minnesota.

Richards, A. G. and Pipa, R. L. (1958). *Smithson. misc. Collns*, **137**, 247.

Roche, J. and Eysseric-Lafon, M. (1951). *Bull. Soc. Chim. biol.* **33**, 1448.

Roche, J. and Michel, R. (1951). *Adv. Protein Chem.* **6**, 253.

Rudall, K. M. (1955). *Symp. Soc. exp. Biol.* **9**, 49.

Rudall, K. M. (1962). *In* "Comparative Biochemistry" (M. Florkin and H. S. Mason, eds). Vol. 4, p. 397. Academic Press, New York and London.

Rudall, K. M. (1965). *In* "Aspects of Insect Biochemistry" (T. W. Goodwin, ed.). Symp. No. 25, p. 83. Academic Press, London and New York.

Rudall, K. M. (1968a). *In* "Treatise on Collagen" (B. S. Gould, ed.). Vol. 2, p. 83, Academic Press, New York and London.

Rudall, K. M. (1968b). *In* "Comprehensive Biochemistry" (M. Florkin and E. H. Stotz, eds). Vol. 26B, p. 586. Elsevier, Amsterdam.

Schmitt, F. O., Hall, C. E. and Jakus, M. A. (1942). *J. cell comp. Physiol.* **20**, 11.

Seifter, S. and Gallop, P. M. (1966). *In* "The Proteins" (H. Neurath, ed.). Vol. 4, p. 155. Academic Press, New York and London.

Simkiss, K. (1965). *Comp. Biochem. Physiol.* **16**, 427.

Sinohara, H. and Asano, Y. (1967). *J. Biochem.* **62**, 129.

Singleton, L. (1957). *Biochim. biophys. Acta*, **24**, 67.
Smith, J. G. and Clark, R. D. (1965). *Nature, Lond.* **207**, 860.
Sognnaes, R. F., ed. (1960). "Calcification in Biological Systems". No. 64. American Association for the Advancement of Science, Washington, D.C.
Stary, Z., Tekman, S. and Öner, N. (1956). *Biochem. Z.* **328**, 195.
Stegemann, H. (1961). *Naturwissenschaften*, **48**, 501.
Tanaka, S. and Hatano, H. (1953). *J. chem. Soc. Japan*, **74**, 193.
Tanaka, S., Hatano, H. and Itasaka, O. (1960). *Bull. chem. Soc. Japan*, **33**, 543.
Tekman, S. and Öner, N. (1963). *Nature, Lond.* **200**, 77.
Travis, D. F. (1963). *Ann. N. Y. Acad. Sci.* **109**, 177.
Trim, A. R. (1941a). *Nature, Lond.* **147**, 115.
Trim, A. R. (1941b). *Biochem. J.* **35**, 1088.
Tsuchiya, Y. and Suzuki, Y. (1962). *Bull. Jap. Soc. Scient. Fish.* **28**, 222.
Van Gansen-Semal, P. (1960). *Nature, Lond.* **186**, 654.
Voss-Foucart, M. F. (1968). *Comp. Biochem. Physiol.* **26**, 877.
Watson, M. R. (1958). *Biochem. J.* **68**, 416.
Watson, M. R. and Silvester, N. R. (1959). *Biochem. J.* **71**, 578.
Weis-Fogh, T. (1960). *J. exp. Biol.* **37**, 889.
Weis-Fogh, T. (1961a). *J. molec. Biol.* **3**, 520.
Weis-Fogh, T. (1961b). *J. molec. Biol.* **3**, 648.
Wigglesworth, V. B. (1957). *A. Rev. Ent.* **2**, 37.
Wilbur, K. M. (1964). *In* "Physiology of Mollusca" (K. M. Wilbur and C. M. Yonge, eds). Vol. 1. Academic Press, New York and London.
Wilbur, K. M. and Simkiss, K. (1968). *In* "Comprehensive Biochemistry" (M. Florkin and E. H. Stotz, eds). Vol. 26A, p. 229. Elsevier, Amsterdam.
Wilbur, K. M. and Watabe, N. (1963). *Ann. N. Y. Acad. Sci.* **109**, 82.
Williams, A. P. (1960). *Biochem. J.* **74**, 304.

CHAPTER 16

Evolution of Mucopolysaccharides and Related Molecules

In the foregoing chapters I have attempted to bring together all that is known at the present time of the distribution, composition, structure, biosynthesis and function of invertebrate mucopolysaccharides, glycoproteins and other related molecules. Despite sparse available information in many cases it has been possible to compare the invertebrates with the vertebrates: in both are mucopolysaccharides, glycoproteins and structure proteins with very similar structures, others with less similar structures but with still recognizable structural affinities, and still others in each group with no immediately obvious counterpart in the other group. What we should like to be able to see is whether we can distinguish any specific relationships between the protein-carbohydrate complexes and related molecules of the various invertebrate groups and also between the invertebrate materials and those found in vertebrate tissues and secretions. We want to know whether we can trace evolutionary pathways back through homologous groups of molecules to the ancestral molecular forms from which a wide range of molecular types have arisen just as it has been possible to suggest structures for the ancestral molecules of modern protein chains (for example, the haemoglobins). One of the difficulties inherent in extending the arguments and procedures which have been used successfully to trace the evolutionary homologies of protein molecules to biopolymers such as polysaccharides is that one is no longer dealing with molecules likely to show direct homologies. Thus the essential feature of the molecular approach to phylogeny is an attempt to trace the history of the genotype by examining changes in molecular phenotypes, which have taken place during the course of evolution. To do this we must make comparisons between the chemical structures of molecules either constituting the genomes of organisms or else directly reflecting the genome. Ideally we should be able to determine evolutionary pathways by the determination of the base sequences in the nucleic acids of the organisms in question, but technically this is at present unfeasible. Since, however, the primary structures of proteins are determined by a definitely-coded arrangement of the nucleotides in the nucleic acids, and since primary structures of proteins are now fairly readily determined, we have in proteins suitable informational macromolecules for the determination of evolutionary pathways. Such informational molecules have been called "semantides" by Zuckerkandl and Pauling (1965) and series of semantides which show high degrees of structural similarity, such as high levels of correspondence in the sequences of proteins chains, are said to be directly homologous and assumed to be related genotypically. Polysaccharides on the other hand present a more difficult problem, since they are more than one step removed from the genome. Being synthesized by enzymes, perhaps more than one protein being involved in their synthesis, their sequences or overall structures are only an indirect result of the genetic code embodied in the nucleic acids. Such molecules are called episemantides. When groups of structurally similar episemantides are found, such that a genotypic relationship between their synthesizing enzymes seems indicated, they are said to be indirectly homologous. Thus while we may observe similar connective mucopoly-saccharides in vertebrates and molluscs and may tentatively suggest that these are indirect homologues and that the two groups may share a common ancestral mucopolysaccharide molecule linking them genotypically, we can only really prove this postulate by isolation of the relevant poly-saccharide synthetases and by demonstration of direct homologies in the sequences of amino acids in the proteins. Alternatively, but possibly less satisfactorily, we might attempt sequence determinations of the protein molecules attached to the respective polysaccharides in the hope of finding homologies there. Such an approach might give misleading results if one considers the possibility of unrelated (genotypically) polysaccharides becoming associated with related or directly homologous protein chains by secondary evolution. We might however consider that where a series of apparently indirectly homologous molecules is found in two separate evolutionary lines in both cases leading back through progressively simpler evolutionary types to a group of animals which seem, on morphological or embryological grounds, to have affinities with both lines, it should provide good ancillary evidence for inferring a genotype relationship between the episemantides and organisms concerned.

In examining indirect homologies we find that in addition to possible groups of episemantides distributed among phyla are also cases where there appear to be groups of polysaccharides closely related in structure, apparently indirectly homologous episemantides, within species. That is to say we may find groups of genotypically, or apparently genotypically, related polysaccharides in a single organism. Thus for example in the vertebrates the group of extracellular connective tissue mucopolysaccharides which includes hyaluronic acid, the two chondroitin sulphates, dermatan sulphate and keratosulphate all have closely similar chemical structures which seem to suggest something more than convergent homology.

Present opinion suggests that the evolution of protein structure, in terms of changes in sequence (deletions, additions or alterations of amino acid

residues) and hence tertiary organization, is a result of changes or mutations in the structures of the nucleic acid molecules constituting the genes
responsible for coding the primary structures of the proteins concerned.
Thus it is possible to see that as the gene for a particular protein, say for
example cytochrome c, has become distributed among the various phyla
during millions of years, so independent evolution or mutation of the gene
will have taken place and differences in the sequences of amino acids in
the protein chains between phyla and even species become apparent.
The more the primary structures of proteins of species or phyla differ the
longer the genes have been evolving independently. It is therefore possible
by comparing numbers of differences in protein amino acid sequences and
by making certain assumptions with regard to the length of time required
for mutations to produce the observed changes, to estimate the time which
has elapsed since the proteins, or rather their genes, have differentiated
from a common ancestor and also to make deductions regarding the
probable form of the ancestral molecule. Thus the cytochrome c chains
of vertebrates and yeasts still have recognizably similar sequences with
correspondences at 58 points suggesting a common ancestry some one to
two thousand million years ago.

Where we find two or more directly homologous protein chains in the
same organism it is envisaged that these have originated from a single
ancestral protein chain whose gene has duplicated and translocated to give
rise to duplicate protein chains; these evolved independently, through
mutations in the duplicate genes, while remaining in the same organism.
Further duplications, translocations and independent mutations will in
the course of events give rise to further homologous chains, all of which
will be phylogenetically related. The alpha, beta, gamma and delta haemo-
globin chains together with myoglobin are believed to have originated in
such a manner.

The structures or sequences of polysaccharides are not direct transcrip-
tions of the genetic code as are the proteins. However, mutations in the
gene specifying the sequence of amino acids in a specific polysaccharide
synthetase are likely to be reflected in the polysaccharide itself, although
less clearly than in the protein. Non-deleterious mutations resulting in
changes either in sequences at or around the enzyme active centre or in
regions affecting the enzyme conformation will be likely to produce
alterations either in specificity for the monosaccharide to be incorporated
into the polysaccharide or for the type of glycosidic bond formed. We
should therefore, on the basis of the principles already described, be able
to postulate mechanisms for the evolution of polysaccharides and draw
some tentative deductions about inter-relationships and possible ancestral
molecules of modern polysaccharides from their structures.

K

The polysaccharides of animals, plants and other organisms fall into two general compositional categories: the homopolysaccharides and heteropolysaccharides. Homopolysaccharides are those consisting of linear or branched chains of a single type of monosaccharide sub-unit, while heteropolysaccharides are those made up of two or more different types of sugar residue, often arranged in regularly repeating sequences. Cellulose and glycogen are representative examples of the first type of polysaccharide while the vertebrate connective tissue group of mucopolysaccharides is representative of the second. If we are to consider polysaccharides as actually having evolved then it seems not improbable that the hetero-polysaccharides have evolved from the structurally simpler homopoly-saccharides. It probably would be easier at this stage to consider the sim-plest case of the evolution of linear polysaccharides.

Polysaccharide synthesis has been established as being mediated by the action of specific enzymes, synthetases, which transfer monosaccharide from a nucleotide-sugar derivative to the end of an extending polysaccharide chain that acts as the sugar acceptor. Thus the nature of the polysaccharide must be inherent in the genome of the organism synthesizing the poly-saccharide as we have already observed. Presumably the specificity of the synthetase determines not only the particular type of monosaccharide incorporated but also the nature of the glycosidic bond formed. The mode of synthesis for homopolysaccharides of the linear type is rather easy to envisage but when we come to heteropolysaccharides with two or more monosaccharides arranged in a specific and repeating sequence, it becomes difficult to see how synthesis is controlled. A system of sequence determina-tion analogous to that involved in the synthesis of proteins seems unlikely. This leaves us the possibility of an elaborate system of control over two or more synthetases which have to be able to carry out such operations as the recognition of the last monosaccharide on the chain. Simplification of the mode of biosynthesis of heteropolysaccharides under the influence of several different specific synthetases might be possible if we were to con-sider all the enzymes involved as having originated from a single ancestral molecule which would be the synthetase responsible for the production of an ancestral homopolysaccharide.

To illustrate this let us take as a rather simple and naive example the evolution of a homopolysaccharide containing monosaccharide (a), in a representative sequence –(a)–(a)–(a)–(a)–, to a polysaccharide which also contains monosaccharide (b) such that the sequence becomes –(a)–(b)–(a)–(b)–.

Initially we can say that the gene "A" will dictate the synthesis of the protein (A) which is the specific synthetase for the production of the homopolysaccharide –(a)–(a)–(a)–(a)– from a nucleotide derivative of the monosaccharide (a). Suppose that we now allow duplication and trans-location of gene "A" to take place. Two duplicate (A) chains will be formed and will be subject to independent evolution. Mutation of one of the duplicates of gene "A" may now produce a gene "B" which will, in turn, dictate the production of synthetase (B) with a different nucleotide sugar specificity, say for sugar (b). Thus this synthetase could, in theory, produce a polysaccharide with the sequence –(b)–(b)–(b)–(b)–. If for the sake of simplification we assume that the other half of the duplicate pair remains unmutated (call this "A*") with the same specificity for sugar (a), the organism will now produce two different homopolysaccharides, –(a)–(a)–(a)–(a)– and –(b)–(b)–(b)–(b)–.

Synthetases (A*) and (B) are likely to have closely similar sequences and tertiary structures, and let us suppose that because of this they are likely to be subject to association into a quaternary structure, just as is the case with the homologous hemoglobins in the active hemoglobin tetramer. Not existing completely independently, the tendency might be for (A*) and (B) not to act independently also. We can then envisage the possible course of events indicated in Fig. 1. Here we have a dimer of one molecule of (A*) and one molecule of (B). In the "relaxed" state the nucleotide-sugar and the "acceptor" binding sites on one synthetase [say (A*)] may be open, while the corresponding sites on (B) are closed or unavailable. Once nucleotide-sugar and acceptor are bound to (A*) the effect is to cause the active sites on (B) to come into the open conformation. Nucleotide-sugar (b) can then bind to (B) while the acceptor chain, now extended by a molecule of (a), detaches from (A*) and binds to (B) where a molecule of (b) is attached. Once the acceptor is bound to (B) the nucleotide-sugar binding site becomes available on (A*) once more and the sequence of events outlined is repeated, the extending chain oscillating

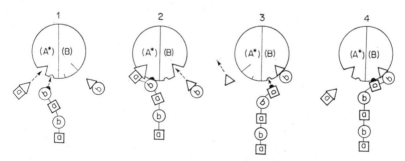

Fig. 1. Mechanism of action of a polysaccharide synthetase composed of two homologous polypeptide chains (A*) and (B) with differing nucleotide-sugar specificities. For details see text. [a] and (b) are sugars while Δ represents nucleotide.

κ*

between (A*) and (B) so that an alternating sequence heteropoly-saccharide –(a)–(b)–(a)–(b)–(a)–(b)– is produced.

Obviously this is a rather naive view and is of course only hypothetical; actual heteropolysaccharides most frequently also alternate the glycosidic bond types as well as the monosaccharides in the chain. Polysaccharides containing more than two sugars in a regular repeating sequence would also present a more complex control problem as would branching structures such as are found in glycoproteins. We might extend the general basis of the hypothesis to include the evolution of a group of indirectly homologous heteropolysaccharides, from a single homopolysaccharide, in one organism. If for example further gene division, translocation and mutation were to take place in the case described above, we might have originating from gene "A", genes "A*", "B" and "C", with capabilities for the coding of synthetases (A*), (B) and (C) whose specificities are for monosaccharides (a), (b) and (c). Supposing trimerization of the three synthetases not possible and formation of hybrids by dimerization the rule, we would obtain hybrid pairs of synthetases (A*B), (A*C) and (BC) producing polysaccharides –(a)–(b)–(a)–(b), –(a)–(c)–(a)–(c)– and –(b)–(c)–(b)–(c)–.

Having stated some of the basic principles and possibilities in the mechanisms by which molecules of the polysaccharide type evolve, what can we say in this regard with respect to the various classes of macromolecules discussed in this book?

It seems difficult at the present stage of our knowledge to make any very deep analyses of the relationship of what we have called mucopolysaccharides to the other major group of protein-carbohydrate complexes, the glycoproteins. As we have noted already, the demarcation line between the two groups is fairly sharp in the vertebrates. There we have on the one hand a group of polysaccharides of a largely linear nature, containing acetyl hexosamine and usually highly-charged by virtue of the presence of uronic acid residues and ester sulphate groups individually or in combination. On the other hand we find in the glycoproteins polysaccharides of far lower molecular weight than the mucopolysaccharides, usually branched to a greater or lesser degree, and containing a different spectrum of sugars to the mucopolysaccharides. These sugars include acetyl hexosamines, but neutral hexoses and methyl pentose are of great importance while uronic acid is absent. Nor is ester sulphate a feature of the majority of glycoproteins, although it is found in some. Instead, where the glycoproteins are highly charged this is by virtue of sialic acid residues. On the face of things, therefore, vertebrate mucopolysaccharides and glycoproteins might seem to be independent evolutionary experiments. We should remember, however, that both classes of molecule constitute covalent associations of carbohydrate and protein. And in the case of the chondroitin sulphate-protein complexes at least it seems likely that there are many polysaccharide chains arranged along the length of a protein backbone in exactly the same way that many glycoproteins have their smaller prosthetic groups distributed evenly along the protein chain. Moreover it seems that one of the two main types of protein-carbohydrate link in vertebrate glycoproteins is of the same character as that found in mucopolysaccharides namely an O-glycosidic bond between one of the hydroxy amino acids and a sugar residue. This may be an example of convergent evolution, but it does lay open the possibility that mucopolysaccharides and glycoproteins may have some distant common origins. In this respect we may note that in the invertebrates the division between mucopolysaccharides and glycoproteins, as we define them in the vertebrate context, becomes far more blurred. Thus we find glycoproteins containing uronic acids while far greater variability of composition than is found in vertebrates occurs in the invertebrate mucopolysaccharides, with neutral sugars carrying ester sulphate groups enjoying a prominent position; invertebrate mucopoly-saccharides frequently contain no hexosamine or uronic acid. Macro-molecular complexes of carbohydrate and protein which fit comfortably into no particular category have been one of the ever-present headaches of writing this book. For example, the horatinsulphuric acids from *Charonia lampas* were included in the chapter on polysaccharide sulphates and other mucopolysaccharides but with its composition of acetyl hexosamines, neutral hexoses, methyl pentose, sialic acid and protein, it would equally well have fitted into the glycoprotein section, were it not for the high ester sulphate content. It may be that we are seeing the tendency towards specialization (and simplification) found in the vertebrates so clearly separating the two classes of complex, at an earlier or only partially-developed level.

As Mathews (1967) has pointed out, we must be careful not to limit ourselves too much in judging chemical homologies by rigid adherence to requirements for embryological homology. Macromolecules of the con-nective tissue type are usually considered as being produced by cells of mesodermal origin, but may in certain instances be synthesized by cells derived from ectoderm. Epithelium and its characteristic secretions may originate in either the mesoderm or the ectoderm, although derivation from the latter is the more usual case.

The most well-known and well-characterized vertebrate glycoproteins are those of epithelial origin. These are the ones usually rich in sialic acid and containing, in addition to amino sugar, one or all of the neutral sugars galactose, mannose and fucose. Glycoproteins of connective tissues also contain these sugars, but in addition glucose seems to be a characteristic constituent. It may be that the presence of glucose distinguishes glycoprotein of mesodermal origin from that derived from extoderm; such deductions must, as I have said, be accepted with considerable reservations. In this respect it is of interest to recall that glucose is also a characteristic constituent of collagens from both vertebrates and invertebrates but that while most vertebrate collagens are supposed to be of mesodermal origin the invertebrate collagens are believed to be ectodermally derived. Vertebrate collagens do not in general have high carbohydrate contents, whereas invertebrate collagens tend to contain large amounts of carbohydrate; *Ascaris* cuticle collagen is rather exceptional in its low carbohydrate content. Vertebrate collagens thought to be of epithelial origin such as kidney basement membrane (Dische *et al.*, 1965; Spiro, 1966) and mammalian lens capsule (Pirie and Van Heyningen, 1956) seem like invertebrate collagens to contain large quantities of carbohydrate.

Epithelial secretions in marine invertebrates at least seem to be rather different from protein-carbohydrate complexes produced by vertebrate epithelium, the latter usually being glycoprotein, as we have seen. In the marine invertebrates the characteristic mucous lubricatory secretions of the outer epithelium seem to be largely sulphated polysaccharides with more affinity to vertebrate connective tissue mucopolysaccharides than epithelial glycoproteins. We have noted in earlier chapters how the polysaccharide sulphates with heparin-like structures seem to be derived from secretions of molluscan tissues with ectodermal origin rather than from the mast cells of connective tissue. Again we must exert caution in considerations involving embryological homologies. Hyaluronic acid observed in the nervous tissues of insects (Chapter 3) seems to be of ectodermal origin rather than being derived from mesodermal tissues, but this again may fit in with the fact that it is associated with collagen and invertebrate collagen also is usually ectodermal rather than mesodermal. The embryology of the hyaluronic acid in the insect peritrophic membrane seems less clear. One would expect this to be derived from the mid gut, which is endodermal; but if, as is thought to be the case with the Diptera, it is secreted by cells in the proventriculous, it may have an ectodermal origin.

The occurrence in both vertebrates and invertebrates of a group of acid mucopolysaccharides with apparently homologous structures presents an interesting problem for the molecular phylogenist. The distribution of hyaluronic acid, the chondroitin sulphates, keratosulphate, chondroitin and heparin in the animal kingdom are shown in Fig. 2. It has been suggested that the appearance of these connective tissue polysaccharides, characteristic of vertebrates, in the tissues of invertebrates of the protostomian line may be due to biochemical convergence. The similarities between these mucopolysaccharides found in the cartilage tissues of molluscs, arthropods and annelids equally may be a manifestation of convergent evolution

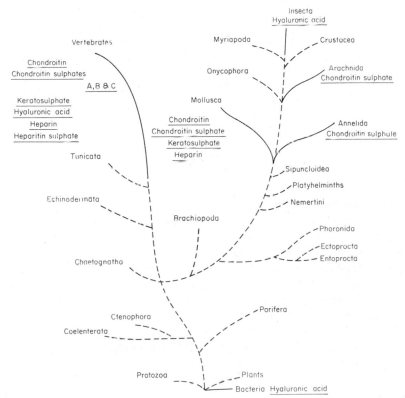

Fig 2. The distribution of chondroitin sulphate and other major connective tissue mucopolysaccharides in the major groups of the animal kingdom. Solid lines indicate the presence of these molecules but dotted lines do not necessarily indicate their absence but rather that no definite conclusions either way can yet be definitely drawn.

(Mathews, 1967). At the present time there is little evidence either for or against this argument. If one could show that the chondroitin sulphate structure was the only possible form of acid polysaccharide suitable for the organization, maintenance or function of a cartilage tissue, this at least would argue the necessity of this biochemical convergence; we have however seen that in one mollusc this role seems to have been quite satisfactorily filled by a glucan sulphate. We know nothing of the very early biochemical origins of the connective tissue acid mucopolysaccharides, unless the occurrence of hyaluronic acid in the bacteria can be taken as an indication. With regard to the mucopolysaccharides present in the connective tissues of phyla predating the origin and separation of the deuterostome and protostome lines, our evidence is largely histochemical. Thus material with the staining properties of acid mucopolysaccharide appears to figure in the coelenterate mesoglea, in association with collagen. Even if we can demonstrate that molecules of the chondroitin sulphate type do not occur in the connective tissues of the more primitive phyla, this need not necessarily argue against a relationship between the chondroitin sulphates of vertebrates and invertebrates. The synthetase producing the ancestral polysaccharide of the connective tissue mucopolysaccharides could have been carried into both evolutionary lines and then in both lines have suffered in the course of time, the gene divisions, translocations and mutations necessary to give rise to the chondroitin sulphates, etc. Thus while these mucopolysaccharides need not have been carried into the chordate or arthropod-mollusc lines in their present form, from a recognizable chondroitin sulphate ancestor, they would still have had a common origin and be genetically related. This form of convergence seems far more probable than one in which the groups of molecules have arisen completely independently with no common origin at all. If we accept the hypothesis that the heteropolysaccharides originate from homopolysaccharides, then the glucan sulphate of *Busycon canaliculatum* cartilage tissue may provide us with some clue to the possible nature of the ancestral mucopolysaccharide molecule. We can predict with some measure of safety that the ancestral polysaccharide had a β-linked structure, arising from the necessity for a linear chain.   Person and Philpott (1967) have listed and discussed many similarities between the supporting (connective) tissues of plants and animals. It is of interest to note that the principal supporting polysaccharide of plants is cellulose, while that of invertebrates is chitin, one a β-linked polymer of glucose and the other the 2-acetamido derivative of the same molecule. Our search for the ancestral form of the connective tissue matrix mucopolysaccharide may perhaps tend towards a glucan-based molecule with charged substituents to provide solubility. Final proof of any hypothesis referring to the relationships of groups of polysaccharides must eventually rest upon isolation and structural characterization of the enzymes involved in their synthesis. It is now possible to predict the ancestral sequences of proteins from the sequences of the contemporary proteins taken from a wide range of species and phyla. Perhaps eventually it will become possible, as our knowledge of the mechanisms of enzyme action is extended, to predict the form of the molecule synthesized by the ancestral enzyme.

   That some of the biochemical pathways involved in the biosynthesis of polysaccharides are very primitive and fundamental indeed seems to be indicated by the ubiquity, over a wide range of life forms, of the activated intermediates of synthesis.   Thus hyaluronic acid has been shown to have the same nucleotide-sugar precursors in both bacteria and mammals

(Schiller, 1966) while the uridine derivatives of 2-acetamido-2-deoxy-D-glucose and 2-acetamido-2-deoxy-D-galactose are found in both vertebrates and invertebrates; UDP-glucose is used for glucan synthesis in plants and animals. Active sulphate, adenosine 3-phosphate 5-phosphosulphate, is the vehicle for polysaccharide sulphation in marine algae, vertebrates and invertebrates.

The importance of sulphate in the structures of the acid mucopolysaccharides of invertebrates and lower vertebrates is a point which has not yet been satisfactorily explained. As we have seen, the chondroitin sulphates of cartilaginous fish and of molluscs, arthropod and annelid are all over sulphated in comparison to their mammalian counterparts. It may be relevant that the polysaccharides of marine plants are also heavily sulphated. It would seem that the reduction in the use of sulphation in polysaccharide synthesis may be connected in some way with the emergence from a marine environment. Marine invertebrates, notably molluscs, make great use of sulphated polysaccharides for purposes other than structural, for example lubrication of the body surface and locomotory surfaces. However, though perhaps we have rather insufficient evidence at present, it would seem that land molluscs make more use of neutral polysaccharides or glycoproteins. It may well be simply a matter of the relative availabilities of sulphate.

The chondroitin sulphates, keratosulphate, heparin, heparitin sulphate and hyaluronic acid seem to be by and large the only mucopolysaccharides of importance occurring in the vertebrates, although further studies of lower vertebrates and the less well characterized tissues of mammals may add to this list. In contrast to the vertebrates the invertebrate phyla can show a far greater diversity of acid mucopolysaccharides, but we must remember that we are dealing there with an equally greater diversity of life forms. Like the vertebrates the invertebrates have their chondroitin sulphates (perhaps not chondroitin sulphate B), keratosulphate, heparin and hyaluronic acid but as we have seen over and above these are a great number of other acid mucopolysaccharides which have no counterparts in the vertebrates. We have too little information about the rest of the evolutionary line leading to the vertebrates to say whether or not many of these mucopolysaccharides were ever part of this line as well and whether they have therefore been quite recently lost when the chordates and vertebrates emerged. In the case of the echinoderms, thought to be members of the line leading to the vertebrates, we do have a fairly considerable body of data regarding the character of the specialized group of mucopolysaccharides coating the eggs. Except for the fact that these substances are ester sulphated polysaccharides, we find nothing comparable with them in the vertebrates. The egg jelly mucopolysaccharides are a homogeneous group

of molecules and one would think are certainly indirectly homologous
episemantides. They seem to be composed largely of either fucose or
galactose or both monosaccharides in conjunction and one might be tempted
by this to conjecture that one type of homopolymer, say the galactan sulphate, has preceded the other in evolution and that the heteropolysaccharides containing both monosaccharides represent a transition in the
evolution. This is however a naive oversimplification and an examination
of the geological record for the appearance of the species whose egg jellies
have been analysed shows no correlation of any particular polysaccharide
type with origin in time. The fact that both echinoid and asteroid egg jellies
contain the fucose-galactose sugar pattern seems indicative of an early
origin for this type of polysaccharide. It is unfortunate that we have little or
no information regarding the occurrence of polysaccharide sulphates in the
eggs of other invertebrate phyla, though we noted in Chapter 5 that the
glandular region of the reproductive tract in one mollusc contains a sulphated mucopolysaccharide made up of galactose, fucose and 2-amino-2-
deoxy-D-glucose, a composition strikingly similar to that of *Asterias
amurensis* egg jelly (Chapter 6). Bayne (1966) found galactose and hexosamine in the egg jelly of a pulmonate gastropod (Chapter 12).

The wide range of other sulphated mucopolysaccharides which have been
found in invertebrates presents too scattered a view for any recognizable
pattern to emerge at the present time. Either too few species have been
sampled or the characterizations of the materials themselves are inadequate.
Even when these criteria have been partially satisfied we find that simple
patterns of relationship may not always be easy to observe. Thus in the
case of the hypobranchial mucins (Chapter 5) we find that those from two
species contain what are essentially cellulose sulphates while one from the
same family has a hypobranchial mucin containing a polysaccharide or
mixture of polysaccharides made up of hexosamine, glucose, fucose and
sulphate. In a fourth case a mollusc from a different family has, in its
hypobranchial mucin, a polysaccharide composed entirely of hexosamine
and here the sulphate, though present, does not even seem to be of the ester
type. Nor is it all clear why one of the cellulose sulphate-containing mucins
should also contain a branched glucan sulphate while the other functions
with apparent equal efficiency without this extra mucopolysaccharide.

What one can say of that remarkable acid mucopolysaccharide onuphic
acid is difficult to see. This glucan phosphate seems without parallel elsewhere in the animal kingdom. The molecule seems to have no evolutionary
antecedents, with the possible exception of the mannan phosphates found
in some yeasts. It will be interesting to see whether the glucose is activated
for the synthesis of onuphic acid as the usual UDP derivative and how the
phosphate is transferred. The teichoic acids and related molecules are not

really comparable with onuphic acid and it seems unlikely that it could have evolved from molecules of this type.

The relationship between glycogen and galactan presents an interesting problem. Glycogen is without doubt eminently successful in its role as a storage molecule, acting as the major energy and carbohydrate reserve material throughout the animal kingdom. The necessity which certain of the mollusca seem to have felt for a second reserve polysaccharide, based on galactose, to be used by embryos and young seems to require explanation, yet the reason is not at all clear. Nor is it clear why both the D and ɪ forms of galactose should figure in the structures of this group of polysaccharides. The galactan found in mammalian lung is probably not genetically homologous with molluscan galactan, nor is the function of the former molecule known with any certainty. Homologies of glycogen and galactan cannot be really discussed with our present knowledge nor do we know whether the glycogens distributed so ubiquitously among the phyla of the animal kingdom are all indirect homologues, although one would think that this was a strong probability. Glycogen is probably a molecule of considerable antiquity; some kind of energy reserve must have been an early requirement, but although it is present in many microorganisms, we do not find it in plants. Instead the carbohydrate reserves of plants contain the linear glucan amylose and the branched glucan amylopectin. Amylopectin has a rather similar structure to glycogen, but the average chain length in amylopectin is about 20 residues as opposed to 12 in glycogen. Glycogen and amylopectin may be indirect homologues but there is no evidence either for or against this at the present time.

Chitin has attracted the attention of molecular phylogenists for some time (Jeuniaux, 1963) and the various aspects of the phylogeny of chitin and chitinolysis have been succinctly reviewed by Florkin(1966). One of the several odd features of the evolution of chitin lies in the manner in which it is distributed (Fig. 3) in the animal and plant kingdoms. It occurs in primitive plants, in the unicellular root of the Metazoa, in the Parazoa, in the Cnidaria and in the major evolutionary line which leads up to and culminates, at its twin peaks, in the arthropod and molluscs, the protostomian line. But strange to say there is no trace of this important structure polysaccharide in the other major branch of the animal evolutionary tree, the Deuterostomia, which leads to the chordates and ultimately to the mammals.

Setting aside functional needs and the efficiency of existing systems there seems, on the face of things, to be no immediate reason why we should not find chitin in vertebrate tissues. The fundamental sub-unit, 2-acetamido-2-deoxy-D-glucose, is present and moreover is activated for polysaccharide biosynthesis in vertebrates as the uridine diphosphate derivative, just as it is

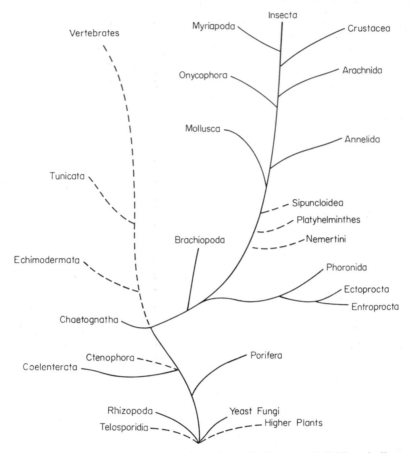

Fig. 3. Distribution of chitin among the major animal groups. Solid lines indicate the presence of chitin and dotted lines its absence.

for chitin synthesis in arthropods and fungi; only the necessary synthetase would seem to be absent. To argue this way is an oversimplification, but it seems valid to ask whether the chordate line lacks chitin because the chitin synthetase gene was lost at the time of separation from what has become the protostomian line, thus causing alternative systems to arise. Rudall (1968a) has observed that as an extracellular cuticle chitin is superb, but it has totally failed to adopt a role as a loose connective tissue. He has put forward various reasons to account for this failure, mainly hinging upon the great stability of chitin fibres being too great a hindrance to growth, regulation and modification. He points out that problems involved in the turnover of an internal chitin-protein connective tissue might be great and might cause drastic changes in neighbouring tissues. Rudall also comments that if any attempt were ever made to use chitin as a connective tissue there has been no survival of such organisms in animals. He does suggest that the connective tissue group of mucopolysaccharides, characteristic of the chordate line but as we have seen present in invertebrates also, are reminiscent of chitin in their structure, although it is not clear whether he suggests an evolutionary homology. He also points out that chitin systems and connective tissue systems differ in that in the former case the polyscaccharide is the fibrous moiety and protein the modifying matrix while in the latter the roles are reversed. In suggesting that contemporary connective tissue systems show no traces of chitin as a constituent molecule Rudall may not be wholly correct; at least we may be able to find a molecule closer to chitin than the connective tissue acid mucopolysaccharides still represented. If we postulate that the chitin synthetase gene was retained by the line leading to the chordates and subsequently suffered mutation, then there would be at least two possibilities. The mutation might have been destructive so that the enzyme was lost for polysaccharide synthesis altogether. Alternatively, the change induced by the mutation in the amino acid sequence, although minor, might have been sufficient to cause a slight change in the specificity of the enzyme. We would then expect to find a new polysaccharide in the place of chitin, with probably a not too dissimilar structure. Examination of the range of structural and connective tissue polysaccharides found in the deuterostomian line does in fact yield one very strong candidate for the role of the chitin homologue, namely cellulose. Cellulose is found abundantly in the connective tissues of the test of tunicates, a protochordate group, much less abundantly in mammalian connective tissue and perhaps also in fossil bone and dentine (Isaacs *et al.*, 1963). In its chemical structure, physical macromolecular organization and physical properties cellulose closely resembles chitin. The presence of cellulose in the tunicates and mammals has been a perturbing evolutionary anomaly; there is nothing to link this cellulose in continuity with the plant celluloses. The hypothesis advanced here would account for its sudden, apparently random appearance in the chordate line and would suggest that its relation to plant cellulose is that of a convergent evoluant. That is to say that the cellulose synthetases of plants and animals are analogues but the cellulose synthetase of animals, the chitin synthetase of animals and possibly the chitin synthetase of fungi are all homologous.

However, having gone so far, we might now consider whether animal cellulose and plant cellulose might not also be indirectly homologous episemantides; chitin having evolved from cellulose originally and animal cellulose having re-emerged by mutation. This would make both animal and plant cellulose synthetases homologous protein semantides.

L

We might go on to suggest that the glucan-based ancestral molecule of the connective tissue acid mucopolysaccharide group discussed earlier might also have been cellulose, the transition having been first via chitin.

Glycoproteins in the invertebrates present a rather ill-defined picture about which it seems possible to say little at the present time; this is because so few have been isolated, satisfactorily purified and characterized. On the whole the carbohydrate compositions of invertebrate glycoproteins seem not to differ too markedly from their vertebrate counterparts; the most profound difference seems to be the absence of sialic acid. Galactose, mannose, fucose and the two amino sugars 2-acetamido-2-deoxy-D-glucose and 2-acetamido-2-deoxy-D-galactose figure importantly in the analyses of invertebrate glycoproteins just as they do for vertebrates but there seems also a significantly greater tendency to find such sugars as xylose or arabinose. Uronic acids also seem to be quite regularly detected in the invertebrate glycoproteins, but we should exert some caution over such identifications to ensure that the criteria of purity are high enough and that acid mucopolysaccharide contaminants are not present.

The attention of biological chemists was first drawn to the field of glycoprotein chemistry by the ubiquity of these molecules in the secretions of the mammalian epithelium. Here the carbohydrate content is often high, with elevated sialic acid levels, and the function would seem to be as a lubricant. However, more recent studies have indicated that a far wider range of vertebrate proteins also carry carbohydrate prosthetic groups, often at a very low concentration, and that while many of these glycoproteins are biologically active, say as enzymes or hormones, the carbohydrate moiety seems to have no effect on the biological activity. It seems possible therefore that the use of glycoproteins as lubricants may be a secondary adaptation while the coupling of a small carbohydrate residue to a protein chain may have an older and more fundamental function.

Eylar (1965) has advanced an interesting and attractive hypothesis which suggests a general role for the carbohydrate in glycoproteins rather than for the glycoproteins themselves. He notes that bound carbohydrate is associated with most of the extracellular proteins but is virtually absent from intracellular proteins. He proposes that the carbohydrate acts as a chemical label which can interact with the cellular membrane and facilitate or promote transport of the protein out of the cell. Eylar also suggests that this method of protein secretion may be a primitive evolutionary device which has been superseded where more rapid secretion of protein is required, as for example from the pancreas by the formation of zymogen granules. Thus the carbohydrate need play no functional role in the activity of biologically-active proteins while glycoproteins which have a mechanical or biological function dependent upon their carbohydrate moiety, such as for example the epithelial mucins or possibly the bone glycoproteins, may have evolved secondarily. Eylar lists an impressive number of extra-cellular glycoproteins, but the list of carbohydrate-free extracellular pro-teins is small. The number of intracellular proteins which almost without exception have nil or negligible carbohydrate contents adds weight to the argument. The glycoproteins from invertebrates described in this book seem to be all extracellular, but it must be admitted that research in this field has been biased towards the examination of invertebrate secretions and body fluids (all extracellular) for bound carbohydrates, with a view to establishing affinities with comparable vertebrate secretions. It is of some interest to note that many of the fibrous and structure proteins secreted by invertebrates and discussed in this book also carry carbohydrate. In this respect it is satisfying to note that the silk proteins, until recently thought to be carbohydrate-free, have now been shown to be glycoproteins.

The evolution of the structure and fibrous proteins from the point of view of the relationships between their polypeptide chains is outside the scope of this book, which is primarily concerned with the association of carbo-hydrates and proteins. In any case, the evolution of such proteins has been adequately dealt with in several reviews by Rudall (1968a, b; Lucas and Rudall, 1968). Rudall considers that collagen has been derived from fibroin type molecules and believes also that elastin and resilin type proteins might be included in such an evolutionary scheme. In his review of the comparative biology and biochemistry of collagen, Rudall (1968a) presents some com-pelling evidence in favour of this argument in the form of a table comparing the amino acid compositions of *Galleria mellonella* silk, elastin, resilin and invertebrate collagen; the correspondence in compositions is remarkable. He also shows how simple modification in the genetic code could change the constitution of the fibroin to correspond very closely to that of elastin. The carbohydrate compositions do little to help us in advancing this hypo-thesis any further. We might note that while collagens universally seem to carry some carbohydrate, however small the quantity, neither resilin nor elastin seem to be glycoproteins, nor does this fit into the hypothesis that such extracellular proteins might be expected to have carbohydrate prosthetic groups. We have already noted that invertebrates generally have collagens with high carbohydrate contents, whereas the vertebrate meso-dermal collagens do not. Rudall (1968a) has suggested that there may be more than one type of association of carbohydrate with the collagen molecule on the basis of the observation that collagens with high carbo-hydrate content tend to have no banding, or indistinct normal banding of the fibre with a narrow core and a special type of banding in the wall of the fibre. However, the carbohydrate-protein links in invertebrate and verte-brate collagens may be rather similar; that of earthworm cuticle collagen is O-glycosidic to serine and threonine while in vertebrate collagen an O-glycosidic link with hydroxylysine seems to be indicated.

The carbohydrate constituents of the other fibrous proteins we have discussed do not seem to argue either for or against any particular affinities with the collagen group of proteins. One rather interesting point is the curious correspondence in the carbohydrate compositions of the operculum, periostracum, and egg case of the mollusc *Buccinum*. The amino acid compositions of these materials seem rather unrelated, yet in all three cases the bound monosaccharides are the same except for the absence of 2-amino-2-deoxy-D-galactose from the operculum. Moreover, the monosaccharides present are sufficiently distinctive to make the compositional correspondence seem even more significant; galactose, mannose and fucose might not together excite comment but the presence of xylose and rhamnose as well becomes noticeable. This again may suggest that the carbohydrate is involved in the process by which the protein is released from the cell. It may also suggest that the organism can exert some sort of genetic economy by using the same carbohydrate prosthetic group for a number of different extracellular proteins of rather similar type. Alternatively, the similarities in carbohydrate composition may indicate affinities between the proteins not immediately apparent from their gross amino acid compositions.

CONCLUSION

The present state of the study of mucopolysaccharides and glycoproteins permits us to make some tentative observations as to the way in which polysaccharides have evolved. But it does not allow us to say with any great certainty that molecules of this type of similar structures, when found in widely-separated phyla, are necessarily evolutionarily related. This is still the case for such well-characterized molecules as the chondroitin sulphates and applies even more so to the immensely diverse group of glycoproteins. The deficiencies which make it impossible for us to map in detail the origins and evolution of connective tissue mucopolysaccharides, mucopolysaccharide secretions, glycoproteins and the other manifestations of carbohydrates in living tissues are many; they include inadequate structural characterization of invertebrate materials, often inadequate purification, the failure to cast the net wide enough by only limited sampling of species or phyla, thereby giving only a limited overall picture, and an inadequate knowledge of the mode of biosynthesis of these molecules and of the actual enzymes mediating the biosynthesis. As to the manner in which the association between carbohydrate and protein originated and has evolved we know nothing. With regard to this problem, we must place ourselves in the position of the computer programmer who has had his programme returned with "further information requested" printed on it by the machine.

REFERENCES

Bayne, C. J. (1966). *Comp. Biochem. Physiol.* **19**, 317.
Dische, R. M., Pappas, G. D., Grauer, A. and Dische, Z. (1965). *Biochem. biophys. Res. Commun.* **20**, 63.
Eylar, E. H. (1965). *J. theor. Biol.* **10**, 89.
Florkin, M. (1966). "A Molecular Approach to Phylogeny". Elsevier, Amsterdam.
Isaacs, W. A., Little, K., Currey, J. D. and Tarlo, L. B. H. (1963). *Nature, Lond.* **197**, 192.
Jeuniaux, C. (1963). "Chitine et Chitinolyse". Masson, Paris.
Lucas, F. and Rudall, K. M. (1968). *In* "Comprehensive Biochemistry" (M. Florkin and E. H. Stotz, eds). Vol. 26B, p. 475.
Mathews, M. B. (1967). *Biol. Rev.* **42**, 499.
Person, P. and Philpott, D. E. (1967). *Clin. Orthop.* **53**, 185.
Pirie, A. and Van Heyningen, R. (1956). "The Biochemistry of the Eye". Charles C. Thomas, Springfield, Illinois.
Rudall, K. M. (1968a). *In* "Treatise on Collagen" (B. S. Gould, ed.). Vol. 2A, p. 83. Academic Press, London and New York.
Rudall, K. M. (1968b). *In* "Comprehensive Biochemistry" (M. Florkin and E. H. Stotz, eds). Vol. 26B. p. 559. Elsevier, Amsterdam.
Schiller, S. (1966). *A. Rev. Physiol.* **28**, 137.
Spiro, R. G. (1966). *Fedn Proc. Fedn Am. Socs exp. Biol.* **25**, 1409.
Zuckerkandl, E. and Pauling, L. (1965). *J. theor. Biol.* **8**, 357.

Addendum

Chapter 2

Person and Philpott (1969) have reviewed the occurrence, nature and biological significance of invertebrate cartilages.

Glycosaminoglycans have been isolated from the skin of the squid *Loligo opalescens* and fractionated (Srinivasan *et al.*, 1969). Highly sulphated chondroitin sulphates were found to be the major components (about 70% of the total mucopolysaccharides). A fourth of the total mucopolysaccharide appeared to be chondroitin and no dermatan sulphate could be detected. Appreciable amounts of other sugars were detected in the extracted and fractionated material, these being mainly 2-amino-2-deoxy-D-glucose and glucose with xylose and traces of galactose and mannose in the chondroitin fractions. The chondroitin sulphate fractions contained traces of 2-amino-2-deoxy-D-glucose and galactose. Amino acid analyses indicated differences in composition among the fractions.

Chapter 5

Ovodov *et al.*, (1969) have isolated polysaccharides from a range of Japan Sea invertebrates and have found remarkable similarities of composition. There were always two components present, demonstrable by gel-filtration, which appeared to be glycogen and a mixed sulphated hetero-polysaccharide fraction which contained galactose, glucose, arabinose, xylose, rhamnose, glucuronic acid and 2-amino-2-deoxy-D-glucose.

Beedham and Trueman (1968) have noted what appears to be a high concentration of acid mucopolysaccharide (identified histochemically) in the calcareous spicule-containing mucoid matrix of the cuticle of the Aplacophoran *Proneomenia*.

Puyear (1969) has detected three substrate specific nucleotide pyrophosphates in 12,000 g supernatants obtained from the hepatopancreas of the blue crab *Callinectes sapidus* Rathbun. The activity of UDP glucose, UDP glucuromic acid and UDP 2-acetamido-2-deoxy-D-glucose nucleotide pyrophosphatase seemed to be related to the stage of the molt cycle. It was suggested that these three enzymes act to regulate the biosynthesis of glycogen, chitin and mucopolysaccharides.

Chapter 8

Study of the chemical composition of the conchiolin of mother-of-pearl from *Nautilus pompilus* Lamark has shown that the insoluble residue which remains after hot alkali treatment (nacroin) is a glycoprotein or muco-polysaccharide in which the carbohydrate moiety is, in part at least, chitin (Goffinet and Jeuniaux, 1969).

Chapter 12

Jaccarini *et al.* (1968) have observed that the secretion of the pallial glands in the rock-boring Lamellibranch *Lithophaga lithophaga* is a neutral glycoprotein with calcium binding ability. This suggests an involvement of this material in chemical boring rather than simple disolution by acid particularly as the pH of the secretion is around 6·5.

Buddecke *et al.* (1969) have further characterized the O-seryl-N-acetyl-galactosaminide glycohydrolase from *Lumbricus terrestris* as an α-glycosi-dase acting on N-acetyl-α-D-galactosaminides but not on the corresponding glucosaminides.

Muramatsu (1968) has isolated an α-N-acetylglucosaminidase, a β-N-acetylglucosaminidase, an α-N-acetylgalactosaminidase and a β-N-acetyl-galactosaminidase from the liver of the gastropod mollusc *Turbo cornutus*.

Chapter 13

Kochetkov *et al.* (1967, 1968) have reported isolating a group of sialic acid-containing sphingoglycolipids from the gonads of the sea-urchin *Strongylocentrotus intermedius*. Zhukova and Smirnova (1969) also found sialoglycolipids in the gonads of the starfish *Patiria pectinifera* and found that in this organism and *S. intermedius* the sialoglycolipids contained glucose in the unusual furanose form.

Chapter 15

Muir and Lee (1969) have shown that over 90% of the D-galactose present in *Lumbricus terrestris* cuticle collagen can be released as reduced di- and trisaccharides with alkaline sodium borohydride. The structure of these oligosaccharides has been shown to be 2-O-α-D-galactopyranosyl-D-galactose and O-α-D-galactopyranosyl-(1 → 2)-O-α-D-galactopyranosyl-(1 → 2)-D-galactose.

Fujimoto (1968) isolated collagens of high hydroxyproline, hydroxylysine and carbohydrate content from the muscle layer of *Ascaris lumbricoides*. Collagen from the muscle layer contained 103, 122, 19 and 40 residues, per 1000 total residues, of proline, hydroxyproline, lysine and hydroxylysine together with 12·6% carbohydrate. This compares with corresponding values of 291, 19, 49, 0 and 0·4% for *Ascaris* cuticle collagen. A similar relationship between pig kidney collagen and pig skin collagen was noted.

Meenakshi *et al.* (1969) have reported the amino acid analyses of the periostraca of twenty-seven species of gastropod and bivalve mollusc.

Neutral polysaccharide, as indicated by the PAS reaction after digestion with diastase, was present in all cases. Glycine accounted for 40% or more of the amino acids of the periostraca of marine and fresh-water bivalves and of fresh-water and terrestrial gastropods. The periostraca of marine gastropods had less glycine and higher ranges of lysine, arginine, aspartic acid and serine than fresh-water species.

A detailed account of the ultrastructure of the egg capsules of the whelk *Buccinum undatum* has been given by Flower *et al.* (1969). The molecular fibrils have an α-helical conformation and these pack into larger, well-defined, ribbon-like units. Ribbons obtained from egg capsules of the two superfamilies Muricacea and Buccinacea have different longitudinal banding patterns and repeat distances although the lateral periodicities are identical. It was suggested that this arises from different types of overlapping of similar molecular fibrils.

REFERENCES

Beedham, G. E. and Trueman, E. R. (1968). *J. Zool., Lond.* **154**, 443.
Buddecke, E., Schauer, H., Werries, E. and Gottschalk, A. (1969). *Biochem. biophys. Res. Commun.* **34**, 517.
Flower, N. E., Geddes, A. J. and Rudall, K. M. (1969). *J. Ultrastruct. Res.* **26**, 262.
Fujimoto, D. (1968). *Biochim. biophys. Acta*, **168**, 537.
Goffinet, G. and Jeuniaux, C. (1969). *Comp. Biochem. Physiol.* **29**, 277.
Jaccarini, V, Bannister, W. H. and Micallef, H. (1968). *J. Zool., Lond.* **154**, 397.
Kochetkov, N. K., Zhukova, I. G., Smirnova, G. P. and Vaskorsky, V. E. (1967). *Dokl. Akad. Nauk SSSR*, **177**, 1472.
Kochetkov, N. K., Zhukova, I. G. and Smirnova, G. P. (1968). *Dokl. Akad. Nauk SSSR*, **180**, 996.
Meenashi, V. R., Hare, P. E., Watabe, N. and Wilbur, K. M. (1969). *Comp. Biochem. Physiol.* **29**, 611.
Muir, L. and Lee, Y. C. (1969). *J. biol. Chem.* **244**, 2343.
Muramatsu, T. (1968). *J. Biochem.* **64**, 521.
Ovodov, Y. S., Gorshkova, R. P., Tomshich, S. W. and Adamenko, M. N. (1969). *Comp. Biochem. Physiol.* **29**, 1093.
Person, P. and Philpott, D. E. (1969). *Biol. Rev.* **44**, 1.
Puyear, R. L. (1969). *Comp. Biochem. Physiol.* **28**, 159.
Srinivasan, S. R., Radhakrishnamurthy, B., Dalferes, E. R. and Berenson, G. S. (1969). *Comp. Biochem. Physiol.* **28**, 169.
Zhukova, I. G. and Smirnova, G. P. (1969). *Carbohydr. Res.* **9**, 366.

L*

Author Index

Numbers in italics refer to pages on which the full reference is given.

A

Subject Index

A

α-chitin, *see* chitin, chitin-protein complexes, X-ray diffraction by chitin-protein complexes

α-galactosaminidase, *see* glycoprotein degradative enzymes in invertebrates

α-D-mannosidase, *see* glycoprotein degradative enzymes in invertebrates

Acid mucopolysaccharides, *see* chondroitin sulphates, chondroitin, keratosulphate, heparin, hyaluronic acid, polysaccharide sulphates, polysaccharide phosphates, echinoderm egg jelly coats

in acanthocephalan epicuticle, 96

in aplacophoran cuticle, 307

in *Asterias forbesi* podia, 95

in *Arion ater* reproductive tract, 96

in *Calinectes sapidus*, 307

in cestodes, 96

in *Convoluta convoluta* mucin, 94

in *Dugesia tigrina* mucin, 94

in *Eisenia foetida* pharyngeal gland, 94

in enteropneust proboscis mucin, 94, 95

in *Helix pomatia* mantle and foot, 96

in helminth cuticle, 96

in *Hemigraspus nudus*, 93, 94

in *Hermodice carunculata* digestive tract, 94

in *Hydra pirardi*, 95

in *Lehmania poirieri* integument, 96

in *Lumbricus* gizzard cuticle, 94

in mollusc egg jelly, 95, 96

in mucous secretions of abareicolid lugworms, 94

in nematocysts, 94

in *Octopus vulgaris* alimentary cuticle, 96

acid mucopolysaccharides

in *Planaria vitta* mucin, 94

in squid liver, 93

in urochordate tunic, 94

2-amino-2-deoxy-D-mannose, as a constituent of sialic acids, 215

5-amino-3:5-dideoxy-D-erythro-L-gulunonulosonic acid, *see* neuraminic acid

Antigen from echinoderm egg surface, 114

Arthropodin, *see also* Insect cuticle

association with carbohydrate, 269

association with chitin, 136, 137

in cuticles of *Sarcophaga falculata*, *Sphinx ligustri*, *Agrianome spinacolis*, 268, 269

Astaxanthin

as a constituent of

Cancer pagurus ovary chromo-glycolipoprotein, 233

ovorubin, 230

ovoverdine, 232

Pecten maximus ovary chromo-glycolipoprotein, 233

Plesionika edwardsii egg chromo-glycolipoprotein, 233

structure, 229

B

β-chitin, *see* chitin, chitin-protein complexes, X-ray diffraction by chitin-protein complexes

β-galactosidase in *Helix*, 196, 210

β-glucosaminidase in *Helix*, 210

β-glucosidase in *Helix*, 210

β-D-mannosidase in *Charonia lampas*, 210

Brachiopod shell matrix proteins, 258, 259

amino acid composition, 260, 261

amino sugar composition, 260, 261

321